Investigation of Microstructural and Corrosion Properties of Steels and Light Alloys

Investigation of Microstructural and Corrosion Properties of Steels and Light Alloys

Editors

Claudio Gennari
Luca Pezzato

MDPI

Basel • Beijing • Wuhan • Barcelona • Belgrade • Novi Sad • Cluj • Manchester

Editors

Claudio Gennari
Department of Industrial Engineering
University of Padova
Padova
Italy

Luca Pezzato
Department of Industrial Engineering
University of Padova
Padova
Italy

Editorial Office
MDPI
St. Alban-Anlage 66
4052 Basel, Switzerland

This is a reprint of articles from the Special Issue published online in the open access journal *Materials* (ISSN 1996-1944) (available at: www.mdpi.com/journal/materials/special_issues/Investig_Microstruct_Corros_Prop_Steels_Light_Alloy).

For citation purposes, cite each article independently as indicated on the article page online and as indicated below:

Lastname, A.A.; Lastname, B.B. Article Title. *Journal Name* **Year**, *Volume Number*, Page Range.

ISBN 978-3-0365-8961-9 (Hbk)
ISBN 978-3-0365-8960-2 (PDF)
doi.org/10.3390/books978-3-0365-8960-2

Contents

About the Editors

Claudio Gennari

Dr. Claudio Gennari is an assistant professor at the Department of Industrial Engineering at the University of Padova. He got his PhD in industrial engineering on the subject of electroplasticity. His main research fields are electroplasticity in metals, metallurgy and corrosion resistance of duplex stainless steels, residual stress measurements on metallic alloys, microstructural investigation of 3D-printed refractory metals (tungsten, molybdenum, and tantalum), and fatigue resistance of 3D-printed metals.

Luca Pezzato

Dr. Luca Pezzato is currently an assistant professor at the Department of Industrial Engineering, University of Padova. His main research activities focus on the production and characterization of coatings for corrosion protection and the study of heat treatments on steels and stainless steels.

materials

Editorial

Investigation of the Microstructural and Corrosion Properties of Steels and Light Alloys

Luca Pezzato and **Claudio Gennari** *

Department of Industrial Engineering, University of Padova, Via Marzolo 9, 35131 Padova, Italy; luca.pezzato@unipd.it
* Correspondence: claudio.gennari@unipd.it

Very few metals can be found in metallic form in nature; the vast majority have to be processed from their ores at a great cost in terms of energy and money. It is therefore energetically favorable for them to reverse to their initial state. This process is commonly known as corrosion or anti-metallurgy, and great efforts are made worldwide to limit this process.

According to the latest NACE estimation (2013), the global cost of corrosion is equivalent to approximately 3.4% of the global GDP (2.5 trillion USD) not considering environmental consequences or safety issues. A reduction between 15% and 35% could be realized if prevention techniques and proper precautions are used, leading to savings between USD 375 and 875 billion. Corrosion involves different sectors, such as industry, military, civilian, services, etc., particularly energy production, transport, chemical and petrochemical industries, the mechanical industry, and the drink and beverage industry. Among these sectors, most of the constituents are made of steel, which is the most-produced metal in the world (1808 million tons in 2018) or light alloys, mainly aluminum (60.1 million tons of consumption in 2018).

A proper alloy design in terms of composition, heat treatment, microstructural features, etc., is mandatory to obtain the best combination of mechanical properties and corrosion resistance during operation, reducing maintenance costs and the overall impact on the global economy. In fact, microstructural features can affect both the corrosion of the material itself and the eventual production of protective layers on their surfaces.

The purpose of this Special Issue is to correlate the key role of the microstructure of steels and light alloys with their corrosion properties.

Maslak et al. [1] studied the impact fracture surfaces as indicators of structural post-fire susceptibility to brittle cracking in different steel grades (i.e., S355J2+N structural steel, X20Cr13 martensitic stainless steel, X6CrNiTi18-10 austenitic stainless steel and X2CrNiMoN22-5-3 duplex stainless steel). They found that various steel grades behave differently after surviving a fire incident, in a manner that is not obvious and difficult to foresee if not supported by appropriate specialized tests. They confirmed the need for a detailed analysis of impact strength to reliably draw conclusions regarding the possible future safe application of material that survived a fire.

Lim et al. [2] investigated a new Al-Mn-Zr alloy by means of microstructure and corrosion behavior after laser welding to a conventional AA3003. Thew found that laser welding caused interdendritic segregation and dissemination of intermetallic compounds throughout the fusion zone, this increased micro-galvanic corrosion sites, destabilizing the passive film. The addition of Zr reduced the size and number of intermetallic compounds minimizing micro-galvanic corrosion and enhancing the corrosion resistance compared to the AA3003.

Pezzato et al. [3] studied the effect of the addition of copper on the corrosion and antifouling properties of PEO-coated zinc-aluminized steel. They were able to successfully embed copper particles inside the PEO coating. The presence of copper particles enhanced

check for updates

Citation: Pezzato, L.; Gennari, C. Investigation of the Microstructural and Corrosion Properties of Steels and Light Alloys. *Materials* **2023**, *16*, 6171. https://doi.org/10.3390/ma16186171

Received: 20 July 2023
Accepted: 31 July 2023
Published: 12 September 2023

the antifouling properties of the sample's surface during the first 20 days of immersion in sea water. However, the corrosion resistance was negatively affected by the higher electrical conductivity of the PEO coating due to the presence of conductive copper particles, even though it remained higher compared to the untreated sample.

Khademzadeh et al. [4] developed a micro-laser powder bed fusion for the additive manufacturing of Inconel 718 alloy. They optimized the process parameters of the additive manufacturing machine to obtain the best microstructural properties for post-printing heat-treated Inconel 718. The result showed that using the optimum input energy density led to the homogenous distribution of nanosized (<10 nm) circular γ' and plate-like γ'' particles in the γ matrix. When uniaxial tensile tests were conducted on heat-treated samples, they showed that ageing temperature is the most determinant factor in the mechanical strength of additively manufactured Inconel 718.

Yeh et al. [5] studied the stress corrosion cracking and crevice corrosion of AISI 304L stainless steel in a saline environment after the deposition of dust on the surface. They used a temperature of 45 °C and different relative humidities (45%, 55% and 70%) for 7000 h. They found that there is a relative humidity threshold (i.e., between 55% and 70%), above which SCC and crevice corrosion appear to have the lowest dust concentration. At the highest dust concentration, the relative humidity threshold is lowered between 45% and 55%.

Liang and Asselin [6] focused their work on the effect of bitumen at temperatures between 60 °C and 120 °C for 30 days on the corrosion of pipeline steel API X100. They analyzed inclusions by means of SEM EDS analysis and observed the effect of the exposure of the steel to bitumen on the matrix and inclusions of the steel. They found that bitumen under these conditions does not affect either the matrix or the inclusion of the steel.

Zong and Liu [7] worked on the microstructure, mechanical properties, and corrosion behavior of ultra-low-carbon bainitic steel with different niobium contents. They performed a tensile test, low-temperature impact toughness, corrosion weight loss method, polarization curves, and electrochemical impedance spectroscopy on four bainitic steels with different niobium contents. They observed that niobium affects the bainite morphology and the size, quantity and distribution of M/A elements. The impact toughness of all the samples at −40 °C is lower than 10 J. Better overall properties were observed in the steel with 0.0692% Nb.

Franceschi et al. [8] studied the effect of different austempering heat treatments on the corrosion properties of high-silicon austempering steel. They performed two sets of heat treatments: one changing the austenitizing temperature and keeping the austempering temperature constant, and the other keeping the austenitizing temperature constant and changing the austempering temperature. They also investigated the corrosion behavior of the treated steel in aqueous borate buffer solution to correlate the corrosion behavior with the microstructure arising from the various heat treatments that were performed. They were able to observe the highest volume fraction of retained austenite when the austenitizing temperature was 850 °C; moreover, the corrosion resistance increased when increasing the amount of retained austenite.

Yeh, Tsai and Huang [9] investigated the influence of chloride concentration on the stress corrosion cracking and crevice corrosion of austenitic stainless steel in saline environments. They analyzed different relative humidity value at 45 °C for 400 h and 10,000 h on the corrosion behavior of AISI 304L after spraying synthetic sea water on its surface. No crack was found on the speciman at 1 g/m^2 of chlorine concentration after 400 h at any relative humidity value; instead, SCC cracks were observed at every relative humidity value for the specimen tested for 10,000 h. The chlorine concentration threshold required to initiate SCC in AISI 304L austenitic stainless steel at 45 °C and 10,000 h is between 0.1 g/m^2 and 1 g/m^2.

Biserova-Tahchieva et al. [10] conducted a comprehensive review of the additive manufacturing processes in selected corrosion-resistant materials. They critically discussed the corrosion resistance of light metallic systems and duplex stainless-steel objects obtained by additive manufacturing, highlighting methodologies that could improve this. Thanks to the data they collected, they concluded that the potentiodynamic corrosion tests are the best methodology to characterize metal additive manufacturing objects.

As can be seen from the papers published in this Special Issue, the collaboration of different institutions from all over the world was made possible, which helps to share the knowledge on, and increase the comprehension regarding, every subject.

As a concluding remark, the Editors want to thank all the authors and the editorial team of *Materials* for their collaboration in the peer review process. We genuinely hope that this Special Issue could improve the understanding of the link between the microstructure and corrosion properties of steels and light alloys.

Acknowledgments: The editors would like to thank all the authors for submitting their papers to the Special Issue "Investigation of the Microstructural and Corrosion Properties of Steels and Light Alloys", as well as all the reviewers and editors for their contributions to improving these submissions.

Conflicts of Interest: The authors declare no conflict of interest.

References

1. Maslak, M.; Pazdanowski, M.; Stankiewicz, M.; Wassilkowska, A.; Zajdel, P.; Zielina, M. Impact Fracture Surfaces as the Indicators of Structural Steel Post-Fire Susceptibility to Brittle Cracking. *Materials* **2023**, *16*, 3281. [CrossRef] [PubMed]
2. Lim, J.-M.; So, Y.-S.; Kim, J.-G. Microstructure and Corrosion Behavior of Laser-Welded Al–Mn–Zr Alloy for Heat Exchanger. *Materials* **2023**, *16*, 6009. [CrossRef] [PubMed]
3. Pezzato, L.; Settimi, A.G.; Fanchin, D.; Moschin, E.; Moro, I.; Dabalà, M. Effect of Cu Addition on the Corrosion and Antifouling Properties of PEO Coated Zinc-Aluminized Steel. *Materials* **2022**, *15*, 7895. [CrossRef]
4. Khademzadeh, S.; Gennari, C.; Zanovello, A.; Franceschi, M.; Campagnolo, A.; Brunelli, K. Development of Micro Laser Powder Bed Fusion for Additive Manufacturing of Inconel 718. *Materials* **2022**, *15*, 5231. [CrossRef] [PubMed]
5. Yeh, C.-P.; Tsai, K.-C.; Huang, J.-Y. The Effect of Deposited Dust on SCC and Crevice Corrosion of AISI 304L Stainless Steel in Saline Environment. *Materials* **2021**, *14*, 6834. [CrossRef] [PubMed]
6. Liang, H.; Asselin, E. Ex Situ Examination of Matrix and Inclusions of API-X100 before and after Exposure to Bitumen at Elevated Temperature. *Materials* **2021**, *14*, 5007. [CrossRef] [PubMed]
7. Zong, Y.; Liu, C.-M. Microstructure, Mechanical Properties, and Corrosion Behavior of Ultra-Low Carbon Bainite Steel with Different Niobium Content. *Materials* **2021**, *14*, 311. [CrossRef] [PubMed]
8. Franceschi, M.; Pezzato, L.; Settimi, A.G.; Gennari, C.; Pigato, M.; Polyakova, M.; Konstantinov, D.; Brunelli, K.; Dabalà, M. Effect of Different Austempering Heat Treatments on Corrosion Properties of High Silicon Steel. *Materials* **2021**, *14*, 288. [CrossRef] [PubMed]
9. Yeh, C.-P.; Tsai, K.-C.; Huang, J.-Y. Influence of Chloride Concentration on Stress Corrosion Cracking and Crevice Corrosion of Austenitic Stainless Steel in Saline Environments. *Materials* **2020**, *13*, 5640. [CrossRef] [PubMed]
10. Biserova-Tahchieva, A.; Biezma-Moraleda, M.V.; Llorca-Isern, N.; Gonzalez-Lavin, J.; Linhardt, P. Additive Manufacturing Processes in Selected Corrosion Resistant Materials: A State of Knowledge Review. *Materials* **2023**, *16*, 1893. [CrossRef] [PubMed]

materials

Article

Microstructure and Corrosion Behavior of Laser-Welded Al–Mn–Zr Alloy for Heat Exchanger

Jeong-Min Lim, Yoon-Sik So and Jung-Gu Kim *

School of Advanced Materials Science and Engineering, Sungkyunkwan University, 2066, Seobu-ro, Jangan-Gu, Suwon-si 16419, Gyeonggi-do, Republic of Korea; soy4718@skku.edu (Y.-S.S.)
* Correspondence: kimjg@skku.edu; Tel.: +82-312907360

Abstract: In this study, an Al–Mn–Zr alloy was designed and its microstructure and corrosion behavior compared after laser welding to that of AA3003. As the results of immersion and electrochemical tests showed, both alloys had a faster corrosion rate in the fusion zone than in the base metal. Laser welding caused interdendritic segregation, and spread the intermetallic compounds (IMCs) evenly throughout in the fusion zone. This increased the micro-galvanic corrosion sites and destabilized the passive film, thus increasing the corrosion rate of the fusion zone. However, Zr in the Al–Mn alloy reduced the size and number of IMCs, and minimized the micro-galvanic corrosion effect. Consequently, Al–Mn–Zr alloy has higher corrosion resistance than AA3003 even after laser welding.

Keywords: laser welding; aluminum alloy; corrosion; zirconium; fusion zone; micro–galvanic; interdendritic segregation; passive film

check for updates

Citation: Lim, J.-M.; So, Y.-S.; Kim, J.-G. Microstructure and Corrosion Behavior of Laser-Welded Al–Mn–Zr Alloy for Heat Exchanger. *Materials* **2023**, *16*, 6009. https://doi.org/10.3390/ma16176009

Academic Editors: Luca Pezzato and Claudio Gennari

Received: 4 August 2023
Revised: 26 August 2023
Accepted: 27 August 2023
Published: 1 September 2023

1. Introduction

A heat exchanger is a core part that directly affects the efficiency of the air conditioner and ventilation system. Fin–tube heat exchangers are generally used [1]. As a material for the fin and tube, AA3003 (Al–Mn alloy) is used due to the advantage of light weight and good formability as well as high corrosion resistance [2,3]. Brazing was used to join the Al fin and tube, but more recently welding has been used. The brazing melts the filler metal and joins the base metals; hence, the use of filler metal is essential. However, the filler metal can cause heat exchanger failure by causing galvanic corrosion between the filler metal, fin, and tube [4–6]. Welding, on the other hand, does not necessarily use filler metal, and can be used as an alternative joining method. In particular, laser welding is attracting attention as a fast and accurate fully automatic process with a narrow heat–affected zone (HAZ) and weld seam [7,8]. However, it has been reported that the microstructural changes caused by laser welding make the fusion zone susceptible to pitting corrosion [9–17]. Therefore, to improve a lifespan of fins and tubes, it is necessary to study the effects of laser welding on the microstructure and corrosion behavior of Al–Mn alloys.

In Al–Mn alloys, the Mn element is precipitated as Al_6Mn IMCs. However, in most Al alloys, the Fe element is contained as an impurity and exists as Al_3Fe and Al_6Fe intermetallic compounds (IMCs). Al_6Fe IMC combines with Al_6Mn to form $Al_6(Fe,Mn)$, which has similar corrosion potential to the Al matrix and reduces the micro-galvanic corrosion between IMCs and the Al matrix [18,19]. However, the remaining Al_3Fe and Al_6Fe have higher corrosion potentials than the Al matrix and still act as cathodic sites, accelerating the micro-galvanic corrosion of Al–Mn alloys. Therefore, in order to increase the corrosion resistance of Al–Mn alloys, studies on the uniform distribution of IMCs by adding alloying elements or nanoparticles are being conducted [20]. Zr is known as an alloying element for refining the grain and IMCs. Zr in Al alloys precipitates as fine Al_3Zr, which prevents the grain growth [21–23]. In addition, the melting temperature of Al_3Zr is very high, thus some Al_3Zr remains unmelted in the molten pool during the welding [24]. These fine Al_3Zr

particles can play the same role as nanoparticles. The addition of nanoparticles reduces the IMC size and spreads the IMC of the welded and base Al alloy [25]. Therefore, the result of increasing the corrosion resistance of the fusion zone can be expected through the adding of Zr.

In this study, an Al–Mn alloy containing the Zr was prepared to estimate the effect of Zr on the corrosion behavior of laser-welded Al–Mn alloy, and the corrosion behavior of Al–Mn–Zr alloy was compared with that of AA3003. The microstructural change during the laser welding was observed using optical microscopy (OM), scanning electron microscopy (SEM), and electron probe microanalysis (EPMA). Then, the corrosion behavior was estimated by immersion test, potentiodynamic (PD) test, and electrochemical impedance spectroscopy (EIS).

2. Materials and Methods

2.1. Materials and Laser Welding Process

The Al alloys used in this research were commercial AA3003 and Al–Mn–Zr alloys (UniCorAl®, Suwon, Republic of Korea, U3003). U3003 is an Al–Mn alloy with added Zr which was developed to improve corrosion resistance. The U3003 alloy was prepared using a 99.9% Al ingot and three master alloys (Al–20Mn, Al–5Fe, and Al–15Zr). An appropriate amount of material was filled into a graphite crucible at 780–800 °C, then the molten metal was injected into a mold and air cooled. Following the casting, hot and cold rolling were performed to produce a 1 mm-thick plate. The chemical compositions of the two alloys were analyzed by inductively coupled plasma–mass spectrometry (ICP–MS, Perkin Elmer (Waltham, MA, USA), NexION 2000S), as shown in Table 1.

Table 1. Chemical compositions of Al alloys.

Alloys	Composition (wt.%)						
	Mn	Fe	Zr	Si	Cu	Mg	Al
AA3003	1.126	0.357	-	0.100	0.120	0.017	Bal.
U3003	0.485	0.086	0.164	0.032	-	0.002	Bal.

The laser-welded specimens were plate shaped with a size of 60 mm × 20 mm × 1 mm, and the center was welded in the longitudinal direction. Figure 1 shows the specimen preparation procedure.

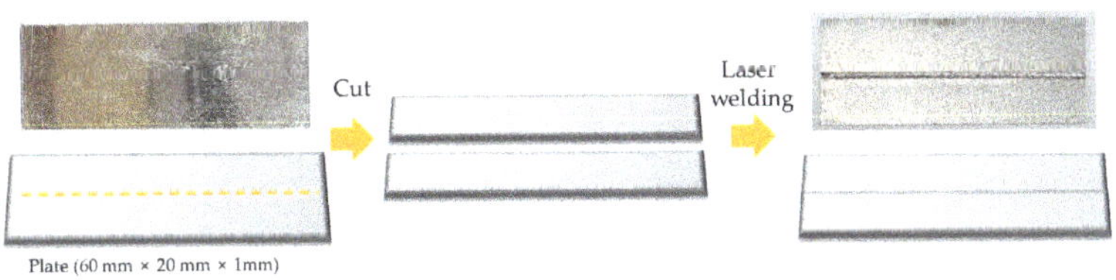

Figure 1. Procedure for laser-welded specimen preparation.

Dual-beam laser welding was performed in a butt joint configuration with no filler metal. Pure nitrogen gas was used as the shielding gas with a flow rate of 10 L/min. The welding parameters are listed in Table 2.

Table 2. Laser welding parameters.

Power Mode	Speed (m/min)	Beam Power (kW)	Beam Size (μm)	Wavelength (μm)	Incident Angle (°)	Defocusing (mm)
Continuous	10	Main 3.0 Ring 0.5	Main 210 Ring 540	1070	0	0

2.2. Surface Analysis

To observe the microstructure of the specimens, the specimens were cut perpendicularly to the welding direction and mounted using epoxy resin. The cross sections of the specimens were mechanically polished up to 4000 grit using SiC paper, followed by polishing to 1 μm using alumina powder. Before observation, the specimens were etched using a Keller reagent (5 mL HNO_3, 3 mL HCl, 2 mL HF, and 190 mL deionized water) according to ASTM E407 [26]. The microstructure of the laser welded specimens was characterized by OM (Witec (Ulm, Germany), Alpha 300S) and SEM (JEOL (Akishima-shi, Japan), JSM 6380). Their chemical compositions were measured by EPMA (JEOL, JXA 8900R) equipped with SEM.

2.3. Corrosion Testing

To evaluate the corrosion behavior of laser-welded Al–Mn alloys, immersion, potentiodynamic (PD), and EIS testing were performed. The test solution was artificial acid rain, and the composition is presented in Table 3 [27]. The ion concentration of the test solution was controlled by H_2SO_4, HNO_3, and NaCl, and the pH was adjusted with 0.1 M NaOH solution.

Table 3. Chemical composition of artificial acid rain.

pH	SO_4^{2-} (mg/L)	NO^{3-} (mg/L)	Cl^- (mg/L)
4	2.212	1.545	400

For the immersion test specimens, the top weld side was polished to a 1000-grit size using SiC paper. Then, the specimen was immersed in the solution for 12 weeks. During the immersion test, the solution was changed every 2 weeks to maintain the chemical composition. After the immersion test, the corrosion depth was measured using a surface profiler (KLA Tencor (Milpitas, CA, USA), Alpha Step 500).

The electrochemical tests were conducted on the bare Al–Mn alloy (before laser welding) and the fusion zone of laser-welded specimens, respectively, using a three-electrode system and a potentiostat (Bio Logic (Seyssinet-Pariset, France), VSP 300). The working electrode was the test specimen, the counter electrode was the pure carbon rod, and the reference electrode was the saturated calomel electrode. Each electrochemical test was performed in triplicate, and before conducting electrochemical tests, the open circuit potential (OCP) was measured for 30 h. The PD tests were performed with a scan rate of 0.166 mV/s, following the ASTM G5 standard [28]. The anodic potential was swept from the OCP to 1 V_{OCP}, while the cathodic potential was swept from OCP to −0.6 V_{OCP}. The frequency range of the EIS tests was from 100 kHz to 10 MHz with alternating current amplitude of 20 mV.

3. Results and Discussion

3.1. Microstructure of Laser-Weld Beads

Figure 2 shows the cross-sectional images of laser-welded AA3003 and U3003.

The red boxes in Figure 2a,c indicate the interface between the base metal (BM) and the fusion zone (FZ), and enlarged images are shown in Figure 2b,d. In the FZ, the grains grow epitaxially from the BM/FZ boundary towards the weld center, which is in a perpendicular direction to the welding direction. Unlike in the BM, the distribution of IMCs in the FZ exhibits directionality, clearly distinguishing BM and FZ. Since the laser welding enables rapid melting and cooling, the effect of heat on the BM is reduced [29,30]. Therefore, both alloys have a very narrow HAZ. Consequently, in this study, the cross sections of the laser-welded specimens were divided into two regions, BM and FZ.

Figure 3 shows the SEM images of the BM/FZ interface (red boxes in Figure 2), and the distribution of Fe and Mn at the same location.

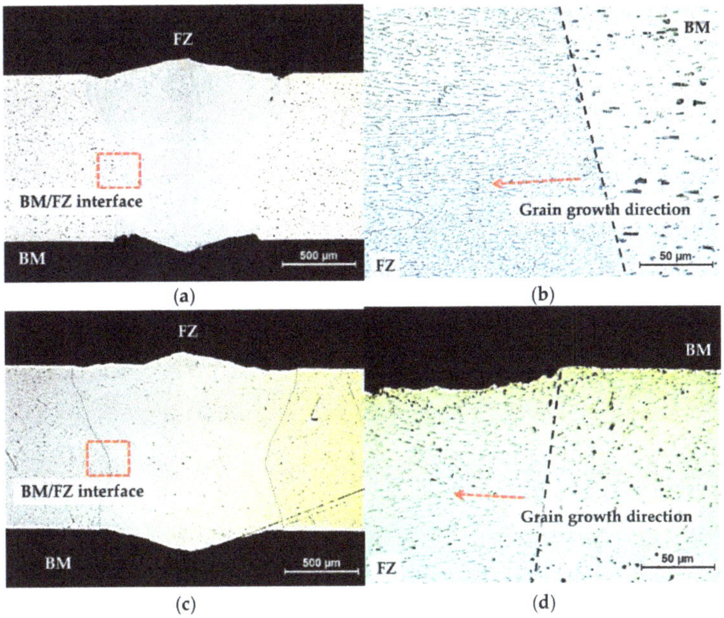

Figure 2. Cross-sectional images of laser-welded specimens. AA3003 (**a**) 50× and (**b**) 500× (BM/FZ Interface); U3003 (**c**) 50× and (**d**) 500× (BM/FZ interface).

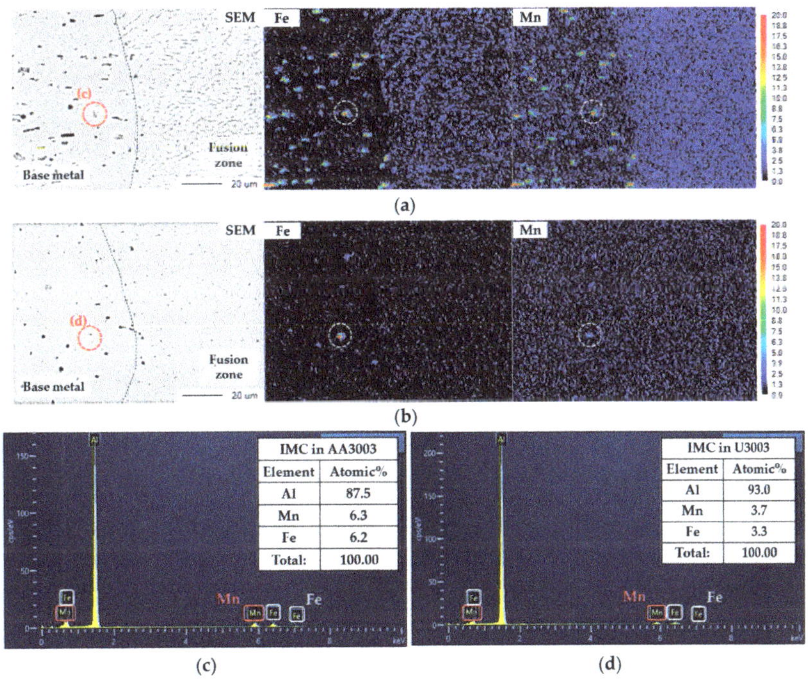

Figure 3. SEM images of BM/FZ interface and EPMA element mapping for Fe and Mn of (**a**) AA3003 and (**b**) U3003. The same position as the red circle in the SEM image is indicated by a white circle in the mapping image. EDS point analysis results of IMCs in red circles of (**c**) AA3003 and (**d**) U3003.

In both alloys, Fe and Mn were concentrated in the IMC, indicated by the white circle in the BM, whereas they were dispersed over the entire area in the FZ. Consequently, the laser welding affected the microstructure of FZ, especially the IMC distribution. Figure 3c,d show the EDS point analysis results of the IMC inside the red circle for each alloy. The IMCs of both alloys contained similar amounts of Fe and Mn, with a significant presence of Al. Therefore, the IMC shown in the SEM image is $Al_6(Fe,Mn)$, which is a representative IMC of Al–Mn alloy [31,32]. Since $Al_6(Fe,Mn)$ is formed by substituting a Fe atom at the Mn position of Al_6Mn or by substituting a Mn atom at the Fe position of Al_6Fe, $Al_6(Fe,Mn)$ has a full composition range from Al_6Mn to Al_6Fe [33,34]. Al_6Mn has a similar potential with the Al matrix, then increasing the corrosion resistance. However, Al_6Fe has a higher corrosion potential than Al matrix, of about 300 mV, thus Al_6Fe acts as a cathode to the Al matrix and accelerates the corrosion rate of the Al matrix by the micro-galvanic corrosion [19,35]. Therefore, the distribution of IMC affects the corrosion behavior of Al–Mn alloy. Since the micro-galvanic corrosion rate increases as the number of micro-galvanic sites and the area ratio of the cathode to the anode increase, the IMC size and distribution should be analyzed.

The microstructures of each region are shown in Figure 4.

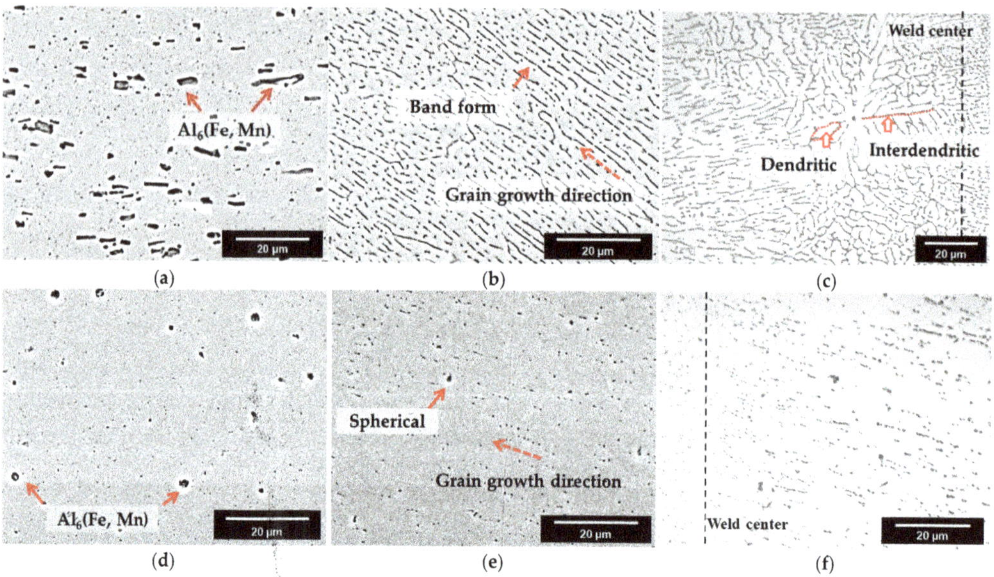

Figure 4. Microstructures of laser-welded specimens. AA3003 (**a**) BM, (**b**) FZ, near BM/FZ interface, and (**c**) FZ, weld center; U3003 (**d**) BM, (**e**) FZ, near BM/FZ interface, and (**f**) FZ, weld center.

In Figure 4a–c, the IMC in the BM (solid arrows) of AA3003 is a short-rod shape with an average size of 1.282 μm^2 and the IMCs are distributed without regularity. The IMCs in the FZ of AA3003 appear as very thin bands along the grain growth direction (dotted arrows). In Figure 4d–f, the IMCs in the BM of U3003 are spherical with an average size of 0.291 μm^2 and are distributed irregularly. However, in the FZ, the size of IMCs is smaller than in BM and distributed along the grain growth direction. This means that during the laser welding, the alloying elements in the molten pool were accumulated among the dendrite arm and interdendritic segregation occurred in both alloys [36,37]. Moreover, in Figure 4c, some equiaxed dendrites appear at the weld center of AA3003, further showing the interdendritic segregation.

Additionally, if the cooling rate of the laser-welding process is fast, then the dendrite arm spacing is reduced [38]. Consequently, the interdendritic IMCs in the dendrite arms are smaller and more evenly spread.

During the solidification, Zr in Al alloy is precipitated as fine Al_3Zr particles, and this precipitation process hinders the nucleation and growth of IMCs [20,39]. Therefore, in BM, the IMC size of U3003 is smaller than that of AA3003. Additionally, after laser welding, since the melting point of Al_3Zr is significantly higher (1500 °C) than that of Al (660 °C), some Al_3Zr particles remain unmelted [24,40]. These fine Al_3Zr particles act as a heterogeneous nucleation site of Al_6(Fe,Mn) IMCs. Consequently, a large number of fine spherical IMCs are formed.

Figure 5 shows the area ratio of IMCs to the Al matrix calculated from the SEM images of Figure 4.

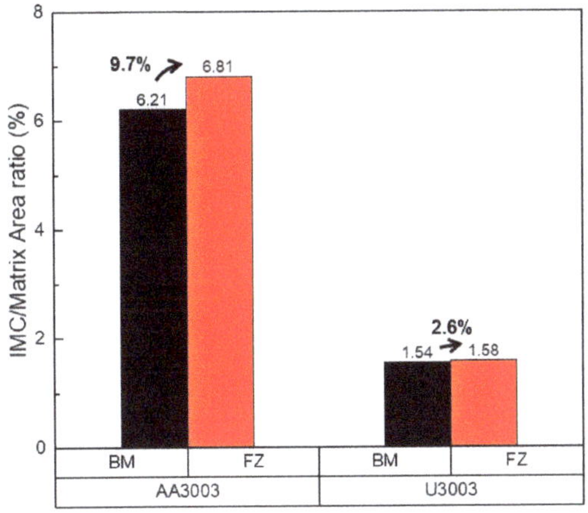

Figure 5. The area ratio of IMCs to Al matrix in BM and FZ calculated from SEM images in Figure 4.

In the AA3003, the area ratio of IMCs to Al matrix is 6.21% in BM, and 6.81% in FZ. In the U3003, the area ratio of IMCs is 1.54% in BM and 1.58% in FZ. Since the rapid cooling of laser welding refines and spreads the IMC, the area ratio of the IMC in the FZ of both alloys increases [41]. In FZ of AA3003, the IMCs are band shaped and are spaced close to each other by the interdendritic segregation; thus, the area of IMC becomes 9.7% wider than that of BM. However, in the case of U3003, Zr in the Al alloy hindered the growth of Al_6(Fe,Mn) and made the IMC spherical, resulting in an increase of the IMC area by only 2.6%. Since U3003 has a lower content of Fe and Mn elements than AA3003, the amount of IMC is less in U3003 in both regions. The galvanic corrosion rate is proportional to the area ratio of the cathode to the anode; thus, the corrosion rate of the Al–Mn alloy is increased as the IMC area increases [42,43]. Therefore, it can be inferred that the corrosion rate of U3003 is lower than that of AA3003, in both regions.

3.2. Corrosion Testing

To investigate the corrosion behavior of laser-welded specimens, immersion testing was performed for 12 weeks. The surface and cross-sectional images of specimens after the immersion test are shown in Figure 6.

In case of AA3003 (Figure 6a,b), the whole region of AA3003 was corroded, and the maximum corrosion depth of FZ was 61.3 μm, which was more severe than that of BM. In Figure 6c,d, it can be seen that U3003 was corroded very slightly in BM, and the partial area of FZ was corroded with a maximum corrosion depth of 18.5 μm. Therefore, the corrosion rate of AA3003 was faster than that of U3003 in both regions.

Figure 7 shows the polarization curves of each specimen and the results are presented in Table 4.

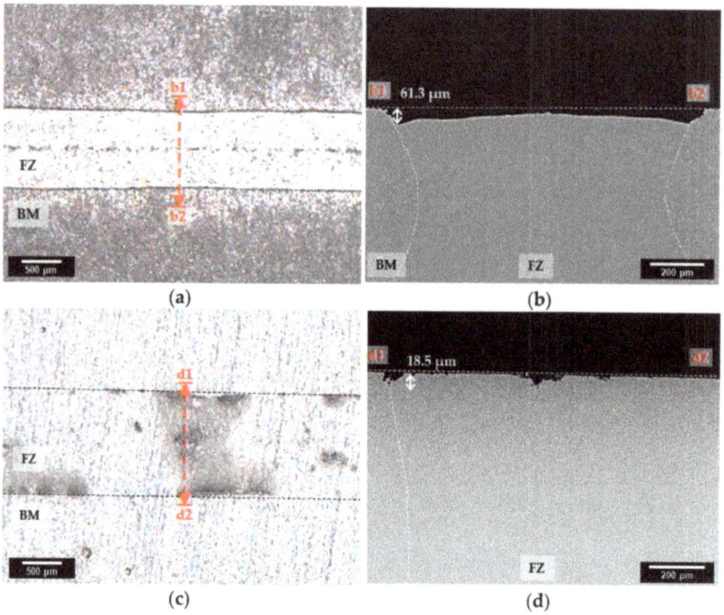

Figure 6. Surface and cross-sectional images of specimens after 12 weeks' immersion test. (**a**) Surface image of AA3003 and (**b**) cross-sectional image at the b1–b2. (**c**) Surface image of U3003 and (**d**) cross-sectional image at the d1–d2.

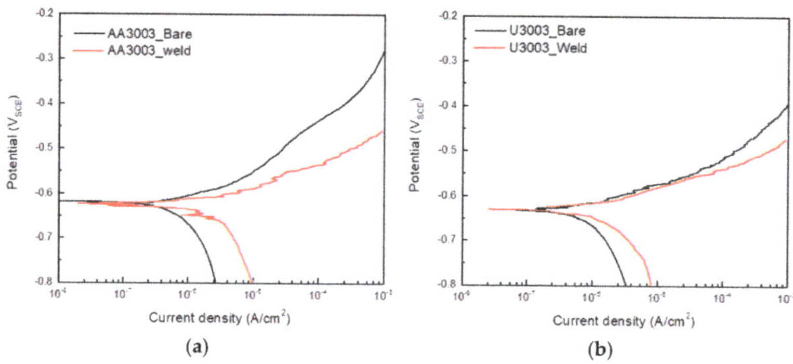

Figure 7. Potentiodynamic polarization curves of (**a**) AA3003 and (**b**) U3003.

Table 4. Potentiodynamic test results for AA3003 and U3003.

Alloy	AA3003-Bare	AA3003-Weld	U3003-Bare	U3003-Weld
Corrosion potential (mV_{SCE})	623.7	621.7	631.2	631.2
Corrosion current density ($\mu A/cm^2$)	0.78	3.15	0.46	0.69

Consistent with the results of the immersion test, the corrosion rate of BM was faster than that of the FZ in both alloys. However, the corrosion potential difference of BM and FZ was less than 5 mV. Therefore, the galvanic corrosion by the microstructure difference between the BM and FZ was not significant. Consequently, the microstructure change caused by the laser welding affected the micro-galvanic corrosion between the IMC and matrix.

Figure 8 shows the high-magnification images of the cross sections after immersion testing. In the BM of AA3003 (Figure 8a), the corrosion of the Al matrix was concentrated around the IMC.

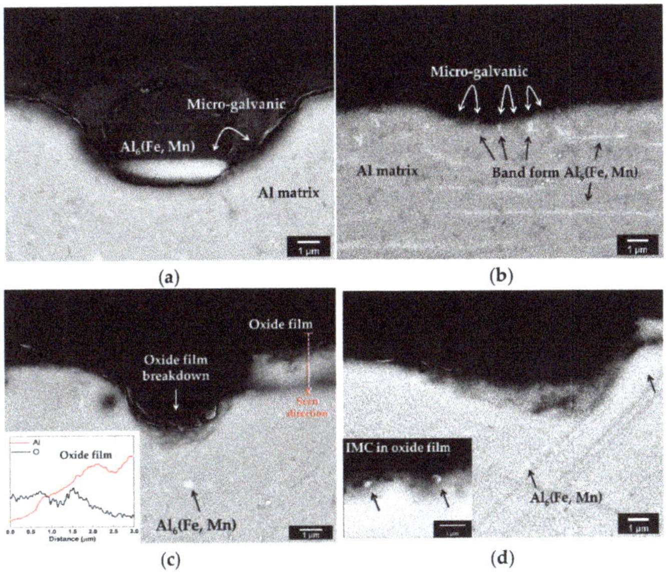

Figure 8. High-magnification images of cross sections of specimens after 12 weeks' immersion testing. AA3003 (**a**) BM and (**b**) FZ; U3003 (**c**) BM and (**d**) FZ.

This means that the micro-galvanic corrosion occurred between the Al_6(Fe,Mn) IMC and Al matrix. In contrast, in FZ (Figure 8b), corrosion occurred uniformly over the entire area. Since the IMCs were refined and redistributed in FZ after laser welding, the number of micro-galvanic sites increased. Therefore, corrosion occurred uniformly over a wide area without being concentrated, and the micro-galvanic rate of FZ was increased compared to that of BM. In the case of U3003 (Figure 8c,d), the IMC size was quite small, and micro-galvanic corrosion next to the IMC did not appear. Since Al_6(Fe,Mn) has a composition between Al_6Mn and Al_6Fe, the corrosion potential also has a value between that of Al_6Mn and Al_6Fe. As the Mn content increased in IMC, the corrosion potential decreased from that of Al_6Fe to that of Al_6Mn, which had a corrosion potential similar to that of the Al matrix [19,35]. The Mn/Fe ratio of AA3003 is 3.15 and that of U3003 is 5.6. Since the Mn content of IMC increases with the Mn/Fe ratio, the corrosion potential of Al_6(Fe,Mn) IMC is lower in U3003 than in AA3003. Therefore, the micro-galvanic effect was not significant in U3003.

In the previous section, the area ratio of IMC to the Al matrix of BM and FZ only slightly increased in U3003. In addition, no significant micro-galvanic corrosion was observed. Nevertheless, the corrosion depths of BM and FZ were quite different, as shown in Figure 6. There is another reason for the corrosion acceleration of FZ in addition to micro-galvanic corrosion between the IMC and Al matrix. In Figure 8c, the result of the EDS line profile performed at the red dashed arrow is shown in the lower left corner. The scan direction was inward from the alloy surface. The oxygen content decreases and Al content increases along the scan direction. Therefore, it can be found that the oxide film was formed on the specimen surface [44,45]. In Figure 8c, the oxide film becomes thin on the corroded surface. Moreover, in Figure 8d, the IMCs are appearing in the passive film. The breakdown of passive film is the main mechanism of pitting corrosion of Al [46]. Therefore, in this study, the stability of oxide film in each region was investigated through the EIS test.

To investigate the corrosion mechanism of laser-welded Al–Mn alloys, the EIS tests were performed on FZ and BM, and the impedance plots were analyzed based on an appropriate equivalent circuit using ZSimpWin Software version 3.21. Figure 9a shows the equivalent circuit used to determine the optimized resistance and capacitance values from the EIS data.

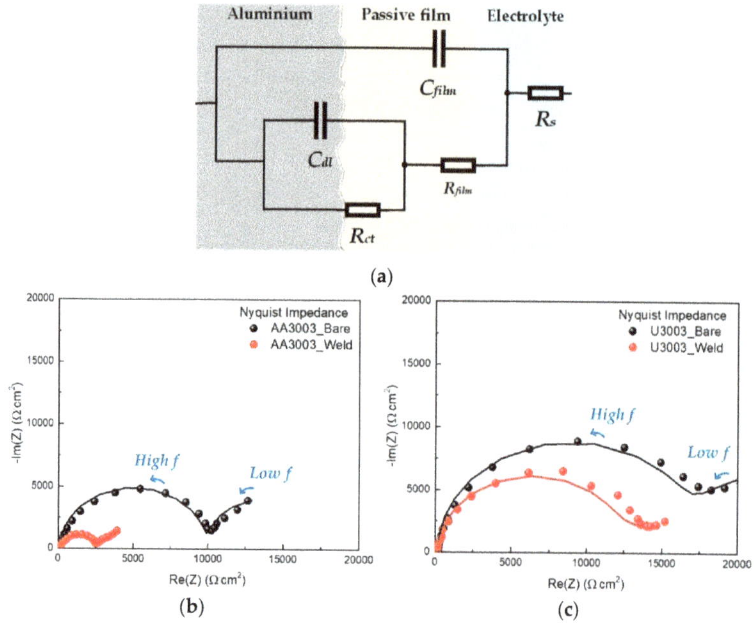

Figure 9. (**a**) Equivalent circuit to fit the EIS data. EIS Nyquist plots of (**b**) AA3003 and (**c**) U3003. Scatter plots are experimental values and line plots are calculated values.

In the equivalent circuit, R_s is the solution resistance, C_{film} is the capacitance of the passive film, R_{film} is the electrical resistance of an ionic conduction path through the passive film, C_{dl} is the double-layer capacitance and R_{ct} is the resistance associated with the metal dissolved in the electrolyte [47]. Figure 9b,c show the experimental Nyquist plots (scatter plots) and calculated Nyquist plots using the equivalent circuit (line plots). The Nyquist plots in Figure 9 show two time constants, and the measurement plots and the calculation plots agree well. The semicircle in the low frequency is correlated with charge transfer resistance, and the high-frequency semicircle indicates the corrosion resistance by the passive film [48–50]. Since the real axis value at the low-frequency intercept is the sum of the polarization resistance and solution resistance, the corrosion resistance increases as the diameter of the semicircle increases [51,52]. The high-frequency semicircle of BM is bigger than that of FZ in both alloys. Therefore, the corrosion resistance of FZ was lower than that of BM due to the effect of passive film stability. Additionally, as a result of analyzing the polarization resistance (R_p), which means the diameter of semicircles, using an equivalent circuit, in the case of BM, that of AA3003 was 17,748 Ω·cm^2 and that of U3003 was 28,692 Ω·cm^2. In the case of FZ, the polarization resistance of AA3003 was 2250 Ω·cm^2 and that of U3003 was 11,993 Ω·cm^2. Therefore, the corrosion resistance of U3003 was higher than that of AA3003 in FZ as well as BM.

The EIS fitting results are presented in Table 5.

Table 5. EIS analysis results for AA3003 and U3003.

Alloy		R_s ($\Omega \cdot cm^2$)	C_{film} (F/cm^2)	R_{film} ($\Omega \cdot cm^2$)	C_{dl} (F/cm^2)	R_{ct} ($\Omega \cdot cm^2$)	i_{corr} ($\mu A/cm^2$)
AA3003	BM	144.12	7.39×10^{-6}	9840	1.7×10^{-3}	7908	0.48
	FZ	86.2	10.3×10^{-6}	404	0.01×10^{-3}	1846	3.80
U3003	BM	181.62	6.6×10^{-6}	17,622	0.3×10^{-3}	11,070	0.30
	FZ	117.42	11.2×10^{-6}	993	0.6×10^{-3}	11,000	0.71

The corrosion rate (i_{corr}) and polarization resistance (R_p) are inversely proportional, as shown in Equation (1) [53],

$$i_{corr} = RT/zFR_p \tag{1}$$

where R represents the gas constant (8.314 J/mol·K), T represents temperature, F represents Faraday's constant (96,500 C/mol), and z represents the number of equivalents exchanged. R_p represents the polarization resistance produced by the sum of R_{film} and R_{ct} because two resistors connected in series [54]. As in the immersion test, the corrosion rate of FZ was faster than that of BM in both alloys. In the case of AA3003, the corrosion rate of FZ (3.80 $\mu A/cm^2$) increased by 7.9 times compared to that of BM (0.48 $\mu A/cm^2$), and in the case of U3003, the corrosion rate of the FZ (0.71 $\mu A/cm^2$) increased by 2.4 times compared to that of BM (0.30 $\mu A/cm^2$).

In both alloys, the R_{film} of FZ was lower than that of BM, and the C_{film} was higher. This means that the thickness of the passive film of FZ became thinner and unstable [55,56]. According to Equations (2) and (3), the thickness of the passive film (d_{film}) is proportional to R_{film} and inversely proportional to C_{film} [48,57].

$$d_{film} = R_{film}A/\rho \tag{2}$$

$$d_{film} = \varepsilon\varepsilon_0 A/C_{film} \tag{3}$$

where A represents the surface area, ρ represents the specific resistivity of the passive layer, ε represents the permittivity of passive film, and ε_0 represents the permittivity of the vacuum (8.85×10^{-12} F/m). After laser welding, the IMCs were evenly distributed on the surface of the FZ, making the passive film thin and unstable [58–60]. As shown in Figure 8d, the IMCs appeared in the passive film, which acted as defects. Consequently, the weakening of the passive film induced an increase of corrosion rate of the FZ of both alloys.

The summary of the corrosion mechanism of laser welded Al–Mn alloys is schematically shown in Figure 10.

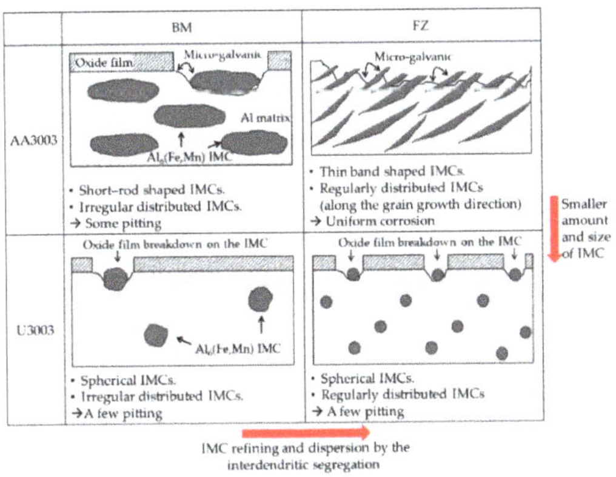

Figure 10. Schematic mechanism of corrosion behavior of laser-welded Al–Mn alloys.

After laser welding, the IMCs were refined and dispersed in the FZ due to interdendritic segregation. This dispersion of IMCs increased the area of IMC to the matrix, and destabilized the passive film. As a result, the corrosion rate of the FZ increased in both alloys. However, in comparison to AA3003, U3003 exhibited a smaller IMC size and reduced quantity, along with a lower Mn/Fe ratio. Therefore, U3003 was more resistant to micro-galvanic corrosion between the IMC and the Al matrix. Consequently, even subsequent to the laser-welding process, U3003 demonstrated higher corrosion resistance than AA3003.

4. Conclusions

In this study, the corrosion behavior of a laser-welded Al–Mn–Zr alloy for a heat exchanger was investigated and compared with that of AA3003. According to the immersion test and the EIS test, the corrosion rate of AA3003 was faster than that of U3003 in both FZ and BM. During the laser welding, interdendritic segregation was occurring in the FZ, and the $Al_6(Fe,Mn)$ IMC was refined and spread more evenly. This change in IMC distribution destabilized the passive film and increased the micro-galvanic corrosion rate. However, the addition of Zr suppressed the precipitates of IMCs, and the remaining Al_3Zr made the IMC spherical in the FZ. Therefore, the IMC area ratio to the Al matrix did not change significantly and the effect of micro-galvanic corrosion was negligible in U3003. Consequently, U3003 had higher corrosion resistance than AA3003 even after laser welding. Therefore, using U3003 as a material for fins and tubes can improve the corrosion resistance of laser-welded joints of the fins and tubes in heat exchangers.

Author Contributions: Conceptualization, J.-M.L. and Y.-S.S.; methodology, J.-M.L.; validation, Y.-S.S.; investigation, J.-M.L.; writing—original draft preparation, J.-M.L.; writing—review and editing, J.-G.K.; supervision, J.-G.K. All authors have read and agreed to the published version of the manuscript.

Funding: This research was supported by the Sungkyunkwan University and the BK21 FOUR (Graduate School Innovation) funded by the Ministry of Education (MOE, Korea) and National Research Foundation of Korea (NRF).

Institutional Review Board Statement: Not applicable.

Informed Consent Statement: Not applicable.

Data Availability Statement: Data is contained within the article material.

Acknowledgments: This research was supported by the Sungkyunkwan University and the BK21 FOUR (Graduate School Innovation) funded by the Ministry of Education (MOE, Korea) and National Research Foundation of Korea (NRF).

Conflicts of Interest: The authors declare no conflict of interest.

References

1. Sinha, A.; Chattopadhyay, H.; Iyengar, A.K.; Biswas, G. Enhancement of heat transfer in a fin-tube heat exchanger using rectangular winglet type vortex generators. *Int. J. Heat Mass Transf.* **2016**, *101*, 667–681. [CrossRef]
2. Jaradeh, M.M.R.; Carlberg, T. Solidification studies of 3003 aluminium alloys with Cu and Zr additions. *J. Mater. Sci. Technol.* **2011**, *27*, 615–627. [CrossRef]
3. Zaffaroni, G.G.B.; Gudla, V.C.; Ambat, R.; Mangelsdorf, A. Microstructural and corrosion issues of embossed and welded aluminium heat exchanger tubes. In Proceedings of the EUROCORR 2017, European Corrosion Congress, Prague, Czech Republic, 3–7 September 2017.
4. Gagliardi, A.; Lanzutti, A.; Simonato, M.; Furlanetto, R.; Magnan, M.; Andreatta, F.; Fedrizzi, L. Failure analysis of a plate heat exchanger used in a blast chiller. *Eng. Fail. Anal.* **2018**, *92*, 289–300. [CrossRef]
5. Hagiwara, M.; Baba, Y.; Tanabe, Z.; Miura, T.; Hasegawa, Y.; Iijima, K. Development of corrosion resistant aluminum heat exchanger, Part 1: Development of new aluminum alloy sheets for sacrificial anode. *SAE Trans.* **1986**, *95*, 477–481.
6. Humpston, G.; Sangha, S.; Jacobson, D. New filler metals and process for fluxless brazing of aluminium engineering alloys. *Mater. Sci. Technol.* **1995**, *11*, 1161–1168. [CrossRef]
7. Mathers, G. *The Welding of Aluminium and Its Alloys*; Elsevier: Cambridge, UK, 2002; pp. 150–155.
8. Zhao, H.; White, D.R.; DebRoy, T. Current issues and problems in laser welding of automotive aluminium alloys. *Int. Mater. Rev.* **1999**, *44*, 238–266. [CrossRef]

9. Coelho, B.N.; Lima, M.S.F.D.; Carvalho, S.M.D.; Costa, A.R.D. A comparative study of the heat input during laser welding of aeronautical aluminum alloy AA6013-T4. *J. Aerosp. Technol. Manag.* **2018**, *10*, e2918. [CrossRef]
10. Haboudou, A.; Peyre, P.; Vannes, A.; Peix, G. Reduction of porosity content generated during Nd: YAG laser welding of A356 and AA5083 aluminium alloys. *Mater. Sci. Eng. A* **2003**, *363*, 40–52. [CrossRef]
11. Cao, X.; Wallace, W.; Immarigeon, J.-P.; Poon, C. Research and progress in laser welding of wrought aluminum alloys II. Metallurgical microstructures, defects, and mechanical properties. *Mater. Manuf. Process.* **2003**, *18*, 23–49. [CrossRef]
12. Sheikhi, M.; Ghaini, F.M.; Torkamany, M.; Sabbaghzadeh, J. Characterisation of solidification cracking in pulsed Nd: YAG laser welding of 2024 aluminium alloy. *Sci. Technol. Weld. Join.* **2009**, *14*, 161–165. [CrossRef]
13. Watkins, K.G.; McMahon, M.A.; Steen, W.M. Microstructure and corrosion properties of laser surface processed aluminium alloys: A review. *Mater. Sci. Eng. A* **1997**, *231*, 55–61. [CrossRef]
14. Padovani, C.; Fratini, L.; Squillace, A.; Bellucci, F. Electrochemical analysis on friction stir welded and laser welded 6XXX aluminium alloys T-joints. *Corros. Rev.* **2007**, *25*, 475–489. [CrossRef]
15. Miyagi, M.; Kawahito, Y.; Wang, H.; Kawakami, H.; Shoubu, T.; Tsukamoto, M. X-ray phase contrast observation of solidification and hot crack propagation in laser spot welding of aluminum alloy. *Opt. Express* **2018**, *26*, 22626–22636. [CrossRef] [PubMed]
16. Zhang, X.; Liu, B.; Zhou, X.; Wang, J.; Hashimoto, T.; Luo, C.; Sun, Z.; Tang, Z.; Lu, F. Laser welding introduced segregation and its influence on the corrosion behaviour of Al-Cu-Li alloy. *Corros. Sci.* **2018**, *135*, 177–191. [CrossRef]
17. Zhang, D.Q.; Li, J.; Joo, H.G.; Lee, K.Y. Corrosion properties of Nd: YAG laser–GMA hybrid welded AA6061 Al alloy and its microstructure. *Corros. Sci.* **2009**, *51*, 1399–1404. [CrossRef]
18. Anderson, W.; Stumpf, H. Effects of manganese on the electrode or free corrosion potentials of aluminum. *Corrosion* **1980**, *36*, 212–213. [CrossRef]
19. Zamin, M. The role of Mn in the corrosion behavior of Al-Mn alloys. *Corrosion* **1981**, *37*, 627–632. [CrossRef]
20. Kim, Y.S.; Park, I.J.; An, B.S.; Park, J.G.; Yang, C.W.; Lee, Y.H.; Kim, J.G. Improvement of corrosion penetration resistance for aluminum heat exchanger by alloying zirconium. *Mater. Chem. Phys.* **2020**, *241*, 122275. [CrossRef]
21. Jia, Z.; Hu, G.; Forbord, B.; Solberg, J.K. Effect of homogenization and alloying elements on recrystallization resistance of Al–Zr–Mn alloys. *Mater. Sci. Eng. A* **2007**, *444*, 284–290. [CrossRef]
22. Mikhaylovskaya, A.; Portnoy, V.; Mochugovskiy, A.; Zadorozhnyy, M.Y.; Tabachkova, N.Y.; Golovin, I. Effect of homogenisation treatment on precipitation, recrystallisation and properties of Al–3% Mg–TM alloys (TM = Mn, Cr, Zr). *Mater. Des.* **2016**, *109*, 197–208. [CrossRef]
23. Kutsuna, M.; Kitamura, S.; Shibata, K.; Sakamoto, H.; Tsushima, K. Improvement of the joint performance in laser welding of aluminium alloys. *Weld. World* **2006**, *50*, 22–27. [CrossRef]
24. Dongxia, Y.; Xiaoyan, L.; Dingyong, H.; Hui, H. Effect of minor Er and Zr on microstructure and mechanical properties of Al–Mg–Mn alloy (5083) welded joints. *Mater. Sci. Eng. A* **2013**, *561*, 226–231. [CrossRef]
25. Ramkumar, K.R.; Natarajan, S. Investigations on microstructure and mechanical properties of TiO_2 nanoparticles addition in Al 3003 alloy joints by gas tungsten arc welding. *Mater. Sci. Eng. A* **2018**, *727*, 51–60. [CrossRef]
26. *ASTM E407*; Standard Practice for Microetching Metals and Alloys. ASTM International: West Conshohocken, PA, USA, 2016.
27. Hong, M.S.; Park, I.J.; Kim, J.G. Alloying effect of copper concentration on the localized corrosion of aluminum alloy for heat exchanger tube. *Met. Mater. Int.* **2017**, *23*, 708–714. [CrossRef]
28. *ASTM G5*; Standard Reference Test Method for Making Potentiodynamic Anodic Polarization Measurements. ASTM International: West Conshohocken, PA, USA, 2021.
29. Cao, X.; Wallace, W.; Poon, C.; Immarigeon, J.-P. Research and progress in laser welding of wrought aluminum alloys I. Laser welding processes. *Mater. Manuf. Process.* **2003**, *18*, 1–22. [CrossRef]
30. Ion, J.C. Laser beam welding of wrought aluminium alloys. *Sci. Technol. Weld. Join.* **2000**, *5*, 265–276. [CrossRef]
31. Denholm, W.; Esdaile, J.; Siviour, N.; Wilson, B. The nature of the $FeAl_3$ liquid-(FeMn) Al_6 reaction in the Al-Fe-Mn system. *Metall. Trans. A* **1987**, *18*, 393–397. [CrossRef]
32. Yan, F.; Kumar, S.; Mckay, B.; O'Reilly, K. Effect of Mn on Fe containing phase formation in high purity aluminium. *Int. J. Cast Met. Res.* **2014**, *27*, 202–206. [CrossRef]
33. Homonnay, Z.; Vértes, A.; Cziráki, Á.; Oszkó, A.; Menczel, G.; Murgás, L. Mössbauer study of Al_6(Fe, Mn) formation in Al-rich Al−Fe−Mn alloys. *J. Radioanal. Nucl. Chem.* **1990**, *139*, 127–134. [CrossRef]
34. Li, Y.; Zhang, W.; Marthinsen, K. Precipitation crystallography of plate-shaped Al_6 (Mn, Fe) dispersoids in AA5182 alloy. *Acta Mater.* **2012**, *60*, 5963–5974. [CrossRef]
35. Li, J.; Dang, J. A summary of corrosion properties of Al-rich solid solution and secondary phase particles in Al alloys. *Metals* **2017**, *7*, 84. [CrossRef]
36. Pakdil, M.; Çam, G.; Koçak, M.; Erim, S. Microstructural and mechanical characterization of laser beam welded AA6056 Al-alloy. *Mater. Sci. Eng. A* **2011**, *528*, 7350–7356. [CrossRef]
37. Almenara, M.D.; Capace, M.C. Microstructure and mechanical properties of GTAW welded joints of AA6105 aluminum alloy. *Rev. Fac. Ing.* **2016**, *25*, 7–19. [CrossRef]
38. Haghdadi, N.; Phillion, A.B.; Maijer, D.M. Microstructure characterization and thermal analysis of aluminum alloy B206 during solidification. *Metall. Mater. Trans. A* **2015**, *46*, 2073–2081. [CrossRef]

39. Kim, Y.S.; Park, J.G.; An, B.S.; Lee, Y.H.; Yang, C.W.; Kim, J.-G. Investigation of zirconium effect on the corrosion resistance of aluminum alloy using electrochemical methods and numerical simulation in an acidified synthetic sea salt solution. *Materials* **2018**, *11*, 1982. [CrossRef] [PubMed]
40. Aidun, D.; Dean, J. Effect of enhanced convection on the microstructure of Al-Cu-Li welds. *Weld. J.* **1999**, *78*, 349–354.
41. Lu, B.; Li, Y.; Wang, H.; Wang, Y.; Yu, W.; Wang, Z.; Xu, G. Effects of cooling rates on the solidification behavior, microstructural evolution and mechanical properties of Al–Zn–Mg–Cu alloys. *J. Mater. Res. Technol.* **2023**, *22*, 2532–2548. [CrossRef]
42. Deshpande, K.B. Numerical modeling of micro-galvanic corrosion. *Electrochim. Acta* **2011**, *56*, 1737–1745. [CrossRef]
43. Mansfeld, F.; Kenkel, J. Galvanic corrosion of Al alloys—III. The effect of area ratio. *Corros. Sci.* **1975**, *15*, 239–250. [CrossRef]
44. Wei, X.; Zhang, B.; Wu, B.; Wang, Y.; Tian, X.; Yang, L.; Oguzie, E.; Ma, X. Enhanced corrosion resistance by engineering crystallography on metals. *Nat. Commun.* **2022**, *13*, 726. [CrossRef]
45. Terachi, T.; Fujii, K.; Arioka, K. Microstructural characterization of SCC crack tip and oxide film for SUS 316 stainless steel in simulated PWR primary water at 320 °C. *J. Nucl. Sci. Technol.* **2005**, *42*, 225–232. [CrossRef]
46. Liu, Y.; Meng, G.; Cheng, Y. Electronic structure and pitting behavior of 3003 aluminum alloy passivated under various conditions. *Electrochim. Acta* **2009**, *54*, 4155–4163. [CrossRef]
47. Yu, M.; Zhao, X.; Xiong, L.; Xue, B.; Kong, X.; Liu, J.; Li, S. Improvement of corrosion protection of coating system via inhibitor response order. *Coatings* **2018**, *8*, 365. [CrossRef]
48. Hirschorn, B.; Orazem, M.E.; Tribollet, B.; Vivier, V.; Frateur, I.; Musiani, M. Determination of effective capacitance and film thickness from constant-phase-element parameters. *Electrochim. Acta* **2010**, *55*, 6218–6227. [CrossRef]
49. Benoit, M.; Bataillon, C.; Gwinner, B.; Miserque, F.; Orazem, M.E.; Sánchez-Sánchez, C.M.; Tribollet, B.; Vivier, V. Comparison of different methods for measuring the passive film thickness on metals. *Electrochim. Acta* **2016**, *201*, 340–347. [CrossRef]
50. Hernández, H.H.; Reynoso, A.M.R.; González, J.C.T.; Morán, C.O.G.; Hernández, J.G.M.; Ruiz, A.M.; Hernández, J.M.; Cruz, R.O. *Electrochemical Impedance Spectroscopy (EIS): A Review Study of Basic Aspects of the Corrosion Mechanism Applied to Steels*; IntechOpen: London, UK, 2020.
51. Wang, L.; Snihirova, D.; Deng, M.; Vaghefinazari, B.; Höche, D.; Lamaka, S.V.; Zheludkevich, M.L. Revealing physical interpretation of time constants in electrochemical impedance spectra of Mg via Tribo-EIS measurements. *Electrochim. Acta* **2022**, *404*, 139582. [CrossRef]
52. Choi, W.; Shin, H.C.; Kim, J.M.; Choi, J.Y.; Yoon, W.S. Modeling and applications of electrochemical impedance spectroscopy (EIS) for lithium-ion batteries. *J. Electrochem. Sci. Technol.* **2020**, *11*, 1–13. [CrossRef]
53. Saitou, M. Determination of a dimensionless fluctuation coefficient using electrochemical impedance spectroscopy. *Int. J. Electrochem. Sci.* **2019**, *14*, 8028–8038. [CrossRef]
54. Le, D.; Yoo, Y.; Kim, J.; Cho, S.; Son, Y. Corrosion characteristics of polyaniline-coated 316L stainless steel in sulphuric acid containing fluoride. *Corros. Sci.* **2009**, *51*, 330–338. [CrossRef]
55. Ma, M.; Liu, H.; Chen, L. Effect of cerium on the initiation of pitting corrosion of 444-type heat-resistant ferritic stainless steel. *High Temp. Mater. Process.* **2020**, *39*, 576–587. [CrossRef]
56. Cramer, S.D.; Covino, B.S., Jr.; Moosbrugger, C.; Sanders, B.R.; Anton, G.J.; Hrivnak, N.; Kinson, J.; Polakowski, C.; Muldoon, K.; Henry, S.D. *ASM Handbook*; ASM International Materials Park: Geauga County, OH, USA, 2003; pp. 491–492.
57. Mohammadi, F.; Nickchi, T.; Attar, M.; Alfantazi, A. EIS study of potentiostatically formed passive film on 304 stainless steel. *Electrochim. Acta* **2011**, *56*, 8727–8733. [CrossRef]
58. McCafferty, E.; Shafrin, E.G.; McKay, J.A. Microstructural and surface modification of an aluminum alloy by rapid solidification with a pulsed laser. *Surf. Technol.* **1981**, *14*, 219–223. [CrossRef]
59. Fares, C.; Belouchrani, M.; Bellayer, S.; Boukharouba, T.; Britah, A. Influence of intermetallic compounds and metallurgical state of the 2017A aluminum alloy on the morphology of alumina films developed by anodic oxidation. *J. Tribol. Surf. Eng.* **2011**, *2*, 239–251.
60. Cirik, E.; Genel, K. Effect of anodic oxidation on fatigue performance of 7075-T6 alloy. *Surf. Coat. Technol.* **2008**, *202*, 5190–5201. [CrossRef]

materials

Article

Impact Fracture Surfaces as the Indicators of Structural Steel Post-Fire Susceptibility to Brittle Cracking

Mariusz Maslak [1,*], Michal Pazdanowski [2], Marek Stankiewicz [1], Anna Wassilkowska [3], Paulina Zajdel [1] and Michal Zielina [3]

[1] Chair of Bridge, Metal and Timber Structures, Faculty of Civil Engineering, Cracow University of Technology, 31-155 Cracow, Poland; mstankiewicz@pk.edu.pl (M.S.); paulina.zajdel1@pk.edu.pl (P.Z.)
[2] Chair for Computational Engineering, Faculty of Civil Engineering, Cracow University of Technology, 31-155 Cracow, Poland; michal.pazdanowski@pk.edu.pl
[3] Department of Water Supply, Sewerage and Environmental Monitoring, Faculty of Environmental Engineering and Energy, Cracow University of Technology, 31-155 Cracow, Poland; anna.wassilkowska@pk.edu.pl (A.W.); michal.zielina@pk.edu.pl (M.Z.)
* Correspondence: mmaslak@pk.edu.pl; Tel.: +48-126282324

Abstract: The results of experimental research on forecasting post-fire resistance to brittle failure of selected steel grades used in construction are presented and discussed in this paper. The conclusions are based on detailed analysis of fracture surfaces obtained in instrumented Charpy tests. It has been shown that the relationships formulated based on these tests agree well with conclusions drawn based on precise analysis of appropriate *F–s* curves. Furthermore, other relationships between lateral expansion *LE* and energy W_t required to break the sample constitute an additional verification in both qualitative and quantitative terms. These relationships are accompanied here by values of the $SFA_{(n)}$ parameter, which are different, depending on the character of the fracture. Steel grades differing in microstructure have been selected for the detailed analysis, including: *S355J2+N*—representative for materials of ferritic-pearlitic structure, and also stainless steels such as *X20Cr13*—of martensitic structure, *X6CrNiTi18-10*—of austenitic structure and *X2CrNiMoN22-5-3* duplex steel—of austenitic-ferritic structure.

Keywords: steel microstructure; post-fire properties; fracture surface; brittle cracking; impact test; shear fracture appearance; toughness

Citation: Maslak, M.; Pazdanowski, M.; Stankiewicz, M.; Wassilkowska, A.; Zajdel, P.; Zielina, M. Impact Fracture Surfaces as the Indicators of Structural Steel Post-Fire Susceptibility to Brittle Cracking. *Materials* **2023**, *16*, 3281. https://doi.org/10.3390/ma16083281

Academic Editors: Luca Pezzato and Claudio Gennari

Received: 27 March 2023
Revised: 17 April 2023
Accepted: 19 April 2023
Published: 21 April 2023

1. Introduction

In order to reliably evaluate the suitability of structural steel for potential service in the bearing structure after undergoing an episode of rapid heating followed by more or less prolonged holding it at raised temperature in fire conditions and, finally, effective cooling down, one has to verify whether, and if so to what extent, such steel has preserved its plastic properties, conditioned by sufficient ductility resulting in post-fire resistance to brittle cracking [1]. Subjecting any material to unintentional action of relatively high temperature of uncontrolled time profile always results in structural changes of weakening and sometimes even degrading character [2,3]. Changes of this kind usually are permanent, so they remain in the material after its complete cool down [4,5]. However, in many situations those changes may remain unnoticed, especially when post-fire technical condition evaluation of the building object is limited to only visual inspection of its bearing structure accompanied by often very cursory inventory of observed deformations. In the Authors' opinion, the resistance of structural steel to brittle cracking is of key importance here. In particular, it is not so much about the a posteriori observation of the complete lack of susceptibility to initiate brittle cracks in the tested material, but about the observed post-fire loss of the capability to effectively arrest the unrestrained growth of such cracks.

This paper constitutes a continuation of previous works published earlier in [6–8]. In our opinion, it has been shown there that a typical set of post-fire tests pertaining to the determination of mechanical properties of structural steel after a fire incident should be mandatorily broadened by adding impact tests. If possible, these tests should be instrumented [9–13], with a hammer of sufficiently high potential energy, using appropriate data transmission and processing following the recommendations of the codes [14–17]. The applicability of these tests to forecasting structural steel resistance to brittle cracking has been proven many times (for example in [18]), though most of such tests did not refer to material cooled down after undergoing a fire incident. However, the results obtained allowed for reliable verification of the risk of initiating new cracks in the material, and then for assessing the degree of limiting their possible propagation [19–22]. Mechanisms of ductile failure under load have been recognized and identified as a result of such research. This seems to be crucial for the deliberations presented in this paper [23–28]. The quantitatively significant risk of brittle failure in prolonged use revealed during research in our opinion should constitute an important factor in the decision on disqualifying analyzed steel from the extended service, or at least on restricting the user-accepted scope and conditions of such service, as this brittle failure in structural steel components usually occurs abruptly, with no previous signs of progressive material degradation over time.

In the deliberations presented in this paper, we will focus on the description and interpretation of various impact fracture types obtained during our research on several steel grades. In our opinion, the morphology of such fractures, observed in practice, seems to clearly indicate the forecast post-fire susceptibility of a given material to brittle fracture under external load. The research on correlations of this type, related to structural steels effectively cooled down after an action of fire temperature, is, so far, relatively infrequent [29–35]. The interest of researchers seems to have been concentrated on registration of permanent changes in mechanical properties of tested steel grades, in particular their yield limits, ultimate strength, and modules of linear elasticity. Detailed insight into the mechanisms affecting the brittle or partially plastic form of initiating and propagating cracks observed after a fire incident, correlated with verification of the fracture surface obtained, has to be unequivocally referred to the conditions accompanying fracture development, as these mechanisms may be different when accompanying quasi static tensile test, and different in the case of the dynamic Charpy impact test.

Several grades representative of various types of structural steels have been selected for detailed analysis. These steel grades differ not only in scope of application but also in chemical composition, and thus in internal microstructure. It has been acknowledged that those characteristics should determine the expected mechanical properties identified on these steels after cooling down in fire episodes. A conventional low alloy *S355J2+N* steel [36], representative for a whole group of steels exhibiting ferritic-pearlitic structure (including among others the Chinese *X80* steel [37–39], and also many high strength steels [40]) was selected at the beginning. The results obtained for this steel were juxtaposed with the results characterizing the post-fire properties of selected stainless steels such as *X20Cr13* steel—of martensitic structure [41], *X6CrNiTi18-10* steel—of austenitic structure, and *X2CrNiMoN22-5-3* duplex steel—of austenitic-ferritic structure [42–44].

2. Preparation of Samples

Prior to the impact tests, the samples were subjected to thermal treatment of a formalized course, corresponding to the steady-state heating regime. This treatment simulated the action of fully developed fire on the sample. During the first phase, the samples were heated at a constant speed of 100 °C/min up to 600 °C (first series) or 800 °C (second series), then kept for 60 min at this temperature (Figure 1). After heating, the samples were cooled down to ambient temperature. For comparative purposes, two cooling scenarios were applied, namely slow cooling in the furnace to simulate the self-extinguishing of a fire and rapid cooling in water mist to simulate the fire extinguishing action of a fire brigade.

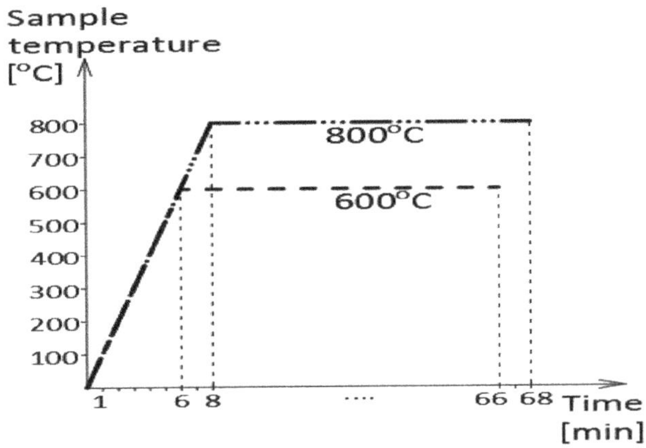

Figure 1. The initial thermal treatment of samples simulating the fire action (according to [7]).

The heating temperature levels indicated above were selected intentionally, as it was preliminarily assumed that the first one during 60 min long exposure time would prove to be too low, while the second one would be sufficiently high, to initiate structural changes of permanent character in the tested steels.

The impact toughness tests were conducted at +20 °C to simulate summer conditions and at −20 °C to simulate winter conditions of the post-fire service of the tested steels.

Thus, the research encompassed 40 quantitatively different cases. For each of the considered steel grades and each of both toughness test temperature levels, four independent test cases related to the steel cooled down after a simulated fire episode (two heating temperature levels multiplied by two different cooling scenarios), and one so-called reference case for the steel in as-delivered condition (i.e., unaffected by the simulated fire action) was considered. Each test case was statistically analyzed on six independent steel samples to assure sufficient reliability of the obtained estimates. This means that $40 \times 6 = 240$ independent impact strength tests were conducted and interpreted in detail.

The final results were archived using a three-digit key to distinguish separate cases as shown in Table 1. The cases denoted by single digits, i.e., 1, 2, 3, and 4, respectively, refer in this key to the six element set of the so-called reference cases, obtained after testing the samples made of the material remaining in as-delivered condition (i.e., the material which was unaffected by the simulated fire action)

Table 1. Description mode of the samples subjected to impact strength tests.

First Digit—Steel Grade	Second Digit—Heating Temperature	Third Digit—Cooling Mode	Additional Information—Testing Temperature
1—*S355J2+N* 2—*X20Cr13* 3—*X6CrNiTi18-10* 4—*X2CrNiMoN22-5-3*	6—600 °C 8—800 °C	0—slow cooling in the furnace 1—fast cooling in water mist	**(+20)**—+20 °C **(−20)**—−20 °C

The juxtaposition of all the tested samples, with their descriptions, is depicted in Figure 2 (only the top layer of a three-tier set is visible there). The noteworthy difference in colors of the scale appearing on the surface of individual groups is induced by the differences in chemical composition of the corresponding steel grades. This scale observed on the surface of samples made of *S355J2+N* steel was only loosely bound to the substrate, while on all the stainless steel grades, it formed a tight surface layer.

Figure 2. Samples prepared for impact strength testing, made of *S355J2+N* steel (1st set), *X20Cr13* steel (2nd set), *X6CrNiTi18-10* steel (3rd set), and *X2CrNiMoN22-5-3* steel (4th set), from left to right.

3. Description of the Impact Strength Tests Conducted

The impact strength tests (Figure 3) of the samples were conducted following the recommendations of the codes [14–17] using instrumented Charpy hammer of the JB-W450E-L type (Figure 4a), of 450 J potential energy. The R8 beater (of American type) was selected, under the assumption that in construction, an impact by an object of more flattened surface is more probable (as compared to the classic R2 beater of European type) [45,46]. The hammer was equipped with a transducer to measure the force applied to the sample, and the accompanying force application point displacement was registered automatically with an encoder. Signals generated by both devices were gathered and processed by a data logger of high sampling frequency and subsequently analyzed by dedicated computer program. The results of each test were illustrated on automatically generated graphs, showing force, breaking energy, and displacement of force application point as functions of time, or alternatively force and breaking energy as functions of force application point displacement. A computer program automatically marked on these graphs the location of characteristic limit points as well.

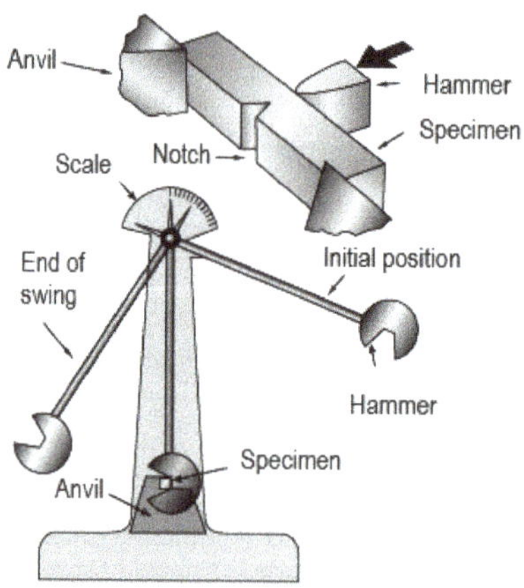

Figure 3. Principle of executing Charpy impact strength test.

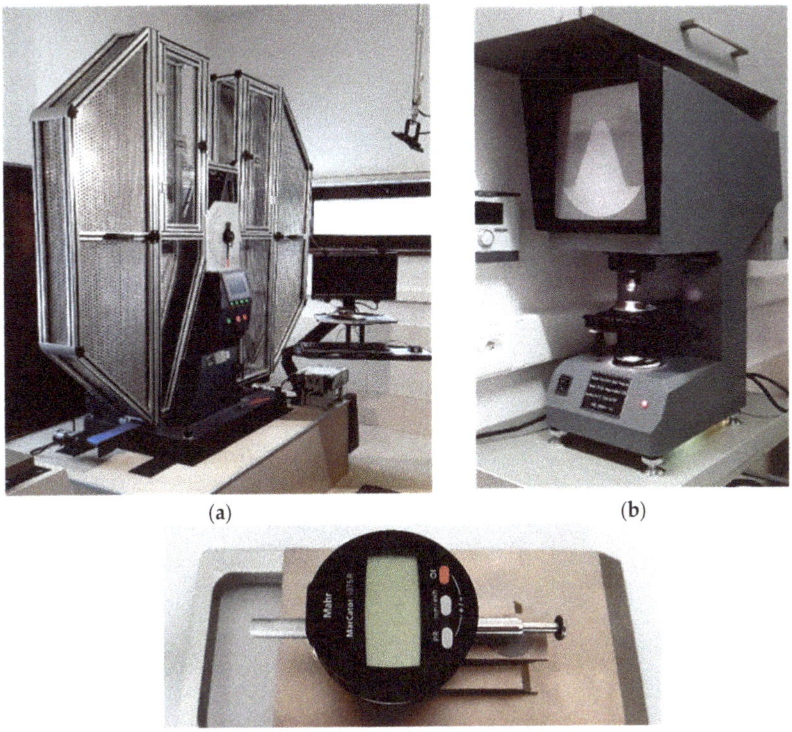

(**a**) (**b**)

(**c**)

Figure 4. Testing equipment used during research: (**a**) instrumented Charpy hammer of the JB-W450E-L type, (**b**) XT-50 notch profile projector, (**c**) the gauge used to measure the lateral expansion of a sample (according to [7]).

For all test cases referring to the *S355J2+N* steel, as well as for those referring to stainless steel grades *X20Cr13* and *X6CrNiTi18-10*, full size impact test samples (ISO Charpy V-10) were used. However, in the cases pertaining to the *X2CrNiMoN22-5-3* stainless steel, the sample size had to be changed to ISO Charpy V-7.5 [47], as the energy of 450 J proved to be insufficient to break the full size sample in its as-delivered condition or a full size sample heated for one hour at 600 °C and subsequently cooled in water mist.

The notch profile testing of reference samples (i.e., of those made of steel in the as-delivered condition, unaffected by thermal treatment simulating the fire action) was conducted on the notch profile projector XT-50 (Figure 4b). It has been shown that these notches were located just below the upper tolerance limit specified in the code [14]. Analogous verification of samples cooled down after prior exposure to fire temperature showed an increase in the size of notches caused by the removal of scale developed as a result of high temperature oxidation resulting in several cases in notches exceeding the size tolerance limits prescribed by the codes. According to the provisions of the code [17], discrepancies of this type may occur in the impact strength tests, resulting in the brittle fracture of a sample overestimating the experimentally determined breaking energy by only 2 ÷ 3 J. However, should the ductile fracture be observed, this overestimation would be negligible. Therefore, in the following considerations, it was assumed that the measurement error induced by this factor would not be taken into account when dealing with post-fire susceptibility of considered steel grade to brittle fracture.

4. Parameters of the Observed Impact Fractures Used to Draw Conclusions

A *F–s* force–displacement graph obtained for a given fracture was used as a basis for the following considerations. In particular, these curves were qualified as category *E* (Figure 5a) or as category *F* (Figure 5b) according to the classification listed in the code [15]. This corresponds to the categories *B* and *C* listed in the code [16], respectively.

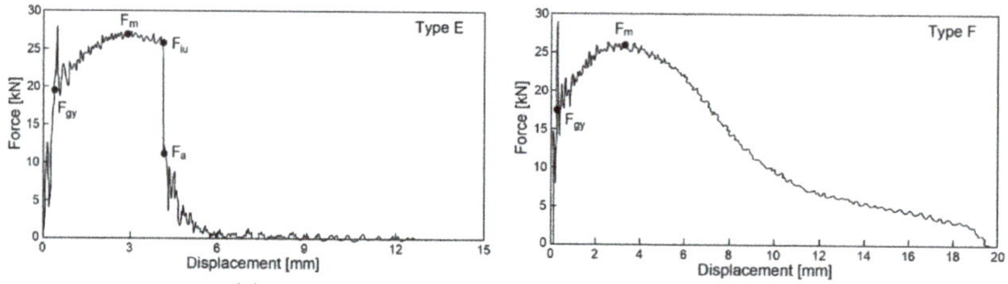

(a) (b)

Figure 5. The force–displacement of force application point graphs as characteristics of typical impact strength fractures including: (**a**) category *E* fracture type, (**b**) category *F* fracture type. Qualification following the recommendations of [15].

The general yield force F_{gy} on these graphs corresponds to the initiation of yield at the developing fracture, while maximum force F_m indicates the global maximum on the *F–s* curve. Unstable growth of the fracture is initiated at F_{in} (force at the initiation of unstable crack propagation). This phase of the process is correlated with abrupt reduction of the applied force accompanied by minimal increase in the displacement of force application point. It is concluded when displacement corresponding to the force F_a (force at the arrest of unstable crack propagation) is reached. At this moment, the effective fracture arrest begins and the sample undergoes plastic fracture. Let us note that the shape of the *F–s* graph in Figure 4b does not reveal the force F_{in}. This means that the graphs of this type, qualified as category *F* (according to [15]), are to be correlated with a fracture exhibiting only completely stable phases. This means, in turn, that self-arrest of the fracture is fully effective. In the case of a *F–s* relationship belonging to the type *E* (according to [15]), as shown in Figure 5a, the capability of the sample to self-arrest an unstable growth of a fracture initiated during the preceding stage is higher the bigger the area under the experimentally obtained curve, provided that such an area is determined only for the displacement larger than the one associated with the limit force F_a. This results in a relatively bigger area of the plastic breakthrough observed on the surface of the fracture. Let us note, also, that when $F_m = F_{in}$ holds, then the given *F–s* relationship is qualified as belonging to the category *C* or *D* in the sense of recommendations contained in [15], according to the character of the final phase of the fracturing process, determined by the way the force F_a is revealed.

In the previous paper [7], we showed how the post-fire brittleness of the material is assessed based on the shape of the *F–s* curve obtained during the impact strength test under the assumption of specific (summer or winter) weather conditions. The evaluated pattern has to be related to the analogous pattern associated with the limit value of the fracturing energy $W_{t,min}$ = 27 J, for which the ductile-to-brittle transition temperature (DBTT) is defined. The steels, for which at a given temperature the experimentally obtained fracturing energy $W_t < W_{t,min}$, would exhibit a tendency to dominantly brittle fractures. Thus, these steels would be incapable of effectively arresting micro-cracks initiated in their microstructure and growing in time. Therefore, such steels may not be recommended for further application in the construction industry. Only the steels for which the relation $W_t > W_{t,min}$ is proven exhibit the capability to self-arrest the micro-crack growth [48]. In each of the tested cases, the surface limited from above by the experimentally obtained *F-s* curve and from below by the horizontal axis of the graph represents the energy W_t required to fracture the sample.

When the experimentally obtained *F–s* curve does not show the steeply sloped part, as for instance happens in the case of the sample classified in the category *F* (Figure 5b), it means in general that the share of plastic fracture area is equal to 100% of the total fracture area. However, should such a segment be identified for a given fracture, then the approximate value of the ratio of ductile fracture surface may be estimated by applying one of the following formulae [15,49,50]:

$$SFA_{(1)} = \left[1 - \frac{F_{in} - F_a}{F_m}\right] \cdot 100\% \quad (1)$$

$$SFA_{(2)} = \left[1 - \frac{F_{in} - F_a}{F_m + (F_m - F_{gy})}\right] \cdot 100\% \quad (2)$$

$$SFA_{(3)} = \left[1 - \frac{F_{in} - F_a}{F_m + 0.5 \cdot (F_m - F_{gy})}\right] \cdot 100\% \quad (3)$$

$$SFA_{(4)} = \left[1 - \sqrt{\frac{\frac{F_{gy}}{F_m} + 2}{3}\left(\frac{\sqrt{F_{in}}}{\sqrt{F_m}} - \frac{\sqrt{F_a}}{\sqrt{F_m}}\right)}\right] \cdot 100\% \quad (4)$$

The abbreviation $SFA_{(n)}$ (%) denotes here the shear fracture appearance. The higher its value, the more resistant the considered material is to brittle failure.

The lateral expansion *LE* (mm) was measured on each sample using the gauge depicted in Figure 4c. The idea of this measurement is depicted in Figure 6. As shown there, this expansion occurs only on at least partially ductile fractures. Polynomial regression formulae relating this parameter to the corresponding sample fracturing energy are commonly known [51,52]. These formulae indicate that in the case of ferritic steels, the capability of the material to absorb dynamic energy, here expressed as the fracturing energy W_t, may not be extended indefinitely. At high impact energy, a significant reduction of ductility measured via the *LE* parameter occurs for such steels. However, in general, a higher *LE* value, at the fracture energy values typical for the structural steels, should be associated with higher resistance to brittle failure exhibited by the test samples. Thus, the *LE* parameter in this sense represents the highest relative increase in sample width after brittle failure, determined in the cross section directly hit by the hammer (annex B to the code [14]). In this cross section, the freedom to realize plastic deformations resulting in permanent deformation of its initial rectangular shape at coincident propagation of the fracture is the highest. Detailed analysis of these deformations exhibited by samples cooled down after exposure to fire temperature and later subjected to impact fracture tests constitutes one of basic premises leading to conclusions regarding post-fire impact strength of a given steel grade. In order to do so, one has to identify particular zones associated with subsequent phases of the fracture on the typical fracture surface of mixed ductile (meaning partly plastic and partly brittle) character. These zones are depicted in Figure 7. As may be seen, the fracture is initiated directly under the notch, in spite of the fact that the hammer hits the other side of the sample, since the local stress concentration zone develops directly adjacent to the notch. On the side surfaces of the sample, the number of the degrees of freedom in strains is higher than in the center, and thus in those areas, plastic fracture may occur. The central part of the sample is dominated by an unstable brittle fracture. This manifests itself by an area of laminated fracture in the center of the sample.

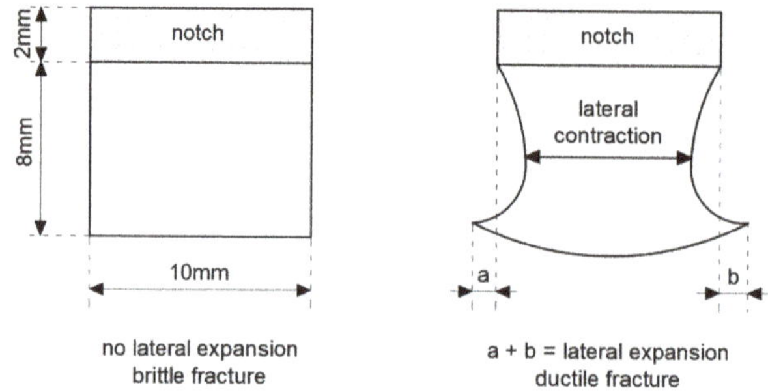

Figure 6. The *LE* (mm) lateral expansion parameter measurement concept.

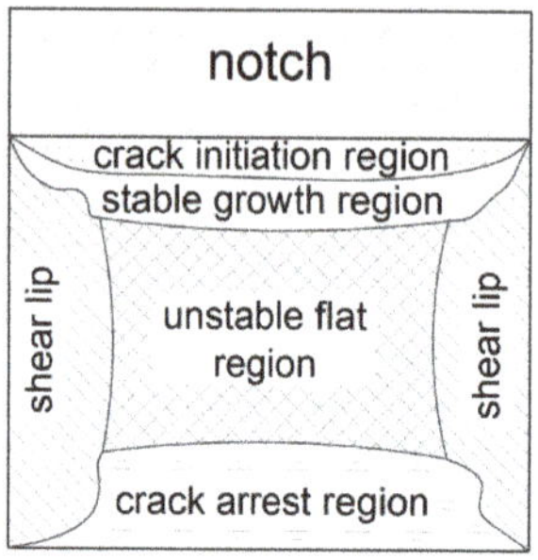

Figure 7. Fracture development stages in an ISO Charpy V sample on the typical fracture surface of mixed ductile character.

Visual observation of the fracture of mixed type allows for simplified determination of the $SFA_{(n)}$ parameter (annex C to the code [14]), yielding the value of $SFA_{(5)}$ (%) as a result. One has to determine the percentile ratio of the laminated fracture area of brittle character (the so-called flat fracture region), indicated by the dimensions A and B in Figure 8, with respect to the full cross section area. The shear area of the fracture surface will therefore complement this value to 100%.

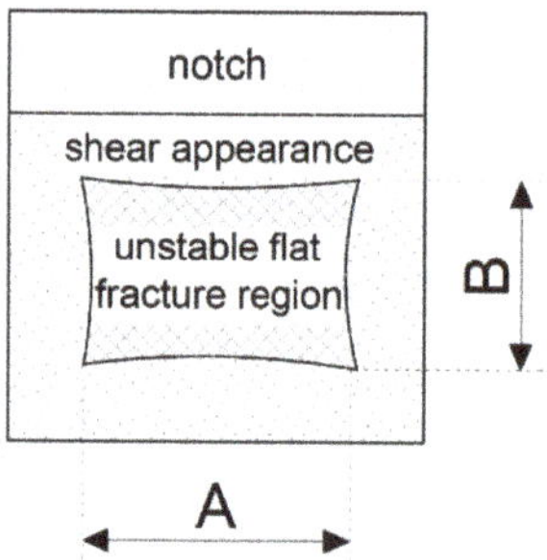

Figure 8. Flat fracture region with partially cleavage cracked zone (partially brittle) and surrounding shear generated fracture area.

More accurate methods used to determine this parameter, based on visual observation of the fracture area, apply advanced image processing approaches [53,54].

5. Microstructural Studies of Impact Fractures

The microstructure of each impact fracture surface obtained in our experiment was analyzed using a scanning electron microscope (SEM) Hitachi S-3400N VP-SEM (Figure 9).

This device may be set up to work under both high vacuum conditions and variable pressure. It typically yields magnifications of 10 to 20,000 times, with a very good surface contrast and depth of field for resolutions of 4 to 10 nm. To perform the chemical analysis, the microscope described above was equipped with an energy dispersive spectroscopy (EDS) *ThermoScientific* Si(Li) detector with a resolution of 135 eV and with the Noran System 7 (NSS) analytical software.

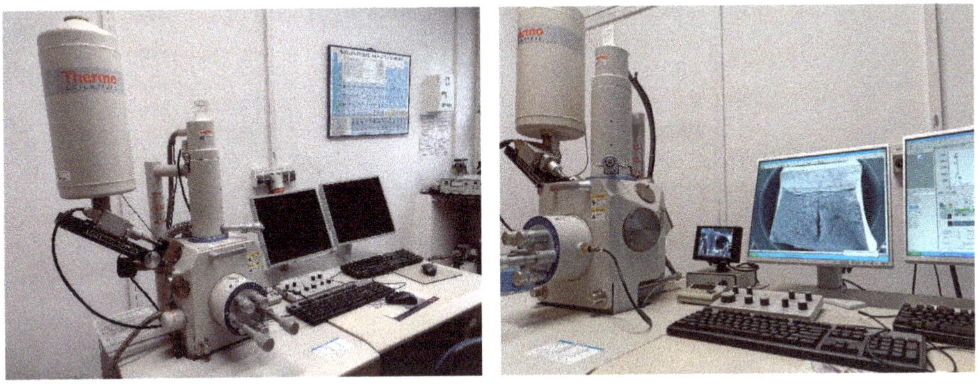

Figure 9. Scanning electron microscope (SEM) used to analyze the microstructure of considered steel.

The SEM works on the principle of generating an electron beam in the thermo-emission gun. This beam, while passing through the electromagnetic lens system, hits the tested sample placed in a vacuum chamber. Depending on the accelerating voltage of the primary beam (from 0.3 kV to maximum 30 kV), the electrons bombarding the sample surface penetrate the tested material to a depth of several micrometers. Secondary (SE) and backscattered (BSE) electrons, as well as characteristic X-rays, are then emitted from the excited volume of a sample. Those signals are used to image the sample surface topography (SE detector), to reveal the contrast of the atomic number of the tested sample (BSE detector), and to identify the micro-sample elemental composition (EDS detector) [55,56].

Detailed results pertaining to the identification and qualitative interpretation of permanent changes observed by the authors of this paper in the microstructure of considered steel grades after action of simulated fire episodes following various development scenarios have been published and discussed in paper [8]. Therefore, these results will be called upon in the following deliberations only in the scope required for proper commenting of the results referring to the same steels, but obtained during the research of their post-fire impact strength, reported below.

In energy dispersive spectroscopy (EDS), determination of the qualitative and then quantitative composition is performed using X-ray excitation parameters adequate to the elements expected in the composition of the sample [57]. In the semi-quantitative EDS, analysis element concentrations are normalized to 100%, seemingly yielding an impression of error-free analysis. Correction of quantitative ratios is executed automatically by application of specialized Noran System 7 software. In particular, the correction by the *Z* factor results from the influence of the atomic number on the X-ray excitation efficiency, the correction by the *A* factor takes into account the probability of X-ray absorption by the sample, the correction by the *F* factor is associated with the possibility of additional emission of energy quantums induced by characteristic and continuous X-rays. It is difficult to determine the percentile concentration of light elements in a sample using the EDS technique. Overlapping of peaks, a result of low spectral resolution, may be counted as an additional disadvantage of EDS. In general, the accuracy of a measurement depends on the mass concentration of heavy elements in the emitting zone whereas the minimal detection limit amounts to 0.1% by weight. Depending on the accelerating voltage and the material type the X-ray emitting zone is in the order of 2–5 μm. This means that the micro-volume of the analyzed specimen is derived only from the subsurface layers of the material. Due to the limitations listed above the chemical composition of the samples tested in the experiment described here was determined independently, in a manner alternative to the classical EDS, and in our opinion yielding more reliable results. The optical emission spectrometer (OES) was applied for this purpose [8].

6. Detailed Results of the Tests Conducted and Their Interpretation

6.1. Results Obtained on Samples Made of S355J2+N Steel

The *S355J2+N* steel is a typical structural low carbon manganese steel of ferritic-pearlitic microstructure. It is weldable and exhibits good impact strength (even at sub-zero temperature values). The material used for the tests underwent normalization. Detailed metallographic examination showed that the structure of this steel contained elongated inclusions of manganese sulfides [8].

The chemical composition of the tested samples made of *S355J2+N* steel was identified with the Foundry–Master optical emission spectrometer (Worldwide Analytical Systems, Uedem, Nordrhein-Westfalen, Germany) and is listed in the Table 2 (according to [8]).

Table 2. Chemical composition of the tested samples made of the *S355J2+N* steel (according to [8]).

Element	C	Si	Mn	P	S	Cr	Ni	Mo
Content by wt. %	0.185	0.164	1.44	0.0066	0.003	0.0298	<0.005	<0.005
Element	**Ti**	**Cu**	**Al**	**Co**	**Nb**	**V**	**W**	
Content by wt. %	<0.002	0.0861	0.0349	0.0032	<0.002	<0.002	<0.015	

Figures 10 and 11 depict impact fractures obtained during our research. Figure 10 refers to the tests conducted at +20 °C, while Figure 11 refers to the analogous test conducted at −20 °C. The zones of stable fracture growth and zones of plastic deformation appearing mainly at the side edges of the tested sample, usually referred to as the plastic lips (Figure 7), are indicated on each picture. Each photo of a presented impact fracture is accompanied by a corresponding *F–s* curve with representative values, indicating the fracture mechanism occurring during the experiment (Figure 5a,b). These values should be interpreted as average values estimated on homogeneous six-element test sample. Each representative average value of the random variable F (kN) (or random forces F_{gy} and F_m, respectively) is accompanied by the averaged displacement of the force application point s (mm). In addition, empirically estimated coefficients of variation, namely the ν_F coefficient—a variation measure of the random variable F (measured along the direction parallel to the vertical axis of the *F–s* graph), and the ν_s coefficient—a variation measure of the random displacement s induced by this force (indicated along the direction parallel to the horizontal axis of this graph) are shown as well. Thus, all the graphs depicted in red are the averaged ones. These graphs fit within the bounds drawn in black and determined as the average value decreased and increased by one standard deviation computed for the random force F (estimated on the statistical sample). The ranges in green, drawn at representative values of the force F indicated in every picture (along the vertical direction) and accompanying displacement s (along the horizontal direction) represent the measure of their random variability determined at the level of a single standard deviation calculated with respect to the proper average value. Sample numbering conforms to the key listed in Table 1.

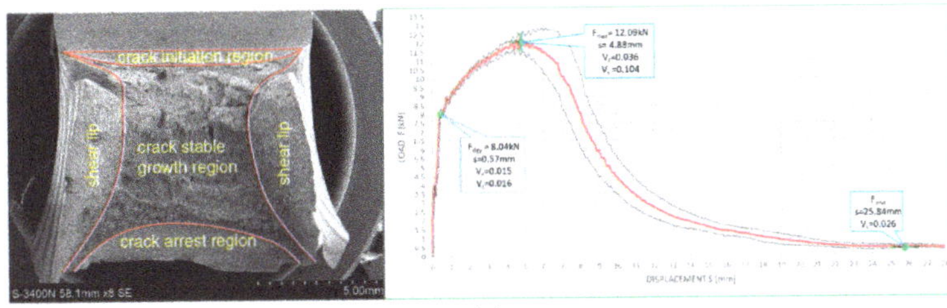

(a)

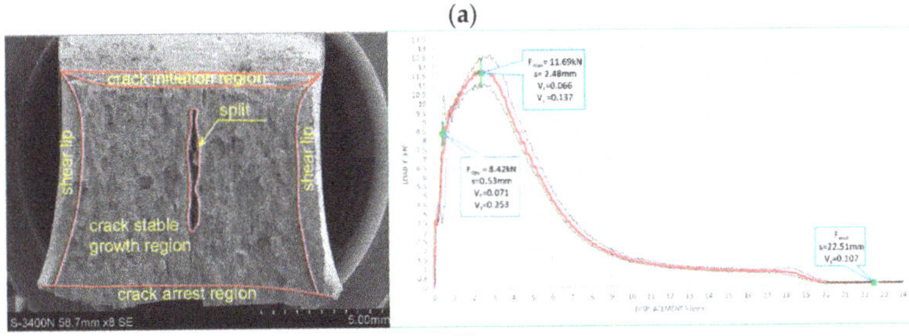

(b)

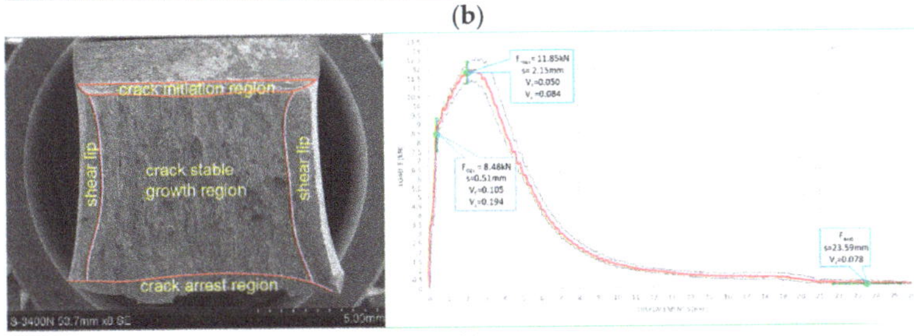

(c)

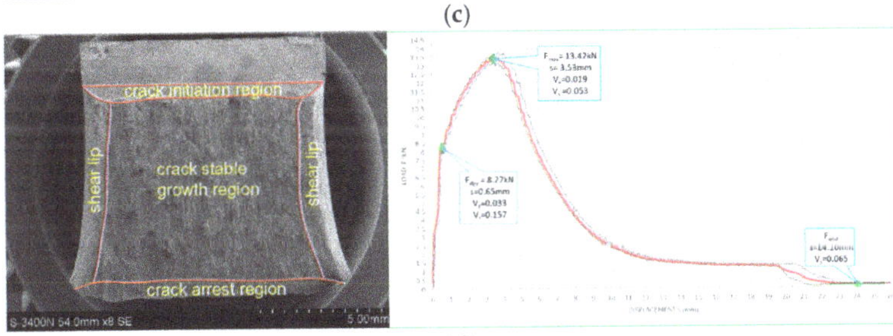

(d)

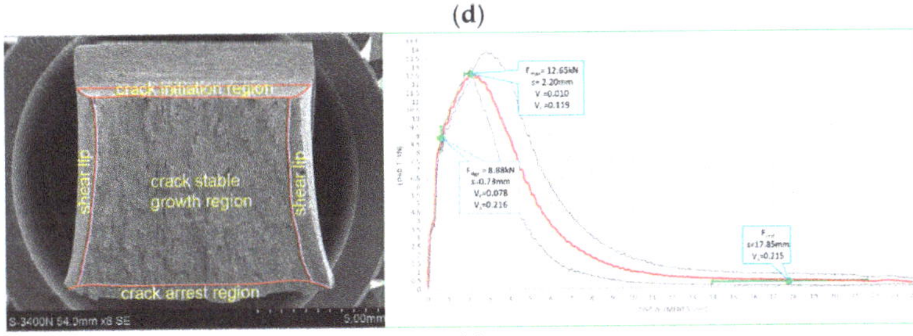

(e)

Figure 10. Morphology of the impact fractures obtained during the tests conducted at +20 °C on the samples made of *S355J2+N* steel—denotations follow the key listed in Table 1. (**a**) *1 (+20)*; (**b**) *160 (+20)*; (**c**) *161 (+20)*; (**d**) *180 (+20)*; (**e**) *181 (+20)*.

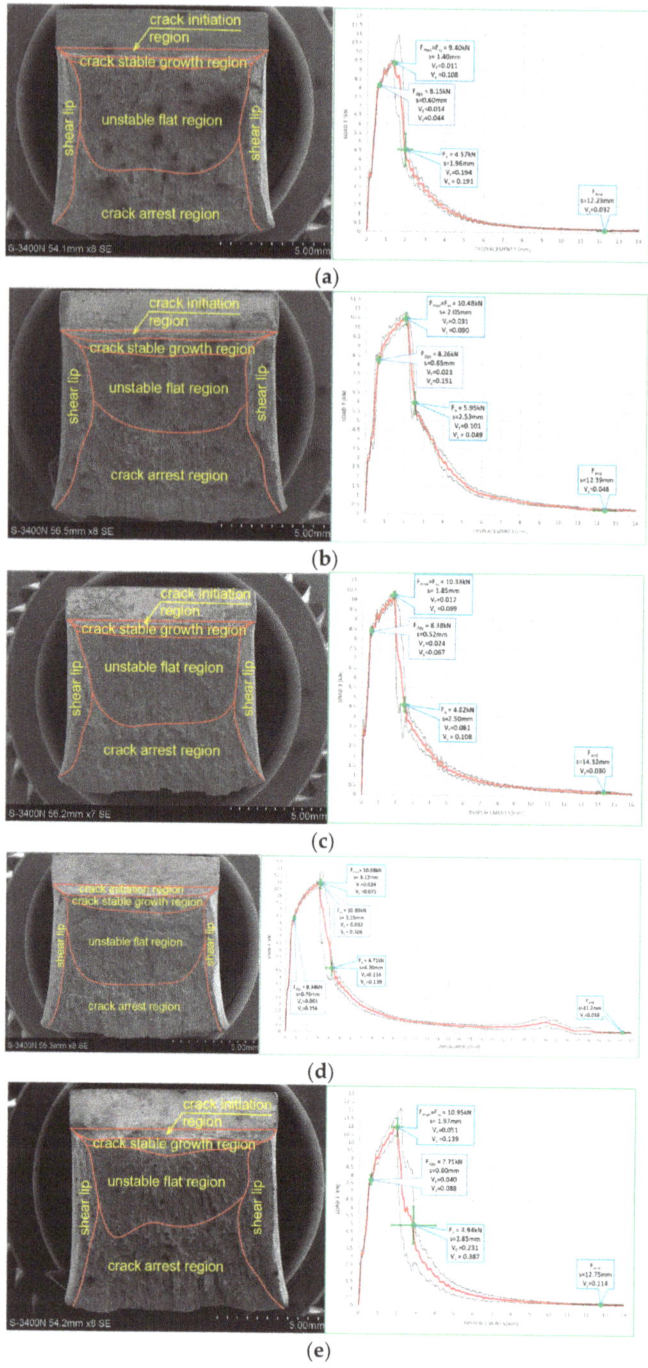

Figure 11. Morphology of the impact fractures obtained during the tests conducted at −20 °C on the samples made of *S355J2+N* steel—denotations follow the key listed in Table 1. (**a**) *1 (−20)*; (**b**) *160 (−20)*; (**c**) *161 (−20)*; (**d**) *180 (−20)*; (**e**) *181 (−20)*.

One may easily notice that in all the cases depicted in Figure 10, the obtained impact fractures exhibit a dominant stable fracture growth zone. This result is confirmed by the accompanying *F–s* curves, which may be assigned to the category *F* (Figure 5b) in each of the considered cases (according to [15]). The force F_{iu} does not manifest itself clearly on any of the graphs. This means that all the stages of the fracture were completely stable, and therefore self-arresting of uncontrolled micro-fractures growing in the affected material was fully successful. Referring to this steel grade, typical for the material of ferritic-pearlitic structure, the steel cooled down after exposure to the simulated fire incident exhibited slightly worse plastic properties than the same steel unaffected by the action of simulated fire. The *F–s* curve related to the sample denoted by *1 (+20)* (Figure 10a) proved to be significantly more "developed in the horizontal direction" than all the remaining curves depicted in a sequence in Figure 10b–e. This is to be associated with significantly higher energy W_t needed to break the sample, and therefore significantly better impact strength. The macroscopically observed plastic deformation of the sample proved to be much bigger in this case as well, thus confirming the above statement. When referring to the sample heated up to 600 °C, no significant influence of the cooling mode applied on the impact strength obtained during the experiment was observed (Figure 10b,c). However, the ductile delamination manifested itself in the material cooled in the furnace (Figure 10b). This has never been observed in analogous scenario of samples cooled down much more rapidly in water mist (Figure 10c). The reduction in impact strength related to the reference value determined earlier on the sample denoted as *1 (+20)* (Figure 10a) and observed on the samples heated at 800 °C (Figure 10d,e) proved to be significantly smaller than the one observed on the samples heated at 600 °C only and depicted in Figure 10b,c. This is a beneficial influence of structural changes generated at such high temperature and associated with austenitic transformation. However, this happened only when the sample after heating was cooled in the furnace (Figure 10d). Rapid cooling in the water mist did not yield such a difference (Figure 10e), as it resulted in local hardening of the material. This in turn induced higher susceptibility to brittle fracture. These conclusions seem to be in complete agreement with results of earlier research conducted by the authors, reported in the papers [6–8].

The impact fractures obtained on samples made of *S355J2+N* steel tested at −20 °C and juxtaposed in Figure 11 exhibit completely different characteristics.

The stable fracture growth zone observed in this testing scenario is very limited in every case. An unstable fracture area with adjoining plastic zone dominates the picture this time. Such morphology of obtained fracture surfaces is consistent with accompanying *F–s* graphs, as one may clearly identify the F_{iu} force on each of these. After the value F_{iu} is reached, the transferred force is abruptly reduced at minimal increase of the accompanying displacement *s*. All the *F–s* graphs presented in Figure 11 should be assigned to category *D* according to the code [15]. In the test scenario considered, simulating the winter conditions, the impact strength of the tested steel is significantly reduced and potential self-arresting of the micro-cracks generated in the material is not very effective. Interestingly, in this case, surviving a fire incident (Figure 11b–e) followed by cooling proves to be beneficial, as it results in significant increase in impact toughness when compared against the sample denoted *1 (−20)* (Figure 11a). This increase was the highest when the sample was heated to the temperature of 800 °C and this was followed by slow cooling down of the sample in the furnace (Figure 11d). Under such conditions, the temperature acting on the steel was sufficiently high to induce in the material changes in its microstructure beneficial from the point of view of impact strength [8]. Importantly, this phenomenon has never been negated by local hardening through rapid cooling of the steel in water mist.

Values of the parameter $SFA_{(n)}$ (%) calculated for the *S355J2+N* steel are listed in the Table 3. Formulas (1)–(4) were used to determine the values listed for $n = 1, \ldots, 4$, while for $n = 5$, a direct approach depicted in Figure 8 had been used.

Table 3. Values of the $SFA_{(n)}$ (%) parameters for the *S355J2+N* steel, determined on impact toughness test fractures obtained during this research.

Sample Number	Temperature of Test (°C)	$SFA_{(1)}$ (%) Average	$SFA_{(1)}$ (%) Standard Deviation	$SFA_{(2)}$ (%) Average	$SFA_{(2)}$ (%) Standard Deviation	$SFA_{(3)}$ (%) Average	$SFA_{(3)}$ (%) Standard Deviation	$SFA_{(4)}$ (%) Average	$SFA_{(4)}$ (%) Standard Deviation	$SFA_{(5)}$ (%)
1	+20	100		100		100		100		100
	−20	47.07	8.21	54.90	7.22	51.30	7.68	69.25	6.11	50
160	+20	100		100		100		100		100
	−20	55.95	5.90	63.20	4.68	59.91	5.15	75.55	3.77	65
161	+20	100		100		100		100		100
	−20	46.45	4.65	54.20	3.54	50.64	3.97	68.99	3.27	52
180	+20	100		100		100		100		100
	−20	43.85	5.90	54.61	5.87	49.81	5.93	67.23	4.38	52
181	+20	91.62	18.74	93.17	15.28	92.47	16.83	95.27	10.57	100
	−20	44.50	7.97	57.27	6.22	51.72	6.98	68.15	5.74	60

These results quantitatively confirm the conclusions drawn in this paper. It has to be noted here that the $SFA_{(1)}$ parameter proved to be the most conservative one in this listing, while the $SFA_{(4)}$ parameter yielded the least conservative results. The approach based on the $SFA_{(5)}$ parameter usually yielded intermediate estimates.

The measured and later averaged values of the lateral expansion parameter *LE* (mm) are juxtaposed below. Figure 12 depicts appropriate average values and coefficients of variation estimated on the statistical sample for each of the samples and test scenarios executed in practice. The presented results confirm the conclusions drawn above.

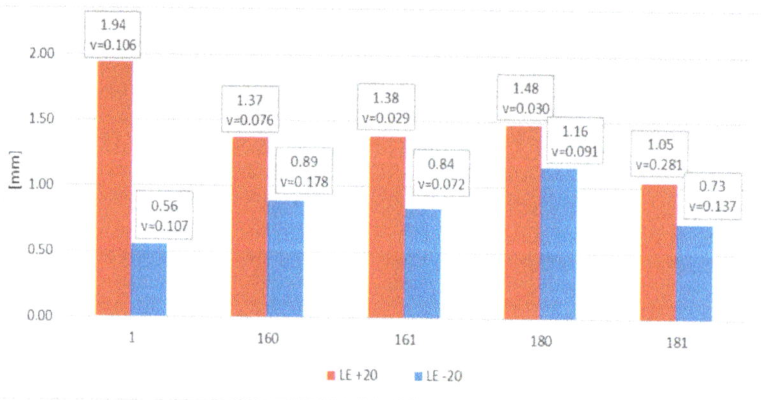

Figure 12. Averaged values of the *LE* parameter and coefficients of variation corresponding to them (estimated on a statistical sample) obtained on impact toughness test samples made of *S355J2+N* steel.

Figure 13 depicts the values *LE* presented previously in Figure 12 in relation to the corresponding values of energy W_t (J) required to fracture the considered sample. This picture indicates that change in the test temperature from +20 °C to −20 °C results in significant decrease in this energy. This means that under those conditions, the impact toughness of given material is much lower.

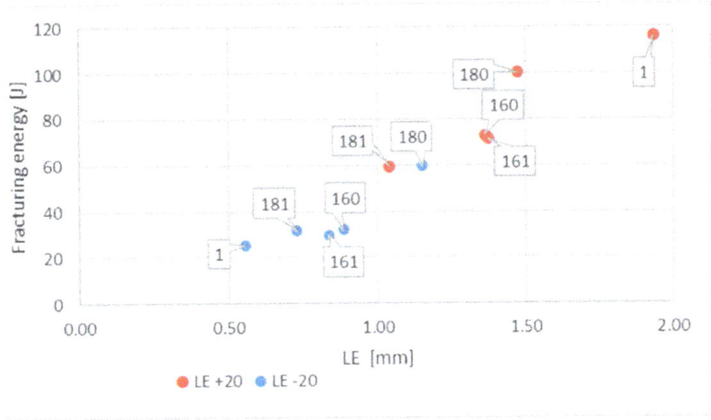

Figure 13. Relationship between averaged values of *LE* parameter and related values of fracturing energy W_t obtained for *S355J2+N* steel at different testing conditions.

6.2. Results Obtained on Samples Made of X20Cr13 Steel

This steel is a single phase stainless high alloy chromium steel, representative for a group of special steels resistant to abrasion. The microstructure of the steel of this type is purely martensitic. This material, characterized by good resistance to corrosion in moderately aggressive environments lacking chlorine content, is usually applied in the tempered and annealed state. It has to be heated to 300–400 °C prior to welding, and after welding it has to be annealed to soften.

The chemical composition of the samples made of this steel, identified using OES [8], is listed in Table 4.

Table 4. Chemical composition of the samples made of *X20Cr13* steel (according to [8]).

Element	**C**	**Si**	**Mn**	**P**	**S**	**Cr**	**Ni**	**Mo**
Content by wt. %	0.247	0.428	0.784	0.0153	0.0166	13	0.1	0.146
Element	**Ti**	**Cu**	**Al**	**Co**	**Nb**	**V**	**W**	
Content by wt. %	0.0045	0.0479	0.0086	0.0132	<0.002	0.009	<0.02	

The impact strength fractures obtained on samples made of this steel are depicted in Figure 14 for the tests conducted at +20 °C, and in Figure 15 for the tests conducted at −20 °C. Each of the presented fracture surfaces is accompanied by a corresponding *F–s* curve. Numbering of the samples conforms to the key listed in Table 1.

One may easily notice that in each of the considered testing scenarios, a brittle fracture was obtained, meaning that the failure of this steel under applied external load would be an abrupt phenomenon, without preceding signs of increasing weakening. Interestingly, the same conclusion may be drawn regarding the sample denoted as *2 (+20)*, which was not affected by prior action of fire temperature (Figure 14a). This statement is confirmed by the *F–s* curves corresponding to respective fractures, as in each of the considered testing scenarios, the limiting force F_{in} reveals itself, initiating the unstable fracture growth phase in the material. Thus, self-arresting of these fractures proved to be ineffective in this case. In addition, on all of the *F–s* graphs presented here, the equivalence $F_m = F_{in}$ holds. Therefore, all these cases may be assigned to category *D* (according to [15]). The results obtained here, due to the brittle behavior of the material revealed during the tests, in general disqualify any possibility for future application of this steel in construction. This brittleness is a result of material hardening by the martensitic structure intentionally induced in it during manufacture.

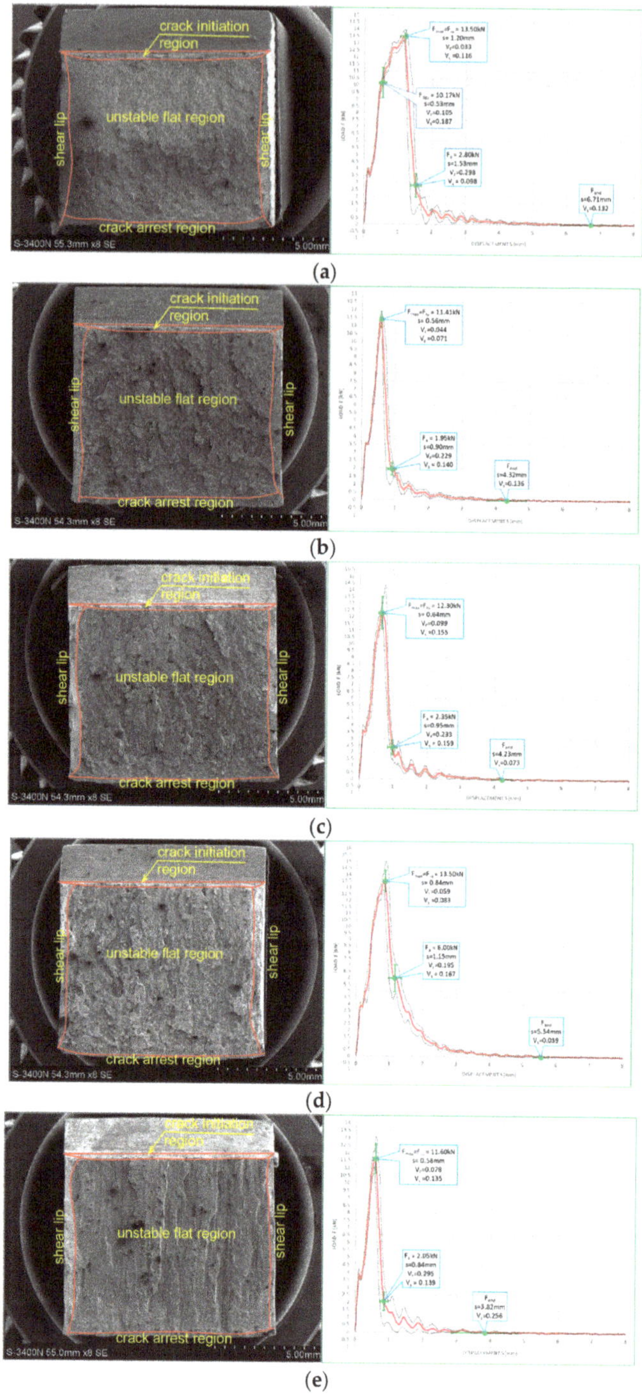

Figure 14. Morphology of the impact fractures obtained during the tests conducted at +20 °C on the samples made of *X20Cr13* steel—denotations follow the key listed in Table 1. (**a**) *2 (+20)*; (**b**) *260 (+20)*; (**c**) *261 (+20)*; (**d**) *280 (+20)*; (**e**) *281 (+20)*.

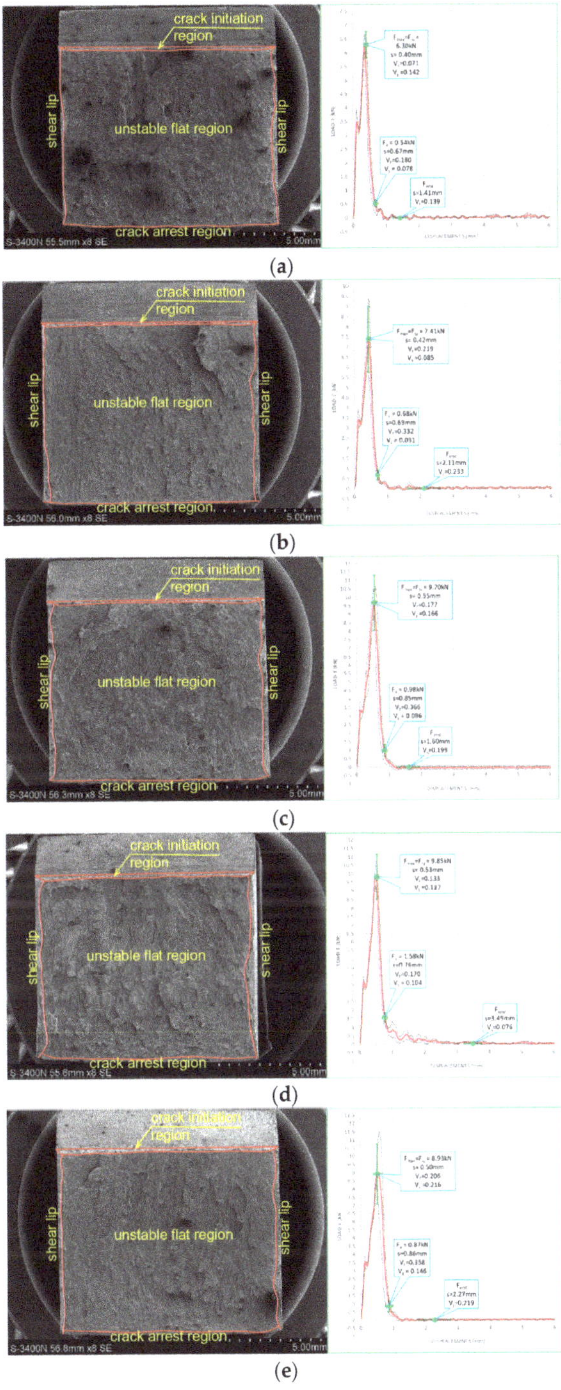

Figure 15. Morphology of the impact fractures obtained during the tests conducted at −20 °C on the samples made of *X20Cr13* steel—denotations follow the key listed in Table 1. (**a**) *2 (−20)*; (**b**) *260 (−20)*; (**c**) *261 (−20)*; (**d**) *280 (−20)*; (**e**) *281 (−20)*.

The results depicted in Figure 15, pertaining to the tests conducted at −20 °C not only confirm but even clearly reinforce the above statement, formulated with respect to analogous samples made of the same steel but conducted at +20 °C. The effective self-arresting of fractures, initiated when the force F_a appears on the F–s graph (Figure 5a), under such conditions becomes negligibly small.

The conclusions drawn above are formally confirmed at first in the values of $SFA_{(n)}$ parameters measured during experiment and juxtaposed in the Table 5 as well as in Figures 16 and 17 depicting the values of the *LE* parameter obtained during each considered testing scenario and its relation to the energy W_t needed to break the sample.

Table 5. Values of the $SFA_{(n)}$ (%) parameters determined for *X20Cr13* steel, determined on impact toughness test fractures obtained during this research.

Sample Number	Temperature of Test (°C)	$SFA_{(1)}$ (%) Average	$SFA_{(1)}$ (%) Standard Deviation	$SFA_{(2)}$ (%) Average	$SFA_{(2)}$ (%) Standard Deviation	$SFA_{(3)}$ (%) Average	$SFA_{(3)}$ (%) Standard Deviation	$SFA_{(4)}$ (%) Average	$SFA_{(4)}$ (%) Standard Deviation	$SFA_{(5)}$ (%)
2	+20	20.19	5.70	36.71	5.55	29.44	5.51	46.98	6.12	11
	−20	7.95	1.28	7.95	1.28	7.95	1.28	28.09	2.39	0
260	+20	14.54	3.20	14.54	3.20	14.54	3.20	37.87	4.39	11
	−20	8.92	1.37	8.92	1.37	8.92	1.37	29.77	2.31	0
261	+20	19.69	4.79	19.69	4.79	19.69	4.79	43.99	5.81	6
	−20	10.54	2.86	10.54	2.86	10.54	2.86	32.19	4.20	0
280	+20	38.99	9.06	38.99	9.06	38.99	9.06	62.04	7.08	6
	−20	16.33	3.68	16.33	3.68	16.33	3.68	40.16	4.53	0
281	+20	16.19	3.92	16.19	3.92	16.19	3.92	39.97	4.54	6
	−20	9.62	2.58	9.62	2.58	9.62	2.58	30.72	4.27	0

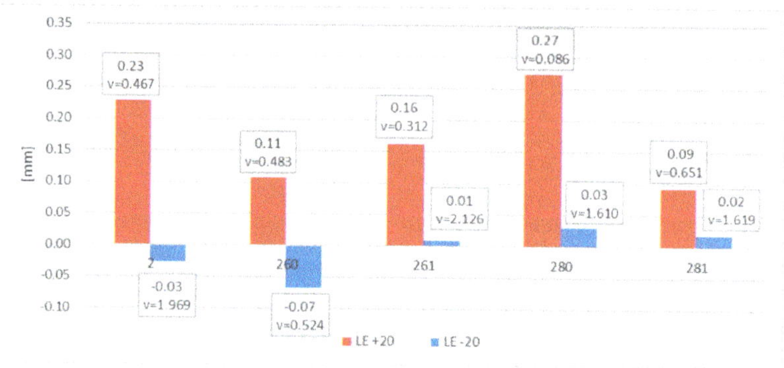

Figure 16. Averaged values of the *LE* parameter and corresponding coefficients of variation (estimated on a statistical sample) obtained on impact toughness test samples made of *X20Cr13* steel.

In the case of this steel, should one assume the values quantified by the parameter $SFA_{(5)}$ to be authoritative, then all the remaining parameters beginning with $SFA_{(1)}$ and ending with $SFA_{(4)}$ clearly overestimate this value.

Negative values of the *LE* parameter obtained for this steel during the tests conducted at −20 °C, accompanied by very low impact strength determined, should be understood as a proof of lateral contraction (*LC*), depicted in Figure 6, occurring in the analyzed cross section.

The data depicted in Figure 15 indicate one more interesting fact. The simulated fire action episodes in the case of tests conducted at +20 °C resulted in a significant reduction in the already very low impact strength of the material. This reduction was slightly smaller when the considered steel was previously heated up to 800 °C, meaning that permanent changes occurring in the microstructure of the material were beneficial from the point of view of this property. Rapid cooling of the material in water mist almost completely eliminated this effect. However, when the tests were conducted at −20 °C (simulating winter conditions), then the preceding episodes of simulated fire action proved to be in general beneficial to the material. This effect was somewhat diminished when the samples were heated up to 800 °C. Nevertheless, the impact strength of this steel related to the testing conditions was clearly negligibly small.

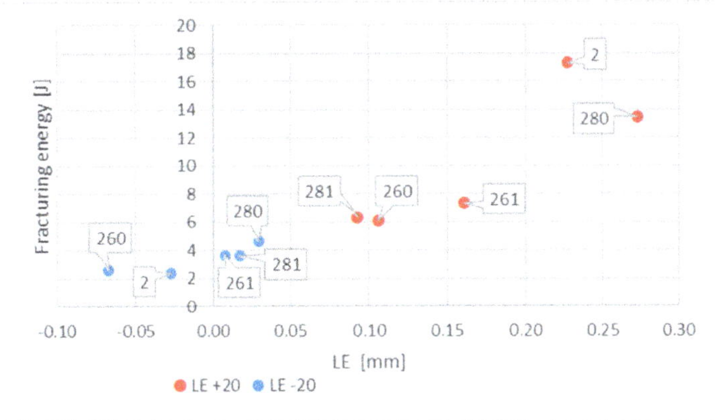

Figure 17. Relationship between averaged values of *LE* parameter and related values of fracturing energy W_t obtained for *X20Cr13* steel at different testing conditions.

6.3. Results Obtained on Samples Made of X6CrNiTi18-10 Steel

This is an acid resistant, nonmagnetic steel, with low yield limit (of about 220 MPa). It exhibits good impact strength even at sub-zero temperature and good mechanical properties when exposed to fire. Furthermore, it exhibits good weldability but simultaneously poor susceptibility to mechanical and electrochemical polishing. The microstructure of such steel consists of austenitic matrix with small amount of titanium carbide precipitates.

The chemical composition of the samples made of this steel identified by the OES spectrometer [8] is listed in Table 6.

Table 6. Chemical composition of the samples made of *X6CrNiTi18-10* steel (according to [8]).

Element	C	Si	Mn	P	S	Cr	Ni	Mo
Content by wt. %	0.0709	0.467	1.84	0.0246	<0.005	18	9.12	0.347
Element	Ti	Cu	Al	Co	Nb	V	W	
Content by wt. %	0.352	0.261	0.0356	0.00983	0.0167	<0.087	<0.02	

The impact strength test fractures obtained on samples made of this steel accompanied by corresponding *F–s* curves are depicted on Figure 18 for the tests conducted at +20 °C, and on Figure 19 for the tests conducted at −20 °C.

All the fractures presented in this group exhibited substantial permanent plastic deformations. This indicates a completely plastic character. The morphology of the examined fractures, in each of the considered scenarios, is dominated by a zone of stable fracture growth. This conclusion is supported by the shape of the *F–s* curve accompanying each case. These curves, according to the classification contained in [15] may be assigned to category *F* (Figure 5b). A large area bounded from above by the *F–s* curve is a measure of adequately high impact strength. It is to be correlated with a high capability to effectively self-arrest micro-cracks induced in the material.

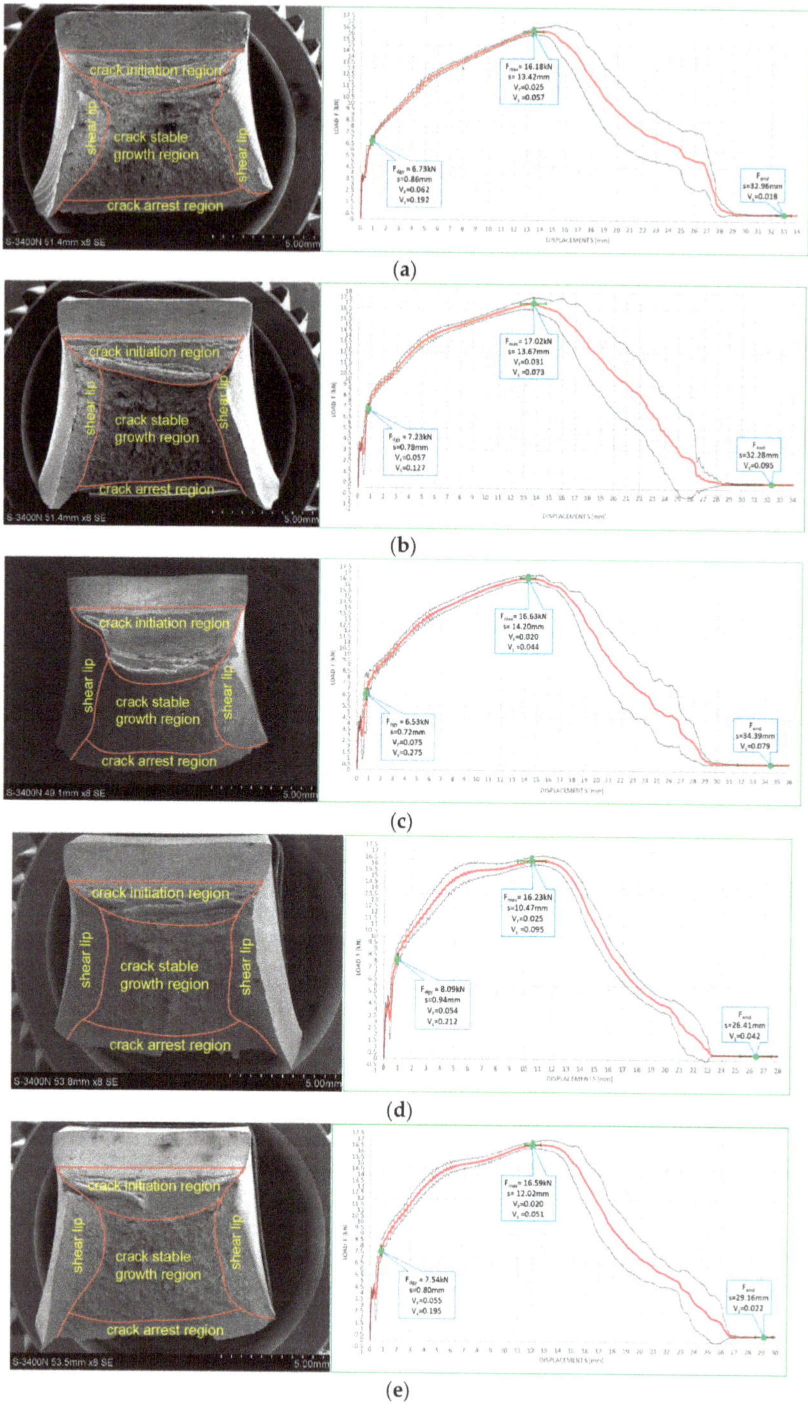

Figure 18. Morphology of the impact fractures obtained during the tests conducted at +20 °C on the samples made of *X6CrNiTi18-10* steel—denotations follow the key listed in Table 1. (**a**) *3 (+20)*; (**b**) *360 (+20)*; (**c**) *361 (+20)*; (**d**) *380 (+20)*; (**e**) *381 (+20)*.

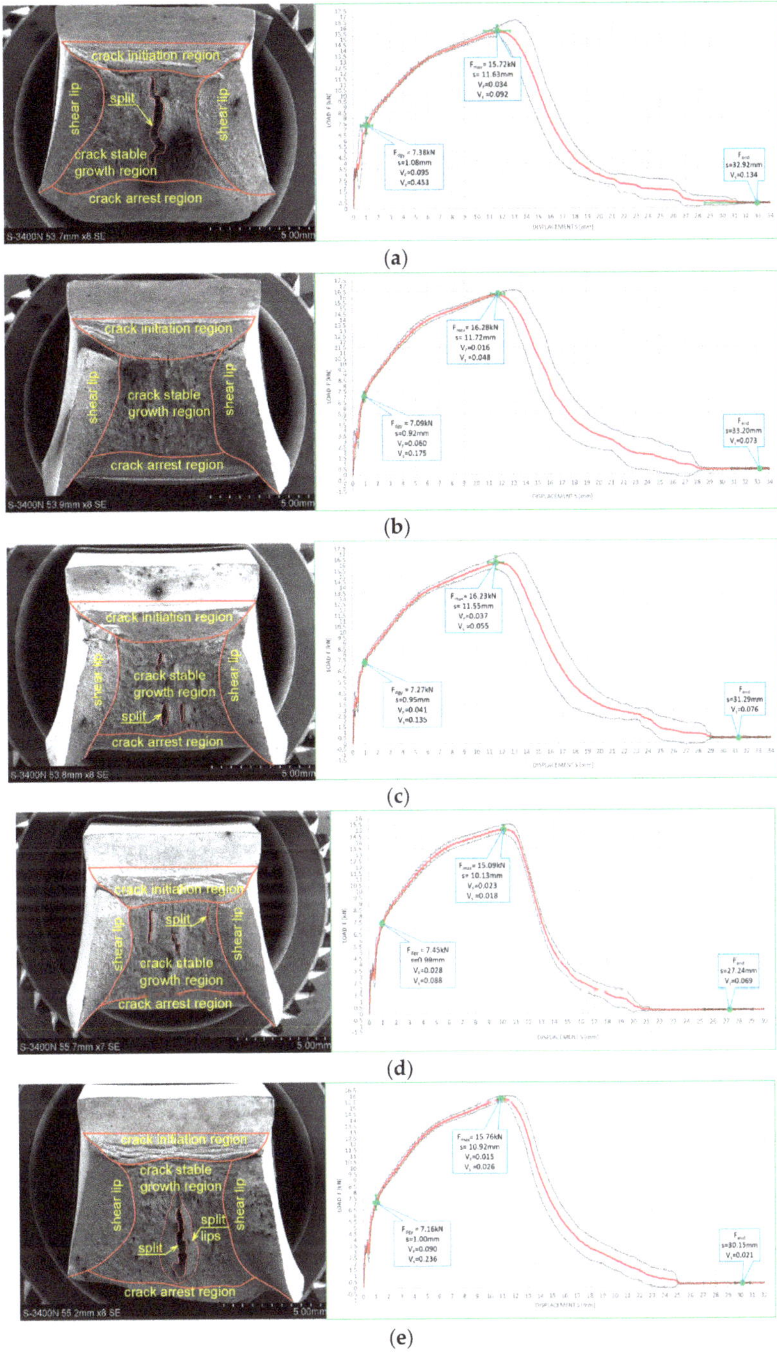

Figure 19. Morphology of the impact fractures obtained during the tests conducted at −20 °C on the samples made of *X6CrNiTi18-10* steel—denotations follow the key listed in Table 1. (**a**) *2 (−20)*; (**b**) *260 (−20)*; (**c**) *261 (−20)*; (**d**) *280 (−20)*; (**e**) *281 (−20)*.

Observation of the impact test fractures obtained during the tests made at −20 °C (Figure 19) leads to analogous conclusions. However, under those conditions, impact strength was proven to be slightly lower. Nevertheless, even in those cases, the obtained *F–s* curves may be assigned to the category *F* of the classification presented in the code [15] (Figure 5b). The ductile delaminations of the material, occurring during the tests conducted in such conditions and indicating its slightly lower resistance to the stresses accumulating in the sample as a result of dynamic loads applied [39], seem to represent a certain qualitative difference here.

The estimates of the $SFA_{(n)}$ (%) parameters conducted by us yielded, regardless of the measurement method applied, unequivocal confirmation of the fully plastic character of the fractures obtained. These parameters are juxtaposed in Table 7.

Table 7. Values of the $SFA_{(n)}$ (%) parameters determined for *X6CrNiTi18-10* steel, determined on impact toughness test fractures obtained during this research.

Sample Number	Temperature of Test (°C)	$SFA_{(1)}$ (%)		$SFA_{(2)}$ (%)		$SFA_{(3)}$ (%)		$SFA_{(4)}$ (%)		$SFA_{(5)}$ (%)
		Average	Standard Deviation	Average	Standard Deviation	Average	Standard Deviation	Average	Standard Deviation	
3	+20	100		100		100		100		100
	−20	100		100		100		100		100
360	+20	100		100		100		100		100
	−20	100		100		100		100		100
361	+20	100		100		100		100		100
	−20	100		100		100		100		100
380	+20	100		100		100		100		100
	−20	100		100		100		100		100
381	+20	100		100		100		100		100
	−20	100		100		100		100		100

The measurements of the *LE* parameter showed a seemingly unexpected effect, that for the considered steel plastic deformations on the impact strength test, fractures were quantitatively larger on the samples tested at −20 °C (Figure 20). This was clearly confirmed by the relation presented in Figure 21.

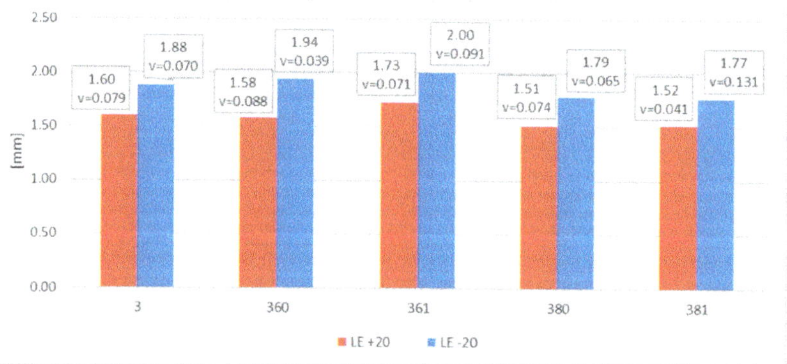

Figure 20. Averaged values of the *LE* parameter and corresponding coefficients of variation (estimated on a statistical sample) obtained on impact toughness test samples made of *X6CrNiTi18-10* steel.

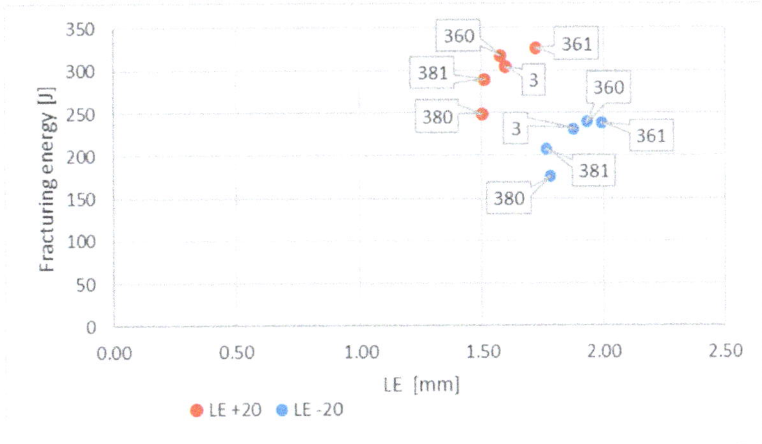

Figure 21. Relationship between averaged values of *LE* parameter and related values of fracturing energy obtained for *X6CrNiTi18-10* steel at different testing conditions.

This relation (Figure 21) in the case of considered steel grade exhibits peculiar characteristics. On one side it shows very high post-fire impact strength of the tested material, a factor not surprising in the case of steels exhibiting austenitic structure. However, this impact strength decreases significantly when tested at simulated winter conditions, nevertheless remaining at sufficiently high level. On the other side, contrary to the results obtained and presented in this paper for other steel grades, the results of tests conducted at −20 °C are located to the right of the corresponding results obtained at +20 °C. Moreover, for this steel, the fire incident simulated at 600 °C proved to be beneficial, as it raised the impact strength tested after cooling, regardless of the cooling mode applied. The higher temperature acting on the steel (800 °C), allowing for the structural changes to occur, finally proved to be detrimental to the tested material (when compared against the test made on samples denoted as *3 (+20)*, as well as *3 (−20)*—Figures 18a and 19a).

6.4. Results Obtained on Samples Made of X2CrNiMoN22-5-3 Steel

The steel of this grade is classified as a typical stainless and acid resistant high alloy steel which is resistant to pitting and also to surface corrosion. It is characterized by two-phase, austenitic-ferritic, chromium–nickel–molybdenum microstructure of duplex type. This steel is recommended for application at temperatures below 300 °C due to the occurrence of the detrimental 475 °C brittleness phenomenon.

The chemical composition of the samples made of this steel identified by the OES spectrometer [8] is listed in the Table 8.

Table 8. Chemical composition of the samples made of *X2CrNiMoN22-5-3* steel (according to [8]).

Element	**C**	**Si**	**Mn**	**P**	**S**	**Cr**	**Ni**	**Mo**
Content by wt. %	0.0507	0.266	1.8	0.027	<0.005	23.7	4.74	2.92
Element	**Ti**	**Cu**	**Al**	**Co**	**Nb**	**V**	**W**	
Content by wt. %	0.0082	0.184	0.0097	0.0622	0.0056	0.0385	<0.02	

The impact strength test fractures obtained on samples made of this steel grade are depicted in Figure 22 for the tests conducted at +20 °C and in Figure 23 for the analogous tests conducted at −20 °C. In all the considered cases, presented fracture surfaces are accompanied by a corresponding *F–s* curve.

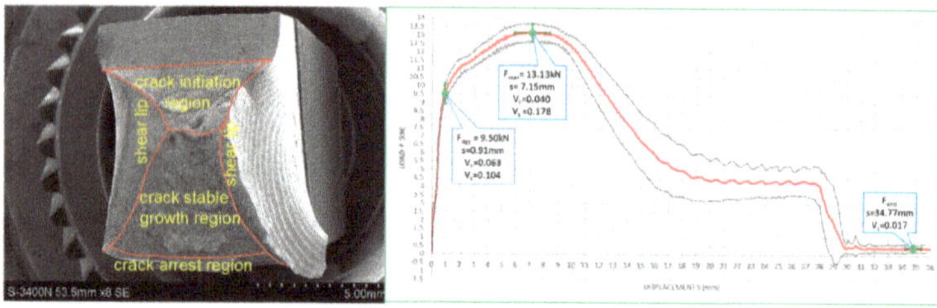

(**a**)

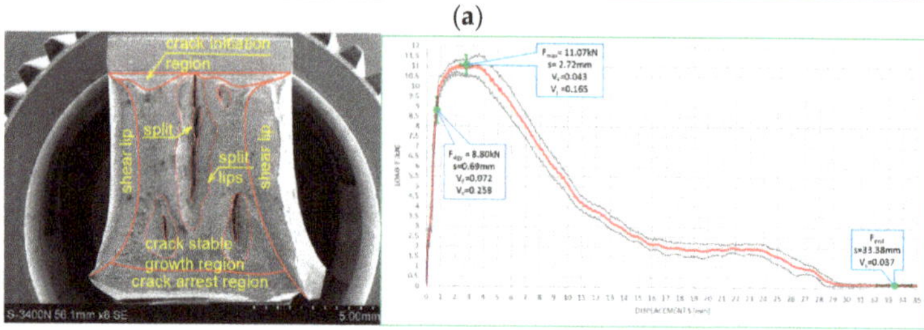

(**b**)

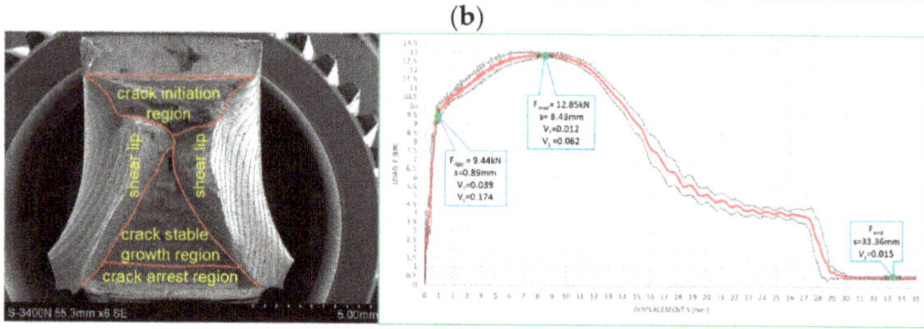

(**c**)

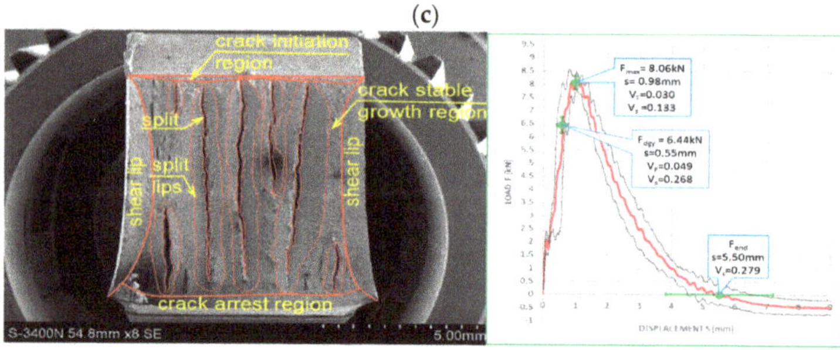

(**d**)

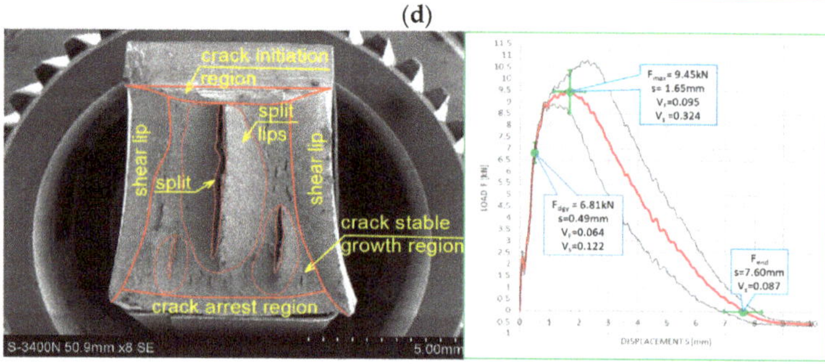

(**e**)

Figure 22. Morphology of the impact fractures obtained during the tests conducted at +20 °C on the samples made of *X2CrNiMoN22-5-3* steel—denotations follow the key listed in Table 1. (**a**) *4 (+20)*; (**b**) *460 (+20)*; (**c**) *461 (+20)*; (**d**) *480 (+20)*; (**e**) *481 (+20)*.

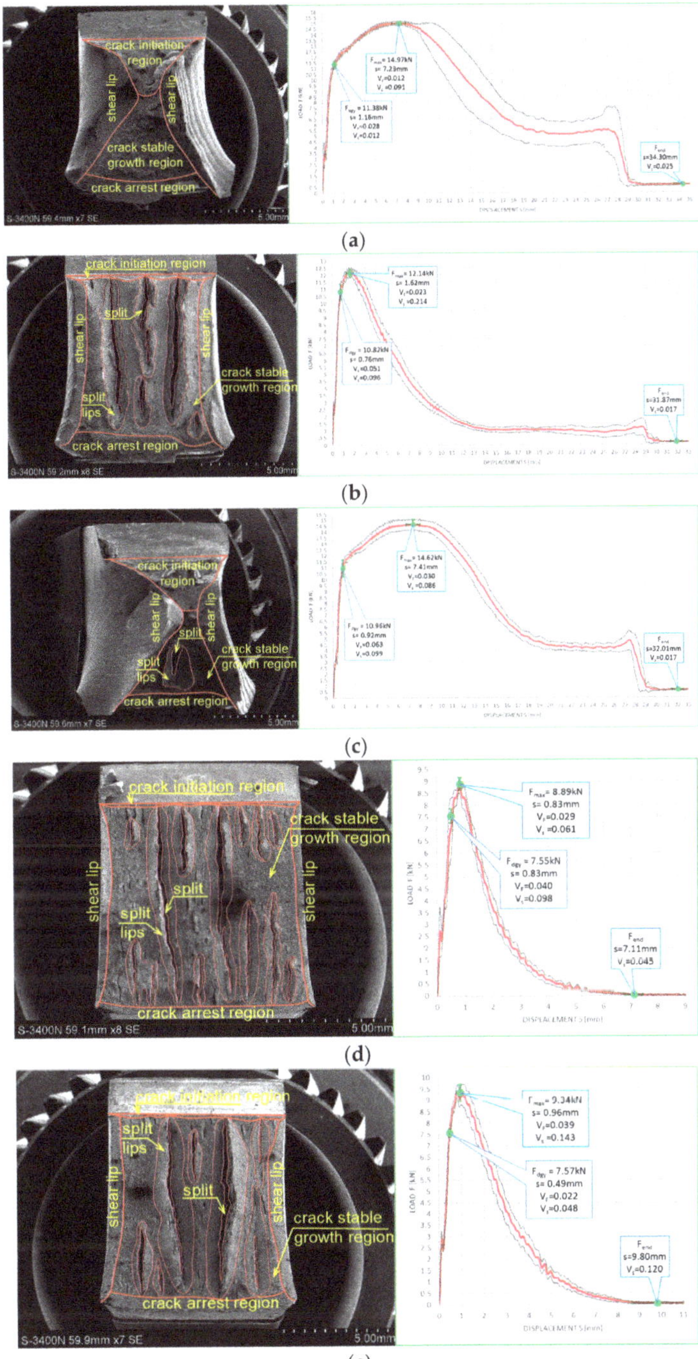

Figure 23. Morphology of the impact fractures obtained during the tests conducted at −20 °C on the samples made of *X2CrNiMoN22-5-3* steel—denotations follow the key listed in Table 1. (**a**) *4 (−20)*; (**b**) *460 (−20)*; (**c**) *461 (−20)*; (**d**) *480 (−20)*; (**e**) *481 (−20)*.

The impact strength fractures obtained on samples made of this steel in +20 °C exhibited completely ductile character. The macroscopically observed changes in shape testify to that. The important qualitative difference with respect to the analogous testing scenarios executed on samples made of *X6CrNiTi18-10* steel appear as multiple transverse delaminations. So far, these delaminations occurred only in samples tested in −20 °C. This destruction mode at completely plastic fracture is usually explained as a result of local heterogeneities in plastic properties of tested material [39]. This phenomenon often causes an incomplete fracture during classical impact strength tests, ending in following quasi static pushing of the sample through hammer clamps. These statements seem to be confirmed by the *F–s* graphs accompanying these fractures, as one may easily identify a horizontal plateau generated during the finishing stage of fracturing the samples (Figure 22a–c). All the *F–s* graphs presented in Figure 22 clearly may be assigned to the category *F* (according to [15]).

This steel exhibited a peculiar reaction to the simulated fire episode. This was especially true when the steel was heated to 800 °C. This temperature level resulted in changes in the microstructure of the affected samples. Under those conditions, regardless of the cooling mode applied, the impact strength test fractures were accompanied by very small displacements *s*, indicating significant deleterious changes in the impact strength of the tested steel. The complementary metallographic analysis conducted by us showed that heating of the steel up to 800 °C resulted in intensive precipitation of numerous carbides, nitrides, and intermetallic phases out of supersaturated ferrite and austenite in its microstructure [8]. Additionally, under such conditions slow cooling of the sample in the furnace (the sample denoted as *480 (+20)*—Figure 22d) yielded longer material transition time through the so-called 475 °C brittle zone. During this phase, usually deleterious brittle phases appear in the material, additionally weakening it, and this in turn obviously decreases its impact strength. Fast cooling in water mist (the sample denoted as *481 (+20)*—Figure 22e) resulted in shorter transition time through this zone. As a result, the material weakening proved to be less significant in this case [8].

Analogous development was observed in the case of samples made of this steel grade and heated up to 600 °C only. During slow cooling in the furnace (the sample denoted as *460 (+20)*—Figure 22b), the transition time through the 475 °C brittle zone proved to be sufficiently long to significantly decrease the post-fire impact strength of the tested steel (as compared against analogous steel not subjected to the simulated fire action—Figure 22a). However, this phenomenon did not appear at all when the samples were cooled sufficiently fast (Figure 22b).

The fractures observed during the tests conducted at −20 °C (Figure 23) proved to be fully plastic as well, exhibiting properties analogous to those described above with respect to the tests conducted at +20 °C. Interestingly, under those conditions, no significant deterioration of the impact strength revealed was observed. Though quantitative differences were observed, their formal importance seems to be negligible. In all considered scenarios, the *F–s* graphs were assigned to the category *F* (according to [15]). This conclusion is clearly confirmed by the juxtaposition presented in Table 9.

Let us note that, as shown in the Figure 24, in the case of this steel grade in all the testing scenarios considered, the values of the *LE* parameter obtained during the tests conducted at +20 °C happened to be higher than analogous values obtained during the tests conducted at −20 °C. However, this does not mean that the impact strength revealed during those tests proved to be better in each of the compared scenarios. A comparison of the samples denoted as *4 (+20)* and *461 (+20)* (Figure 25) indicates a not-so-obvious statement that lowering the test temperature to −20 °C resulted in increased impact strength.

Table 9. Values of the $SFA_{(n)}$ (%) parameters determined for *X2CrNiMoN22-5-3* steel, determined on impact toughness test fractures obtained during this research.

Sample Number	Temperature of Test (°C)	$SFA_{(1)}$ (%) Average	$SFA_{(1)}$ (%) Standard Deviation	$SFA_{(2)}$ (%) Average	$SFA_{(2)}$ (%) Standard Deviation	$SFA_{(3)}$ (%) Average	$SFA_{(3)}$ (%) Standard Deviation	$SFA_{(4)}$ (%) Average	$SFA_{(4)}$ (%) Standard Deviation	$SFA_{(5)}$ (%)
4	+20	100		100		100		100		100
	−20	100		100		100		100		100
460	+20	100		100		100		100		100
	−20	100		100		100		100		100
461	+20	100		100		100		100		100
	−20	100		100		100		100		100
480	+20	100		100		100		100		100
	−20	100		100		100		100		100
481	+20	100		100		100		100		100
	−20	100		100		100		100		100

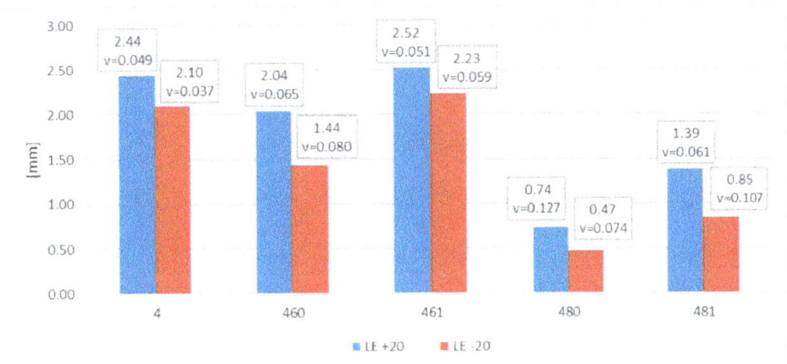

Figure 24. Averaged values of the *LE* parameter and corresponding coefficients of variation (estimated on a statistical sample) obtained on impact toughness test samples made of *X2CrNiMoN22-5-3* steel.

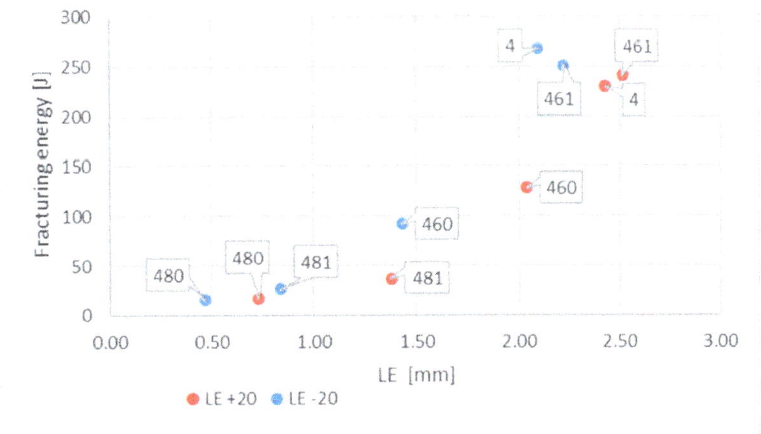

Figure 25. Relationship between averaged values of *LE* parameter and related values of fracturing energy obtained for *X2CrNiMoN22-5-3* steel at different testing conditions.

7. Concluding Remarks

In the authors' opinion, the results presented in this paper confirm the need for detailed analysis of impact strength test fractures obtained experimentally in order to reliably draw conclusions on the possible future safe application of material which has survived a fire incident. In particular, one has to determine whether the material affected by high temperature during fire has retained the capability to safely resist the external loads applied to it with sufficiently high probability. The considered fractures were analyzed a posteriori, i.e., after finished simulated fire incidents, on the samples made of a given steel grade cooled down successfully after prior exposure to high temperature. In that sense, the quantitative and qualitative estimation of given steel grade post-fire impact strength is an action of fundamental importance preceding the decision whether and to what extent the material may remain in continued service after fire. Therefore, it is postulated to include this test in the engineering practice of traditionally performed visual inspection of the deformed structure performed on site, accompanied by experimental verification of the basic mechanical properties of affected material. Obtained results should warrant that during prolonged service after a fire, no brittle fractures would occur in bearing steel structural components, as these fractures during unstable development may result in structural failure. Sufficiently high impact strength of the material warrants that the possible brittle fractures will be effectively suppressed.

The reliability of the conclusions drawn based on the approach described in this paper is enhanced by the fact that the results obtained are confirmed by several methods applied. Regardless of the testing scenario applied to check a given steel grade, similar results were arrived at for *LE* parameter, the LE–W_t relationship, the *F*–*s* curve, or only a simple observation of the morphology of experimentally obtained fracture surface.

It was proven that various steel grades behave differently after surviving a fire incident, in a manner that is not obvious and difficult to foresee if not supported by appropriate specialized tests. Steel grades varying in intentionally induced microstructure have been selected for analysis, as qualitatively different response of tested materials to fire exposure occurring under the same controlled conditions was expected. The a priori formulated thesis on the differences in the impact strength exhibited post-fire by various steel grades under different thermal conditions was experimentally confirmed. Therefore, should a prospective user expect a practically relevant and useful information on the impact strength of tested steel, this information has to be accompanied by the knowledge whether it concerns winter or summer service conditions of affected facility. The differences in this aspect, identified depending on such conditions, may prove to be very significant.

The tests conducted by us have shown that for some steel grades considered in this paper, lowering the test temperature from +20 °C to −20 °C resulted in improved impact strength of the material. Previous fire incidents did not always weaken the tested material as well. In many computational cases important in practice, especially those related to simulation of winter service conditions, the impact strength determined post-fire proved to be higher than that exhibited by the same material not exposed to fire action. However, under those circumstances in general, the final conclusions proved to be the opposite, when the impact strength tests were conducted in a manner simulating summer conditions (that means at +20 °C).

The results obtained were strongly affected by the mode applied to cool the hot steel. In most cases, rapid cooling of samples in water mist, simulating an action of firefighters, resulted in material hardening. This was equivalent to lowering its impact strength. However, this conclusion was not confirmed in the case of a duplex steel, as for this steel, fast reduction of temperature meant fast transition through the deleterious 475 °C brittle zone, and therefore yielded lower degree of material degradation in effect.

The key influence of temperature level reached during the considered fire incident on post-fire properties of analyzed material, and especially its impact strength, was confirmed by this research. Reaching a temperature level initiating structural changes in the considered steel seems to be the key here, as these changes are usually accompanied by developments associated with rebuilding of crystalline lattice and often with numerous precipitates of weakening character.

Continuing this research, we intend to find out how the material properties verified after a fire incident are affected by the extended heating time. It seems that the prolonged effective fire exposure occurring at the same temperature level should in practice result in additional structural changes significantly affecting the post-fire durability of the analyzed steel. These changes occurred so slowly that they did not have sufficient time to reveal themselves during the study scenario analyzed in this article and associated with one-hour heating time.

We also intend to verify how the change in heating speed while keeping the same total heating time affects the post-fire properties of given steel grade. It seems, in this context, that steel heated at a lower temperature increase rate may exhibit a better capacity to accommodate a higher temperature at the same safety level.

Author Contributions: Conceptualization, M.M., M.S. and P.Z.; methodology, M.M., M.S., A.W. and M.Z.; software, M.S., A.W., M.Z. and P.Z.; validation, M.M. and M.S.; formal analysis, M.M., M.S. and P.Z.; investigation, M.M., M.S. and P.Z.; resources, A.W., M.Z. and P.Z.; data curation, A.W., M.Z. and P.Z.; writing—original draft preparation, M.M. and M.P.; writing—review and editing, M.P. and M.S.; visualization, P.Z.; supervision, M.M. and M.S. All authors have read and agreed to the published version of the manuscript.

Funding: This research received no external funding.

Institutional Review Board Statement: Not applicable.

Informed Consent Statement: Not applicable.

Data Availability Statement: The data presented in this article are available within the article.

Conflicts of Interest: The authors declare no conflict of interest.

References

1. Hahn, G.T. The influence of microstructure on brittle fracture toughness. *Metall. Trans. A* **1984**, *15A*, 947–959. [CrossRef]
2. Bednarek, Z.; Kamocka, R. The heating rate impact on parameters characteristic of steel behaviour under fire conditions. *J. Civ. Eng. Manag.* **2006**, *12*, 269–275. [CrossRef]
3. Maslak, M.; Zwirski, G. Changes in structural steel microstructures following heating and cooling episodes in fires. *Saf. Fire Tech.* **2017**, *48*, 34–52. [CrossRef]
4. Maslak, M. Tests of structural steel after a fire in the context of assessing the possibility of its further use in the load-bearing members of building structures. *Przegląd Bud.* **2012**, *6*, 48–51. (In Polish)
5. Peng, P.C.; Chi, J.H.; Cheng, J.W. A study on behavior of steel structures subjected to fire using non-destructive testing. *Constr. Build. Mater.* **2016**, *128*, 170–175. [CrossRef]
6. Maslak, M.; Pazdanowski, M.; Stankiewicz, M.; Zajdel, P. The impact strength of selected steel types after fire. Experimental tests related to simulated fire conditions. In Proceedings of the 7th International Conference "Applications of Structural Fire Engineering" (ASFE), Online, 10–11 June 2021.
7. Maslak, M.; Pazdanowski, M.; Stankiewicz, M.; Zajdel, P. Post-fire susceptibility to brittle fracture of selected steel grades used in construction industry—Assessment based on the instrumented impact test. *Materials* **2021**, *14*, 3922. [CrossRef]
8. Pancikiewicz, K.; Maslak, M.; Pazdanowski, M.; Stankiewicz, M.; Zajdel, P. Changes in the microstructure of selected structural alloy steel grades identified after their simulated exposure to fire temperature. *Case Stud. Constr. Mater.* **2023**, *18*, e01923. [CrossRef]
9. Server, W.L. Instrumented Charpy test review and application to structural integrity. In *From Charpy to Present Impact Testing*; François, D., Pineau, A., Eds.; Elsevier Science and ESIS: Amsterdam, The Netherlands, 2002; pp. 205–212. [CrossRef]
10. Alar, Ž.; Mandić, D.; Dugorepec, A.; Sakoman, M. Application of instrumented Charpy method in characterization of materials. *Interdiscip. Descr. Complex Syst.* **2015**, *13*, 479–487. [CrossRef]
11. Pereira, L.C.; Garcia de Blas, J.C.; Griza, S.; Darwish, F.A.I. Use of instrumented Charpy testing on the fracture toughness characterization of metallic materials. *Tecnol. Em Metal. Mater. E Min.* **2021**, *18*, e2469. [CrossRef]
12. Ali, M.B.; Abdullah, S.; Nuawi, M.Z.; Padzi, M.M.; Zakaria, K.A. Experimental analysis of an instrumented Charpy impact using statistical study based data analysis. *Int. J. Mech. Mater. Eng. (IJMME)* **2011**, *6*, 260–268.
13. Zajdel, P. A suitability assessment using an instrumented impact test of the use of selected structural steel grades on the basis of their changes in response to exposure to fire. *Tech. Trans.* **2021**, *118*, e2021007. [CrossRef]
14. *EN-ISO 148-1*; Metallic Materials—Charpy Pendulum Impact Test. Part 1: Test Method. ISO: Geneva, Switzerland, 2006.

15. *EN-ISO 14556*; Metallic Materials—Charpy V-Notch Pendulum Impact Test. Instrumented Test Method. ISO: Geneva, Switzerland, 2015.
16. *ASTM E 2298-18*; Standard Test Method for Instrumented Impact Testing of Metallic Materials. American Society for Testing and Materials: West Conshohocken, PA, USA, 2018. [CrossRef]
17. *ASTM E 23-92*; Standard Test Methods for Notched Bar Impact Testing of Metallic Materials. American Society for Testing and Materials: West Conshohocken, PA, USA, 2016. [CrossRef]
18. Schmitt, W.; Varflomeyev, I.; Böhme, W. Modelling of the Charpy test as a basis for toughness evaluation. In *From Charpy to Present Impact Testing*; François, D., Pineau, A., Eds.; Elsevier Science and ESIS: Amsterdam, The Netherlands, 2002; pp. 45–56. [CrossRef]
19. Balyts'kyi, O.I.; Ripei, I.V.; Protsakh, K.A. Reliability of steam pipelines of thermal power plants in the course of long-term operation. *Mater. Sci.* **2006**, *42*, 461–465. [CrossRef]
20. Tronskar, J.P.; Mannan, M.A.; Lai, M.O. Measurement of fracture initiation toughness and crack resistance in instrumented Charpy impact testing. *Eng. Fract. Mech.* **2002**, *69*, 321–338. [CrossRef]
21. Kharchenko, V.V.; Kondryakov, E.A.; Zhamaka, V.N.; Babutskii, A.A.; Babutskii, A.I. The effect of temperature and loading rate on the crack initiation and propagation energy in carbon steel Charpy specimens. *Strength Mater.* **2006**, *38*, 535–541. [CrossRef]
22. Müller, K.; Push, G. Use of Charpy impact testing to evaluate crack arrest fracture toughness. In *From Charpy to Present Impact Testing*; François, D., Pineau, A., Eds.; Elsevier Science and ESIS: Amsterdam, The Netherlands, 2002; pp. 263–270. [CrossRef]
23. François, D. Micromechanisms and the Charpy transition curve. In *From Charpy to Present Impact Testing*; François, D., Pineau, A., Eds.; Elsevier Science and ESIS: Amsterdam, The Netherlands, 2002; pp. 21–31. [CrossRef]
24. Cubides-Herrera, C.S.; Villba-Rondón, D.A.; Rodríguez-Baracaldo, R. Charpy impact toughness and transition temperature in ferrite–pearlite steel. *Sci. Et Tech.* **2019**, *24*, 200–204.
25. Mengqi, Z.; Shanwu, Y.; Farong, W. Competition mechanism of brittle–ductile transition of metals under tensile condition. *Mech. Mater.* **2019**, *137*, 103138. [CrossRef]
26. Curry, D.; Knott, J.F. Effect of microstructure on cleavage fracture stress in steel. *Met. Sci.* **1978**, *12*, 511–514. [CrossRef]
27. Chen, L.; Liu, W.; Yu, L.; Cheng, Y.; Ren, K.; Sui, H.; Yi, X.; Duan, H. Probabilistic and constitutive models for ductile-to-brittle transition in steels: A competition between cleavage and ductile fracture. *J. Mech. Phys. Solids* **2020**, *135*, 103809. [CrossRef]
28. Lin, T.; Evans, A.G.; Ritchie, R.O. Stochastic modeling of the independent roles of particle size and grain size in transgranular cleavage fracture. *Metall. Trans. A* **1987**, *18*, 641–651. [CrossRef]
29. MacKenzie, S. Overview of the mechanisms of failure in heat treated steel components. In *Failure Analysis of Heat Treated Steel Components*; Canale, L.C.F., Mesquita, R.A., Totten, G.E., Eds.; ASM Technical Books; ASM International: Materials Park, OH, USA, 2008. [CrossRef]
30. Mirmomeni, M.; Heidarpour, A.; Xiao-Ling Zhao Hutchinson, C.R.; Packer, J.A.; Chengqing, W. Fracture behaviour and microstructural evolution of structural mild steel under the multi-hazard loading of high-strain-rate load followed by elevated temperature. *Constr. Build. Mater.* **2016**, *122*, 760–771. [CrossRef]
31. Pantazopoulos, G.A. A short review on fracture mechanisms of mechanical components operated under industrial process conditions: Fractographic analysis and selected prevention strategies. *Metals* **2019**, *9*, 148. [CrossRef]
32. Kai, Y.; Ozaki, F. Impact fracture energy of steel welded connections in fire and post-fire. *J. Constr. Steel Res.* **2020**, *170*, 106120. [CrossRef]
33. Kai, Y.; Ozaki, F.; Knobloch, M. Impact fracture energies of cold-formed steel square hollow section in and after fire. *J. Constr. Steel Res.* **2021**, *183*, 106740. [CrossRef]
34. Zhu, Y.; Huang, S.; Sajid, H.U. Micro-mechanisms and modelling of ductile fracture initiation in structural steel after exposure to elevated temperatures. *Metals* **2021**, *11*, 767. [CrossRef]
35. Zhu, H.; Zhang, C.; Chen, S.; Wu, J. A modified Johnson-Cook constitutive model for structural steel after cooling from high temperature. *Constr. Build. Mater.* **2022**, *340*, 127746. [CrossRef]
36. Dzioba, I.; Lipiec, S. Fracture mechanisms of S355 steel—Experimental research, FEM simulation and SEM observation. *Materials* **2019**, *12*, 3959. [CrossRef]
37. Joo, M.S.; Suha, D.W.; Baec, J.H.; Bhadeshia, H.K.D.H. Role of delamination and crystallography on anisotropy of Charpy toughness in API-X80 steel. *Mater. Sci. Eng. A* **2012**, *546*, 314–322. [CrossRef]
38. Kim, J.-S.; Kim, Y.-J.; Lee, M.-W.; Kim, K.-S.; Shibanuma, K. Fracture simulation model for API X80 Charpy test in ductile–brittle transition temperatures. *Int. J. Mech. Sci.* **2020**, *182*, 105771. [CrossRef]
39. Kantor, M.; Vorkachev, K.; Bozhenov, V.; Solntsev, K. The role of splitting phenomenon under fracture of low-carbon microalloyed X80 pipeline steels during multiple Charpy impact tests. *Appl. Mech.* **2022**, *3*, 740–756. [CrossRef]
40. Tong, L.; Niu, L.; Jing, S.; Ai, L.; Zhao, X.-L. Low temperature impact toughness of high strength structural steel. *Thin-Walled Struct.* **2018**, *132*, 410–420. [CrossRef]
41. Samuel, K.G.; Sreenivasan, P.R.; Ray, S.K.; Rodriguez, P. Evaluating of ageing-induced embrittlement in an austenitic stainless steel by instrumented impact testing. *J. Nucl. Mater.* **1987**, *150*, 78–84. [CrossRef]
42. Laukkanen, A.; Uusikallio, S.; Lindroos, M.; Andersson, T.; Kömi, J.; Porter, D. Micromechanics driven design of ferritic-austenitic duplex stainless steel microstructures for improved cleavage fracture toughness. *Eng. Fract. Mech.* **2021**, *253*, 107878. [CrossRef]

43. Topolska, S.; Labanowski, J. Impact-toughness investigations of duplex stainless steels. *Mater. Technol.* **2015**, *49*, 481–486. [CrossRef]
44. Sieurin, H. Fracture Toughness Properties of Duplex Stainless Steels. Ph.D. Thesis, Department of Materials Science and Engineering, Royal Institute of Technology, Stockholm, Sweden, 2006.
45. *NISTIR 8043*; Determination of the Compliance of NIST Charpy Impact Machines. National Institute of Standard and Technology, U.S. Department of Commerce: Gaithersburg, MD, USA, 2015. [CrossRef]
46. Klepaczko, J.R. Impact loading on specimens of different geometries, test methods and results. In *Crack Dynamics in Metallic Materials*; Klepaczko, J.R., Ed.; Springer: Wien, Austria, 1990; pp. 396–398. [CrossRef]
47. Strnadel, B.; Matocha, K. Testing samples size effect on notch toughness of structural steels. *Metalurgija* **2009**, *48*, 253–256.
48. Zajdel, P. Interpretation of the results of the impact test performed with an instrumented Charpy pendulum for the purposes of assessing the properties of structural steels. *Inżynieria I Bud.* **2020**, *7*, 341–344. (In Polish)
49. Nanstad, R.K.; Sokolov, M.A. Charpy impact test results on five materials and NIST verification specimens using instrumented 2-mm and 8-mm strikers. In *Pendulum Impact Machines. Procedures and Specimens for Verification*; Siewert, T.A., Schmieder, A.K., Eds.; ASTM STP 1248; ASTM: West Conshohocken, PA, USA, 1995; pp. 111–139.
50. van Walle, E. Evaluating material properties by dynamic testing. In *ESIS Publication 20*; Mechanical Engineering Publications Limited: Suffolk, UK, 1996.
51. Sreenivasan, R. Charpy energy–lateral expansion relations for a wide range of steels. *Int. J. Press. Vessel. Pip.* **2006**, *83*, 498–504. [CrossRef]
52. Stankiewicz, M.; Holloway, G.; Marshall, A.; Zhang, Z.; Slazak, B. Charpy impact test and lateral expansion parameter in welding consumables evaluation for cryogenic application. *Przegląd Spaw.* **2012**, *11*, 2–11. (In Polish)
53. Wang, C.; Tong, Z.; Zhong, W.; Lin, H.; Ning, G.; Zhang, C.; Yu, B.; Xu, S.; Yang, W. A method for directly measuring fracture toughness and determining reference temperature for RPV steels by Charpy impact test. *Eng. Fract. Mech.* **2021**, *243*, 107526. [CrossRef]
54. Park, T.C.; Kim, B.S.; Son, J.H.; Yeo, Y.K. A new fracture analysis technique for Charpy impact test using image processing. *Korean J. Met. Mater.* **2021**, *59*, 61–66. [CrossRef]
55. Wouters, R.; Froyen, L. Scanning electron microscope fractography in failure analysis of steels. *Mater. Charact.* **1996**, *36*, 357–364. [CrossRef]
56. Goldstein, J.; Newbury, D.; Joy, D.; Lyman, C.; Echlin, P.; Lifshin, E.; Sawyer, L.; Michael, J. *Scanning Electron Microscopy and X-ray Microanalysis*, 3rd ed.; Kluwer Academic Publishers: New York, NY, USA, 2003; pp. 1–673. [CrossRef]
57. Wassilkowska, A.; Czaplicka-Kotas, A.; Zielina, M.; Bielski, A. An analysis of the elemental composition of micro-samples using EDS technique. *Tech. Trans. Chem.* **2014**, *18*, 133–148.

materials

Review

Additive Manufacturing Processes in Selected Corrosion Resistant Materials: A State of Knowledge Review

Alisiya Biserova-Tahchieva [1], Maria V. Biezma-Moraleda [2], Núria Llorca-Isern [1,*], Judith Gonzalez-Lavin [2] and Paul Linhardt [3]

[1] Departament Ciència de Materials i Química Física, Universitat de Barcelona, 08028 Barcelona, Spain
[2] Departamento Ciencia e Ingeniería del Terreno y de los Materiales, Universidad de Cantabria, 39004 Santander, Spain
[3] Institute of Chemical Technologies and Analytics, Vienna University of Technology (Technische Universität Wien), 1040 Vienna, Austria
* Correspondence: nullorca@ub.edu

Highlights:

- A full review article on the corrosion behaviour of additive manufacturing (AM) of metallic parts is presented with the aim of covering the significant lack of information on this subject.
- Corrosion resistance of light metallic systems and duplex stainless steels objects obtained by additive manufacturing (AM) processes is critically discussed, based on the synergism between variables linked with chemical composition, manufacturing processes, and service conditions.
- Methodologies for improving the corrosion resistance of metallic parts produced by additive manufacturing (AM) are highlighted.
- The available data allow us to state that potentiodynamic corrosion tests are the best methodology to characterize metal additive manufacturing (MAM) corrosion and this unique test should be selected for the evaluation of this phenomenon. There is a gap in knowledge and test procedure to properly identify the relationship between defects and corrosion behaviour of the most studied metal additive manufacturing (MAM) systems.

Abstract: Additive manufacturing is an important and promising process of manufacturing due to its increasing demand in all industrial sectors, with special relevance in those related to metallic components since it permits the lightening of structures, producing complex geometries with a minimum waste of material. There are different techniques involved in additive manufacturing that must be carefully selected according to the chemical composition of the material and the final requirements. There is a large amount of research devoted to the technical development and the mechanical properties of the final components; however, not much attention has been paid yet to the corrosion behaviour in different service conditions. The aim of this paper is to deeply analyze the interaction between the chemical composition of different metallic alloys, the additive manufacturing processing, and their corrosion behaviour, determining the effects of the main microstructural features and defects associated with these specific processes, such as grain size, segregation, and porosity, among others. The corrosion resistance of the most-used systems obtained by additive manufacturing (AM) such as aluminum alloys, titanium alloys, and duplex stainless steels is analyzed to provide knowledge that can be a platform to create new ideas for materials manufacturing. Some conclusions and future guidelines for establishing good practices related to corrosion tests are proposed.

Keywords: corrosion; additive manufacturing (AM); titanium alloys; aluminum alloys; duplex stainless steels; defects

Citation: Biserova-Tahchieva, A.; Biezma-Moraleda, M.V.; Llorca-Isern, N.; Gonzalez-Lavin, J.; Linhardt, P. Additive Manufacturing Processes in Selected Corrosion Resistant Materials: A State of Knowledge Review. *Materials* **2023**, *16*, 1893. https://doi.org/10.3390/ma16051893

Academic Editors: Claudio Gennari and Luca Pezzato

Received: 23 January 2023
Revised: 21 February 2023
Accepted: 22 February 2023
Published: 24 February 2023

1. Introduction

AM is continuously growing and helping to expand the options for materials used in manufacturing. Although thermoplastics have been the most attractive materials in the early days of 3D printing, metallic systems have been driving an important revolutionary change in AM during the last decade, addressing fundamental challenges for engineers and researchers.

AM has opened a huge new research field and prompted the investigation of alloys and new techniques in order to optimize the whole process, from the raw material properties to the final component characterization. This is due to the advantages that the process can provide compared to the traditional manufacturing techniques. Additive manufacturing development is exponentially accelerating cost adjustment and efficiency in producing different components for a wide range of applications. The reduction in energy of up to 25% and cutting of waste and materials costs up to 90% are some impactful achievements, in addition to the main attractiveness of the AM processes.

Furthermore, a variety of complicated shapes can be easily designed through the AM technique and, thus, specific components can be obtained for biomedical applications [1–3] and the aeronautical industry [4–6], with titanium and aluminum alloys being, respectively, the most-used metallic materials in such applications due to their mechanical properties and lightness which are essential to both industries. Fe-based materials, and more specifically, stainless steels, as the best substitutes in applications where corrosion resistance in combination with mechanical properties is required, are frequently used in AM, as well. Understanding the AM process conditions, postprocessing, and structure-performance of stainless steels should be considered to create more corrosion-resistant stainless steels [7,8]. There is also growing interest in the use of AM for the production of jewelry and luxury watch components, as well. This interest is driven not only by the potential design innovation offered by Additive Manufacturing, but also from the environmental and economic point of view, as the recovery of precious metals such as gold is a complex process, whereas this step is circumvented by applying gold as an alloying element. The main precious metal that benefits nowadays from AM is platinum and its alloys, which are used to create highly-value added metallic objects [9]. Nickel-aluminum bronze (NAB) alloys are extensively used in the marine industry and there has been a recent increase in interest in their being additively manufactured. Their microstructure can vary in comparison to conventionally casted NAB alloys and some studies have shown improvements in their mechanical and corrosion properties [10].

Such applications require high-quality control and specific corrosion resistance properties besides mechanical resistance. The influence of the conventional techniques on such properties in metallic components produced by AM has been extensively studied. However, it has been observed in the literature that there is a lack of information and data, specifically on the corrosion properties of AM metallic materials in comparison to traditionally fabricated objects. For these reasons, the main purpose of this paper is to obtain a state of the art of the most significant problems related to all kinds of defects occurring during and after the main AM processes and to correlate these issues to the corrosion resistance of the as-built AM metallic parts.

2. Additive Manufacturing Processes Classification

2.1. Introduction

In order to understand the impact of additive manufacturing processes, a general definition and description, as well as classification of the various additive manufacturing processes, will be covered in the following section.

Additive manufacturing (AM) is defined by the ISO/ASTM 52900 [11] as the process of joining materials to make objects from 3D model data, usually layer upon layer, as opposed to subtractive manufacturing methodologies, such as the traditional machining of building components. AM refers to a variety of processes in which material is deposited, joined, or solidified [12]. Metal additive manufacturing (MAM), also known as metal 3D printing, consists of complex metal parts fabrication with improved functionalities. The advantages of using AM of metals over traditional manufacturing methods, in some cases, have led to an enormous increase in its use; for example, between 2020 and 2024, metal AM growth is estimated to be around 14% per year [13], alongside a simultaneous increase in the number of studies on additively manufactured steels. For instance, 5.861 articles were found in the Web of Science, considering both additive manufacturing and steels, that were published in the last five years. Since its inception in 1987, AM has shown significant progress in the general understanding of the processes related to it and the impact on the structures and properties of the fabricated metallic components.

AM technologies are based on the principle of modelling a body, loading the data into the equipment, and building the component layer upon layer until the creation of the complex 3D model. AM can be categorized according to the energy source provided during the process (e.g., laser or electron beam). 3D printing deposition is carried out by means of a print head or a nozzle (or multiple) that deposits the material in a controlled flow and temperature.

This layer-by-layer forming saves costs because it does not generate large amounts of waste, as the material is only added where it is really needed, thereby permitting near-net shape manufacture with powder or wire form as the starting feedstock. Virtually, this process does not create any residual material waste if all of the un-melted material can be fully recycled as has been demonstrated recently, focusing attention on the environmental impacts of additive manufacturing vs. traditional machining via life-cycle assessment in a comparative way [14]. Also, almost any geometry can be created, in contrast to conventional machining, the walls of the workpiece can be very thin (which is not possible through casting), and manufacturing is possible in a fairly short period of time. The freedom of design is particularly relevant in biomedicine, as dental implants or orthopedic parts and can be adapted to the specific needs of each patient [15]. However, the main sector driving the increase in the use and development of AM is the automotive sector, due to reductions in weight and, especially, in the weight/cost ratio.

In any case, we must not forget that there are also problems associated with this general technique; therefore, it is essential to study these issues in order to understand and solve them, with the aim of obtaining an even more versatile technique. To this end, numerous studies have been carried out comparing the mechanical properties, corrosion behaviour, and microstructure of metal parts produced by AM and by traditional manufacturing methods [16,17]. It is known that the special conditions of AM produce a fine microstructure with unique directional growth characteristics in non-equilibrium zones. This distinctive microstructure, fine α' martensite in as-built Ti6AlV4 [18,19], together with microstructural defects, originated from the additive manufacturing process, strongly influences the corrosion behavior of metal additive manufacturing materials. Several studies have already been carried out in this respect. However, issues concerning the corrosion and corrosion protection of these materials are still poorly understood [20].

Some AM processes have evolved from conventional welding processes, while others, such as powder bed fusion processes, have been developed with the specific intent of enabling the manufacturing of complex 3D geometrical objects. Two main groups can be distinguished in this technique:

- Powder bed technologies. Within this group, a classification can also be made according to the energy source used for deposition. Here, we find Selective Laser Melting (SLM), Electron Beam Melting (EBM), or precision inkjet printing in which the metal powder is mixed with a binder, so that after deposition the piece is sintered, resulting in the final model.
- Blown powder technologies, also known as Laser Metal Deposition, LMD, or Laser cladding, wherein both are based on the availability of the metal powder to blow coaxially to the laser beam, which melts the powder onto a metal substrate to form a metallurgical bond upon cooling to room temperature.

Depending on the material used, a distinction can be made between metal in the form of 'wire' and metal in the form of 'powder.' For the former, the metal is placed in the form of a coil which is connected to the extruder nozzle. For the latter, several requirements must be met. For additive manufacturing, metal powder should have a spherical shape to ensure good flow and coating ability. A particle size range between 50 μm and 150 μm, depending on the machine type and its distribution, must be adapted to the application, as well as the chemical composition of the metal and the gas content, because these may be responsible for defects within the structure [21,22].

There are also other causes associated with AM processes that can produce defects in the structure. One example is the thermal cycling inherent to manufacturing, resulting in different microstructures and types of defects related to the process parameters and the geometry of the final object, as well as the local environmental conditions during processing.

If the manufactured part has a low static loading, the microstructure itself can determine the average mechanical properties. However, if it must withstand cyclic loading, as in aero engine or turbine components, defects limit the lower threshold of the mechanical properties and, therefore, are a major concern, as they restrict the loading conditions during operation. In view of the above, post-manufacturing treatments such as Hot Isostatic Pressing (HIP) are essential, as they can minimize certain types of defects, such as porosity, depending on the material and the AM process used. Other treatments include hybrid solutions between in-situ and post-manufacturing treatments, such as the induction of compressive residual stresses in the material to reduce the influence of surface topography, defects, and residual stresses.

Sander et al. [23], in 2018, published a deep, extended, and critical review of corrosion knowledge of Additively Manufactured alloys, pointing out that this revision practice should be done each year, since AM is being introduced in all industrial sectors in which severe and critical service conditions are present. Indeed, this is the motivation of this paper: to advance the corrosion behavior knowledge of the AM metallic systems, paying special attention to the deep relationship between the existence of defects resulting from each particular process as well as the final microstructure of the produced material. The following sections present a description and classification of the main AM methods and tools that are currently carried out to produce metallic components.

2.2. *Additive Metallurgy by Direct Energy Deposition Processes (DED)*

During direct energy deposition, focused thermal energy is used to melt the material as it is deposited. This is the main difference of Powder Bed Fusion (PBF), where thermal energy is used to selectively fuse regions of a powder bed. DED processes are typically used on existing parts of arbitrary geometry with a relatively high deposition rate; however, they allow for little complexity in shape. Therefore, their main use is in the repair or improvement of other pre-formed parts.

2.2.1. Additive Metallurgy by Laser Melting Deposition (LMD)

The main advantage of this method is the lack of Heat-Affected Zones (HAZ), or hydrogen cracking, which is characteristic of repairs by conventional arc welding, or cold or plasma spraying. In fact, LMD can reduce distortion and microcrack formation by optimizing the parameters, resulting in metallurgical-bonded and high-quality coatings obtained in a short period of time. When used to repair parts, it is done with highly satisfactory results. With respect to the disadvantages, a typical example is when LMD is used with 316 L stainless steel, in which the composition has a great effect, for instance with any redistribution of Mo in printed 316 L alloys. If this happens, their corrosion resistance could be significantly affected, as well as the grain size, altering local degradation mechanisms which have a strong dependence on the corrosive media (decreasing grain size in austenitic-type 304 stainless steel below 2 μm increases the general corrosion in sulfuric acid solution due to passive film destabilization at grain boundaries) [24,25].

2.2.2. Additive Metallurgy by Wire Arc Additive Manufacturing Processes (WAAM)

WAAM uses arc welding tools and makes use of wire as feedstock for additive manufacturing. The deposition rate is high, the equipment is not expensive, and it has a good structural integrity, making it a promising method to replace the current manufacturing methods for components with low and medium complexity [23]. During the WAAM process, every single layer has a heterogeneous grain structure (as well as melt pool boundaries, matrix supersaturation, segregation, phase transformation, new textures, and oxide formation), as in other types of MAM; however, in this case, the former layer is in a solid-state, and, therefore, holds a lower temperature, increasing the effect. In addition, for each individual layer, three zones can be distinguished (Figure 1): the Melt Pool Zone (MPZ), Melt Pool Border (MPB), and heat affected zone (HAZ). The MPZ is produced because the metal wires are completely melted during WAAM. Between different MPZ areas, it is possible to find the MPB, with a completely different grain structure and intermetallic distribution. Finally, HAZ is the previously deposited layer region where the temperature is high enough to alter the microstructure but below liquidus temperature. The region which is the furthest away is total or partially remelted, and belongs to the MPB on the previously deposited layer side [26].

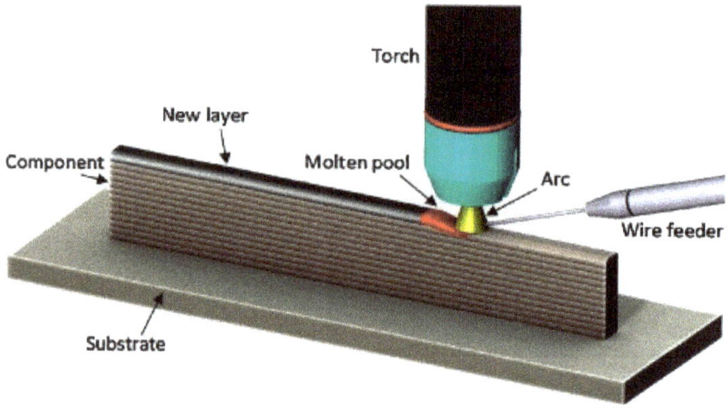

Figure 1. Illustrative diagram of WAAM process and Schematic representation of different zones in WAAM part, being MPB: melt pool border, MPZ: melt pool zone, and RM: remelted, HAZ: heat affected zone [27].

2.3. Additive Metallurgy by Powder Bed Fusion Processes (PBF)

Different scan parameters should be considered when the PBF process is carried out. An inert atmosphere or partial vacuum is necessary to provide shielding of the molten metal. Currently, the most-used practice is laser powder-bed fusion (L-PBF) followed by vacuum heat treatment to produce alloys with controlled microstructures. An energy source is used to selectively melt each layer of the powder, which are already spread according to the required cross-section of the part in the digital model. Once the layer has been scanned, the piston in the building chamber moves down and the piston in the powder chamber moves up according to the defined layer thickness. The coating mechanism or roller deposits the powder in the building chamber, which is scanned again by the energy source. This cycle is repeated layer by layer until the complete part is formed. The time required to complete a part is longer than that for DED technologies; however, a higher complexity and better surface finish can be achieved and minimal post-processing is required. Multiple parts can be built together to make maximum use of the build chamber [28]. If the process is carried out at reduced pressure, a more stable melt pool and reduced porosity are obtained. Part-specific process settings are often controlled to reduce thermally-induced residual stresses and defects.

2.3.1. Additive Metallurgy by Laser Powder Bed Fusion (L-PBF)

Powder-Bed category techniques involve laser powder bed fusion (L-PBF), also known as selective laser melting; SLM; selective laser sintering, SLS; and EBM technologies. The cooling rate for SLM is usually >105 K/s, which is higher than that of direct laser deposition (DLD) (usually from 103 to 105 K/s) and much higher than that of traditional casting methods, which have lower solidification rates [29]. The local rapid heating and, especially, fast cooling rates, coupled with thermal cycling, induce the formation of unique microstructures with refined grain structures, dislocation cells, and internal residual stresses. These conditions also cause the formation of metallurgical defects, including un-melted powder particle microcracks, entrapped gas pores, balling, and rough surfaces.

One of the most common nucleation sites for corrosion problems in three-dimensional printed materials is pores. They reduce the passivation property in the presence of sulphuric and phosphoric acid solutions [30]. There are two types of pores: one type exists around the un-melted powder particles, and another is caused by the trapped gas inside the powder during gas atomization [31]. The porosity can be reduced to a certain extent by optimizing the printing conditions, including laser energy, scanning rate, and scanning direction. It has been found that increasing the laser power or properly decreasing the scanning rate can reduce the porosity of different metals such as nickel- or aluminium-based alloys, or some type of stainless steel such as 316 L by SLM [23]. This will be detailed in the next section.

Selective laser sintering, SLS, was the first PBF-AM process, invented in 1981 by Ross Householder [32] who patented it under the name of "Molding process". A few years later, Carl Deckard and Joe Beaman patented a similar method, under sponsorship of DARPA [33], which was more alike to what is known today as SLS.; however, it would take some years before this technique could be used with metallic materials, as it was initially used with amorphous or semi-crystalline polymer powder or semi-crystalline powder. Usually, this technique is related to a final step that consists of applying thermal treatment in order to remove the polymer binding particles, sintering, and improving the base material microstructure [34–36].

In any case, there are examples of the strength of this technique, such as the seven degree-of-freedom dual-arm hydraulic which was built using the SLM method for undersea use. The hydraulic system was produced with titanium to obtain the necessary properties for naval applications [37].

2.3.2. Additive Metallurgy by Electron Beam Melting (EBM)

Many parameters can be modified and corrected using the EBM technique so that the final part does not require further treatments. Precursor and pre-alloyed powders are selectively preheated and melted by varying the beam sweep, its current, and even the cooling rate, this being one of the main causes of defects in AM due to creating stresses in the microstructure. If the microstructure and mechanical properties of these systems are analyzed and compared to conventional forged and cast products, a columnar microstructure resulting from layer-by-layer melting solidification phenomena can be distinguished. This is because solidification is directional, which implies a continuous melting/solidification front [38].

Some of the major defects that appear during this process are unavoidable to some extent and can severely degrade the mechanical properties of the material. An example of this is Alloy 718, obtained by EBM, which is often post-processed to improve the material properties. Although hot isostatic pressing (HIP) is commonly used to close defects, it cannot close open-to-surface ones. Therefore, a possible solution could be that, if the surface of the EBM-manufactured specimen is suitably coated to encapsulate the EBM-manufactured specimen, then HIPing can be effective in healing such surface-connected defects [39]. It is remarkable that the residual properties of EBM-fabricated components are usually as good as or better than conventional cast or wrought products, even after post processing. Table 1 summarizes the main advantages and disadvantages of both the DED and PBF AM techniques.

Table 1. Main advantages and disadvantages of DED and PBF AM techniques.

Main Process		Advantages	Disadvantages	References
DED	LMD	Less micro-cracking Improved thermal control	Require post-processing Low process rates	[40]
	WAAM	Good structural integrity High deposition rate	Lower accuracy Different microstructure obtained	[24,40,41]
PBF	L-PBF(SLM)	High heat and process speed No support structure required	Surface roughness Powder particle size	[42,43]
	EBM	High process rates Good accuracy Fully dense parts	High surface roughness Requires high-quality powder	[42,44,45]

2.4. Additive Metallurgy by Hybrid-AM Techniques

As already mentioned, AM processes offer numerous advantages, especially in terms of free design in a reduced time; however, this is irrelevant if the part created is full of defects that can affect the mechanical, physical, or chemical properties in such a way as to render the part unusable. To avoid this, in situ or serial AM processes and secondary energy sources that are capable of modifying the resulting material or part properties must be combined. These modifications can be due to both in-situ secondary procedures and process chains, and is anchored in multiphysics mechanisms, so that new hybrid-AM processes can be applied very selectively to the problem which was generated during AM (Figure 2). The desired material properties determine the mechanism that is used and, in turn, the energy source that is applied, which ultimately defines the hybrid-AM process. Some of the properties that can be changed using this mechanism are melt pool dynamics, microstructure development, stress state, surface evolution, and thermal gradients [46].

Overall, additive manufacturing processes for metal systems promise significant revolution in the industry, assuring advantages but also many challenges that must be faced in the upcoming century.

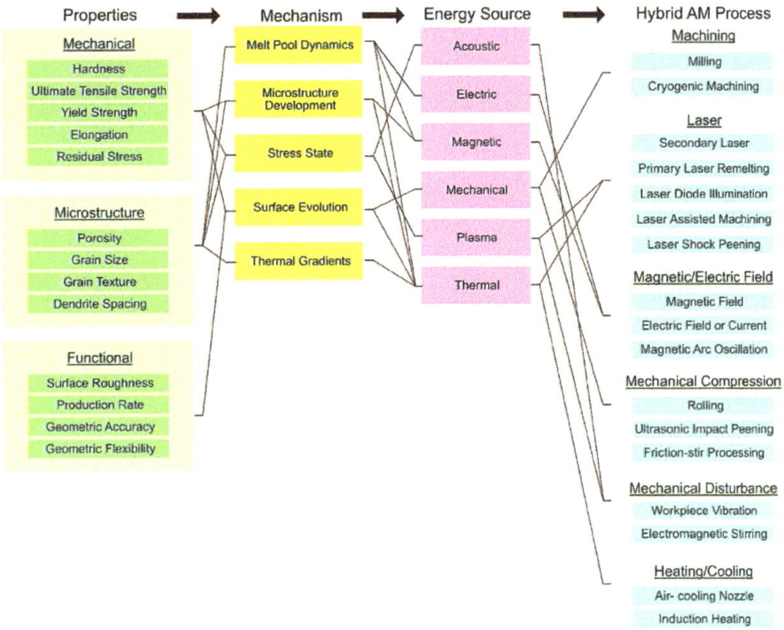

Figure 2. Property—mechanism—energy source—hybrid-AM process framework for hybrid-AM classified by mechanism and energy source utilization. Attending to the scheme, it is easy to understand why such a problem-oriented selection of the sample is possible [46].

3. Defects in Metallic Systems Due to AM Processes

The control of defects in all processing of metals is of crucial importance, and in the same is true of additive manufacturing processes, as these can significantly deteriorate mechanical properties. Additionally, it is even more important to minimize or prevent processing defects when corrosion resistance is required from the processed components. Therefore, in this section, we will discuss the appearance of different defects during the processing of metals and their influence on the loss of corrosion properties.

Despite the great advances and improvements provided by AM techniques, as has already been mentioned in the previous section, the appearance of defects in the manufactured parts can significantly worsen the physico-chemical properties of the components and, with this, their functionality. Since a defect is a discontinuity of a material that penalizes the properties of that material, they must be avoided during the processing of metallic systems. The key issue is to define the concept of defect in the material processed by additive manufacturing due to the particularities of each process concerning the material to be processed, such as chemical composition and geometry, as well as the variables of the process: temperature, flux heat speed, vacuum system, etc. Therefore, it is highly dependent on the process itself.

Defect formation is a common problem in all types of AM, being commonly identified in WAAM and SLM; nevertheless, the defects generated during WAAM are the most studied [47], wherein residual stresses and distortion parameters, among others, can influence the fracture behavior and as a result it can be difficult to achieve the required tolerances of the components. In this sense, a defect is a discontinuity; however, in general, larger and more heterogeneously distributed defects would imply a higher negative impact related to the same situation in other classical processing methodologies. In this way, the critical defect criteria, which are type, size, orientation, and distribution, may be known. Hence, it is also important to understand the properties of AM components and the associated effects of defects, mainly porosity and cracks, to understand their impact. Previous studies have defined the term as critical defect criteria [48].

Corrosion could be controlled by the existence of defects that are caused during the additive manufacturing process; therefore, it is important to know the defects present in a material processed by AM and their influence on the final properties. In this section, the description of the main defects found in the literature and their direct correlation to the properties of several components of alloys obtained by the AM process will be shown. The main defects are not strictly categorized by a unique type; nevertheless, the most frequently found are cracks, porosity, residual stresses, non-homogenous microstructure, unbalanced chemical composition, brittle phases nucleation, properties anisotropy, and microgalvanic cells. Some of these defects are shown in the schematic illustration in Figure 3a, which represents their formation during the layer-by-layer building procedure of AM processes; for example: porosity, lack of flow parts, keyhole defects, microcracks, and the formation of intermetallic phases that are not visible macroscopically (not represented in the scheme). Figure 3b details the appearance of additional defects due to the influence of a corrosive environment on the as-built AM component. For instance, intragranular corrosion and pitting corrosion defects could be present in the microstructure along with the AM process defects, causing significant microstructural damage. Figure 3c offer a general view of a component after post AM processes such as solution annealing treatment at a certain temperature and during a specific time, as well as the surface pickling passivation process, which improves the homogenization of the microstructure and protects, superficially and internally, the material from corrosive environments, respectively.

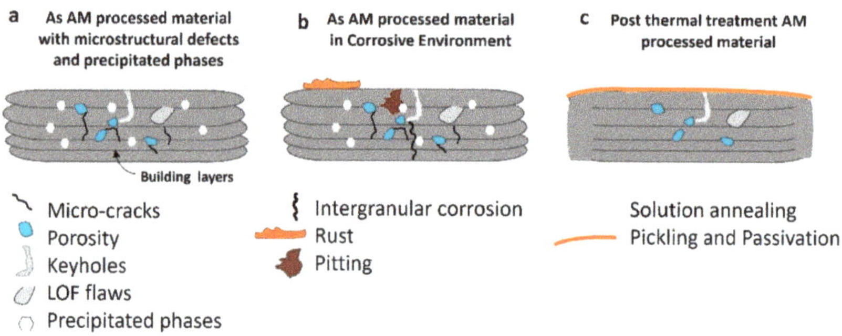

Figure 3. (**a**) Schematic representation of an AM-processed material and the presence of several microstructural defects (micro-cracks, porosity, keyholes, and LOF flows, as represented); (**b**) the same AM processed material in a corrosive environment where intergranular corrosion, pitting corrosion, and rust on the surface can appear; (**c**) the same AM-processed material after being thermally treated (solution annealing) to remove residual stresses and precipitated secondary phases after pickling and passivation as a preventive measure for corrosive environments.

Wu et al. [47] have published a state of the art focused on the defects found in different alloys used in WAAM in order to control the quality improvement requirements for specific industrial applications. An algorithm has been optimized to identify the correct process parameters by Busachi et al. [49] in order to identify the main defects generated during the WAAM process: porosity, undercutting, and humping. B. Zhang et al. [50] reviewed defect formation mechanisms and their classification in SLM for common defects such as porosity, incomplete fusion holes, cracks, surface roughness, and delamination. However, it is necessary to point out that the type of defects is very closely dependent on the chemical composition of the material to be processed, as has been observed by Xu et al. [51] in AlSi10Mg alloy used in SLM, with lack of fusion and porosity being identified as the most common defects. Residual stresses are usually induced in SLM process due to high thermal gradients occurring in the process that can also cause cracks and distortions in the manufactured material, as has been observed by Siddique et al. [52].

Sander et al. [23] conducted an exhaustive state of the art review, linking the chemical composition with the corrosion behavior of numerous materials in multiple adverse conditions. They have pointed out the necessity of a deeper fundamental understanding of MAM and have called attention to the necessity of standards and standardized assays, which can be, for example, specimen preparation. Actually, it is critical to approach an unified understanding of the complex relations between all of the variables involved in each process. Nevertheless, the increasing development of additive manufacturing techniques as an alternative to conventional processing, having an impact on new metallic materials, needs to proceed with comprehensive periodical revision, which is one of the main purposes of this paper.

Classification and Identification

Since the different processes can induce different defects, it is not easy to establish a close classification of them. Moreover, some hypothetical defects could be considered as desirable variables during an AM process, such as porosity, in order to produce lighter components despite the fact that they are cavities within the weld bead. In particular, its presence reduces the overall density of the parts and thus reduces the mechanical properties of the manufactured components. In this study, we will only be considering defects which introduce a penalizing issue to the material microstructure and/or macrostructure. The formation of keyholes is the main defect that appears in WAAM and SLM build components [53]. In research conducted by the LLNL [54], a previously unknown dynamic within LPBF was revealed, in which the process produces small particles or clusters of powder particles that are ejected from the laser's path and which can land back on the parts. This phenomenon can lead to the formation of pores or defects in final parts. After discovering these spatter interactions, LLNL researchers set out to better understand them and find a solution for reducing or even eliminating defects in parts built through a common, laser-based metal 3D-printing process. They show a simulation of a laser interaction, where laser-power was above a threshold that expelled the spatter away from the scan track, preventing the formation of defects due to "laser shadowing".

Porosity refers to cavities within the weld bead and is considered a defect as it affects the performance of the weld. In particular, its presence reduces the overall density of the parts and thus reduces the mechanical properties of the manufactured components. Undercutting is related to the concavity of the weld bead, which compromises the tolerance requirements. Humping refers to an uneven material deposition. LOF defects are the result of an interaction of spatter particles of the material with the incident laser beam. Kim et al. [39] classified two porosity types in AM-produced parts: lack-of-fusion (LOF) porosity, resulting from the interaction of spatter particles of the material with the incident laser beam, and gas porosity, (GP). LOF does not necessarily come from the interaction of spatter particles but can also result from an insufficient energy input which does not completely melt the powder bed.

The various types of microstructural features or defects, their generation mechanisms, their effect on bulk properties, and the capability of existing characterization methodologies for powder-based AM parts were considered by Collins et al. [55], who pointed out the importance of their in-situ detection by non-destructive tests. Recently, Snow et al. [56] found that these flaws in AM-produced material are an important defect that penalize fatigue resistance, by the application of machine learning approaches they distinguish different types by attending to geometry and size: gas porosity and keyhole pores, (~50 μm or less) relatively spherical, and lack of fusion pores with an irregular morphology, often much larger,. Nassar et al. [57] showed that stochastic collisions occur both between particles which are nearly-simultaneously expelled from the laser interaction zone and between particles which are ejected from distant locations to support the existence of LOF defects. The effect of ejected melt and its spatter in PBFAM specimens has been studied, the occurrence of cracks being the main associated defect. Monzon [58] defined three

Table 3. Associated problems to MAM and their solutions.

Problem	Cause	Solution
Refrigeration	Materials are not able to withstand the heat generated during AM.	Add external part that cools the equipment as an electronic chassis [77].
Material cost	Lightness of the materials makes it difficult to measure them accurately until there are not enough layers.	Creation of different gadgets to resolve this issue [78].
Abruply deposition	Control of deposition is not accurate due to the usually lower melting point of light metals	Add intermediate piece on which the molten material accumulates before being deposited or a first laser that pre-fuses the material [79].

4.1.1. Aluminium Alloys

Aluminium alloys are one of the main systems in which AM is used, especially in the SLM process, and they can be used for a variety of applications, from medical to aerospace. In the latter case, aerospace-grade aluminium is of great interest and is still under investigation [80]. However, processing these alloys is challenging due to the difficulties associated with laser-melting aluminium, where parts may show various types of defects. In recent times, several studies have developed approaches to avoid these defects and reported the successful SLM of various Al alloys [81]. For instance, porosity or high surface roughness, residual stresses, and remaining unmolten powder in the printed parts are some of the defects formed that affect the corrosion resistance properties of Al alloys [82–84]. There are several parameters that can influence the final printed parts microstructure and surface roughness, and thus, the corrosion performance. Therefore, post-AM treatments are applicable to reduce such roughness [85,86]. Another example is the application of post-processing electro-polishing, shot peening, or mechanical and electro-polishing, which have shown better corrosion resistance than as-printed parts or sunblasted samples, as reported by Fathi et al. [87].

On the other hand, the effect of the alloying elements such as Si and Mg in AlSi and AlSiMg alloys have shown different corrosion behaviours. Some studies [88,89] have reported that better corrosion behaviour is found in alloys with a higher amount of Si, in comparison to the AlSiMg alloys with a higher amount of Mg. This is mostly due to the formation of Mg_2Si precipitates that affect pitting corrosion. However, in the as-built AM components, the formation of Mg_2Si particles have not yet been confirmed. Revilla et al. [20] mentioned many issues in a detailed review about the corrosion resistance of AM Al alloys and have suggested further studies about Mg- and Fe-containing precipitates in Al.

Moreover, solution treatments (T6) on as-built SLM AM parts have been performed in different studies [90,91] to compare the corrosion behaviours within untreated parts. It has been observed that, after exposure to temperatures between 300 and 550 °C for 2 h and water quenching afterwards, separated Si particles in the matrix resulted in higher corrosion densities and lower corrosion potentials. The larger the Si precipitates are, the more restricted is the formation of the compact oxide layer.

A recent review has pointed out the complex and numerous scenarios for cast aluminum alloys and the relationship between microstructure and corrosion resistance [92]. Moreover, aluminum alloys exhibit elongated grains that are aligned with the extrusion direction. This radically changes with treatments leading to thioxoformed high-performance alloy with a regular equiaxial grain microstructure [93].

4.1.2. Titanium Alloys

There are two main reasons why the use of titanium alloys in AM is relevant. The first is the fact that these alloys are biocompatible; therefore, if there is also the possibility to create almost any shape adapted to specific needs, this fabrication is essential. On the other hand, titanium alloys have seen their activity reduced due to their high cost, meaning that it is essential to reduce waste, which is another advantage of AM [94]. Therefore, it seems that the combination of this material with this technique is a great advantage.

In general, it is necessary to properly characterize the mechanical properties of any material to be used; however, due to the medical use of Ti alloys, it is essential to do so after the production by AM. The rapid solidification during the AM process leads to the formation of highly stressed phases, which results in the deterioration of its mechanical properties and the appearance of corrosion on the parts. In addition, Laquai. et al. detected small inhomogeneities such as voids or cracks whose sizes are below the spatial resolution of optical microscopy for Ti-6Al-4V parts produced by selective laser melting [95]. This issue supports the need to use higher resolution techniques to control AM objects.

Zhang et al. [96] presented an extended review of Ti alloys fabricated by the EBM process in which several important issues are explained that are related to the problems within the AM technique, the defects detected afterwards, and the mechanical properties compared to conventionally fabricated components. However, little information is reported about the corrosion resistance properties of AM-fabricated Ti alloys. For instance, they show the results of EBM-fabricated Ti-6Al-4V alloy that exhibits a transformation from the stable prior β phase to stable α phase. Corrosion resistance enhancing has been observed after a higher fraction of β phase and refined lamellar α/β phases with different behaviour compared to traditionally fabricated Ti-6Al-4V alloys [96].

In pure titanium alloys, the microstructure predominately consists of equiaxed prior-β grains, and a body centered cubic crystal structure. As aluminum and vanadium are added, the transformed β phase is presented, which is a varierty of α-morphologies such as widmanstätten; however, vanadium radically alters the microstructure of Ti-Al, since it predominantly leads to a very fine acicular α phase and some are delineated by an intergranular β phase [97].

Chiu et al. [98] have compared the corrosion assessment of Ti-6Al-4V alloy that is conventionally fabricated and a L-PBF build in a body fluid simulation solution. The results showed similar E_{corr} values with a small improvement in the AM-fabricated Ti alloy. Performing a heat treatment adopted before the final application of the as-built AM pieces has been seen as a solution to improve corrosion behaviour. It has been found that the corrosion rate of as-built AM Ti alloys is almost 16 times worse than cold-rolled Ti components, mainly as a result of the non-equilibrium phases formation during AM processes. If the heat treatment is carried out properly, this loss of corrosion property can be reduced or even eliminated. For Ti-6Al-4V alloy produced by the laser-based powder bed fusion AM technique, the ideal process is a post annealing heat treatment at 800 °C for 2 h, making it so that the corrosion behaviour of the piece can be comparable to the commercial sample ones, due to the stress relief of the martensitic phase and formation of the BCC phase of β Ti-6Al-4V, which has a higher corrosion resistance.

In the case of Ti-Al alloys, there are three major intermetallic compounds, gamma Ti-Al, alpha 2-Ti 3 Al, and Ti-Al3. Among the three, gamma TiAl has received the most interest and applications. A particular method of improving the stability of the passive layer is adding platinum group metals. Therefore, the corrosion resistance increases, and these alloys are not expected to suffer a relevant corrosive effect, as they increase the corrosion potential to nobler values. Additions of Pt group metals facilitate cathodic depolarisation by providing low hydrogen surge sites, which shifts the alloy potentials in the positive direction, where passivation of the oxide film is possible. Relatively small concentrations of certain precious metals are sufficient to significantly increase the corrosion resistance of titanium in reducing acidic media. A very clear example of how this contributes to corrosion enhancement is in a 25%w HCl solution environment (Figure 4). The addition shifts the cathodic process in the active region of the single Ti-Al alloy. The cathodic modification of Ti-Al alloyed with Pt group metals occurred as a result of these metals accumulating on the surface of the Ti-Al alloys, which simultaneously enhanced the hydrogen evolution efficiency and inhibited metal dissolution [99].

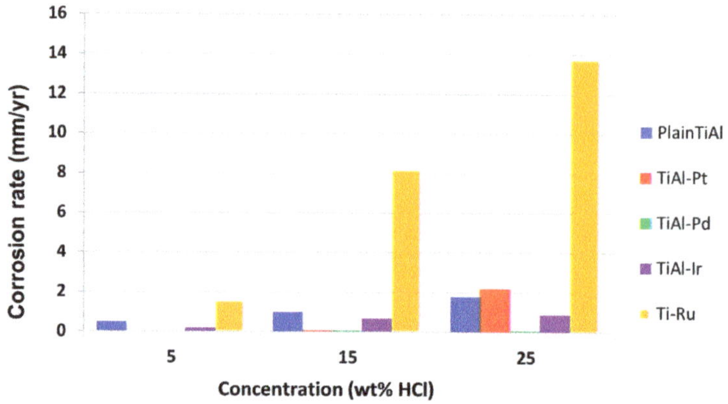

Figure 4. Corrosion rates of Ti-Al alloys with Pt-group metal additions in 5, 15, and 25 wt-%HCl solutions [99].

Overall, light metallic systems such as Ti and Al alloys are essential in the biomedical and aeronautical industries. Additively-manufactured components require controlled processing in order to avoid non-equilibrium phase precipitation and prevent the decreasing of corrosion resistance. Further research is required to understand the influence of some AM processes parameters on the corrosion resistance of Al and Ti components, as well as their improvement in comparison to conventional processed parts.

5. Duplex Stainless Steels and Additive Manufacturing

DSS are widely investigated due to both their mechanical and corrosive properties, although only 1.558 articles were published for the last five years considering their processing through AM techniques and their influence on corrosion resistance. Therefore, in this section, the main features of AM fabrication and problems related to corrosion resistance of duplex stainless steels (DSS) are presented, these being the most widely fabricated through AM processes among stainless steel types.

Duplex stainless steels have extensive applications as structural materials in many industries and different environments, as these steels present both high mechanical and corrosion resistance properties provided by their dual-phase microstructure (ferrite δ and austenite γ) [100]. Some of their applications involve chemical, oil, gas, food, and marine industries, where their good performance and attractive economical cost can substitute alternative materials such as nickel-based alloys. These steels may present precipitation of undesired intermetallic phases, carbides, and nitrides at different temperatures. Particularly, from 600 °C to 1000 °C, the sigma phase (σ), chi-phase (χ), carbides ($M_{23}C_6$, M_7C_3), or nitrides (CrN, Cr_2N) can precipitate and deteriorate their mechanical properties as well as decrease their corrosion resistance [101–103]. Moreover, the intermetallic phases present higher content in Fe, Cr, and Mo and, thus, can be the main factor responsible for the decrease in toughness and pitting corrosion. Due to the ferrite-formation elements (Fr, Cr, and Mo), these secondary phases are considered to precipitate from the ferrite phase and, in some cases, can lead to its total consumption.

To show the complexity of the microstructure that can appear when processing DSS by AM, it is worth noting that there is some controversy around the transformation from delta ferrite during heat treatment after the conventional casting processing which has been just recently solved [104]. The usual microstructure observed in a typical wrought microstructure is composed of ferrite (47 ± 4)% and austenite (53 ± 4)%. Hot-rolling materials provided numerous potential void nucleation sites, which accelerated the propagation of cracks along the ferrite/austenite interface and caused serious cracking [105]. Meanwhile, the weld material has a typical microstructure of ferrite (42.1 ± 3.0)% and Widmanstätten austenite (57.9 ± 3.0)% as reported by Tavares, et al. [106], with massive plates of σ phase precipitated in the ferrite phase.

Because of the attractive properties of the ferritic-austenitic microstructure, and, therefore, the wide range of applications that DSS can perform, industry is increasingly processing DSS through AM techniques. Such processing not only allows for obtaining different shapes and dense materials, but can also be an alternative process to the conventional techniques. However, some of the main problems with these processing techniques are related to the precipitation of undesired secondary phases and the balance control of the two main phases in order to maintain desirable mechanical properties and, more importantly, the corrosion resistance properties, because of the severe conditions in which they are used. In order to simulate these conditions, most of the corrosion studies used (mainly polarization tests) in the literature are carried out under similar conditions; for instance, 3.5 wt% NaCl solutions at room temperature. Thus, in the following section, the corrosion results are presented for some typical grades of duplex stainless steels.

5.1. Main AM Processes and Associated Problems

Additively manufactured steels can be more susceptible to corrosion due to several defects that may occur during processing. Not only the presence of intermetallic phases but also microstructural defects after the processes such as porosity because of incomplete fusion between successive molten layers or localized defects can be of importance when considering the decrease in corrosion resistance. However, in some cases, for example during SLM processing, porosity is highly improved [107,108]. A refined grained microstructure can be achieved and, thus, pitting corrosion resistance is increased. In addition, a higher number of grain boundaries promotes the diffusion of elements which could form stable passive films. Research has shown that different AM stainless steels perform better with regards to corrosion resistance than the same wrought stainless steels. Moreover, as seen above, most of the AM processes for stainless steels require controlled cooling rates and heat treatments in order to obtain an optimal microstructure and balanced austenite-ferrite ratio to maintain the desirable corrosion resistance properties. The general behavior and pitting corrosion resistance of some common duplex stainless-steel grades after being AM processed are discussed in the following.

5.1.1. Wire Arc Additive Manufacturing Processes

One of the most-used AM processes with DSS is the wire arc additive manufacturing process (WAAM), which is a direct energy deposition process, where filler wire and arc are the heating source. This process is cheaper and more readily implemented in industries, as well as enabling higher productivity. However, as mentioned above, one of the main problems associated with its performance is the maintaining of an optimal dual ferritic-austenitic microstructure. Because of the continuous reheating process by forming consecutive layers, it is of great importance to control the heat input, the interpass temperatures, and the high cooling rates in order to avoid undesired intermetallic phases and to maintain the balance between ferrite and austenite.

There are several studies describing the material properties and process effects such as temperature on phase precipitation, performance, and phase balance in DSS [41,109–113] after WAAM. For instance, Eriksson et al. [109] studied the effect of heat input on the mechanical properties of superduplex stainless steel UNS S32507, considering that harmful secondary phases could be formed during the process. They have shown that the base metal consisted of 48 vol% ferrite and 52 vol% austenite; however, during welding, the microstructure transforms to ferrite. Moreover, the precipitation of austenite from ferrite during cooling depends exclusively on the cool rate. The heat-affected zone showed a higher content of ferrite than the welded zone, between 49 vol% and 56 vol%, depending on the heat input. In addition, the highest content of ferrite was found in the top layer, as there was no subsequent layer, thus no reheating. By taking care of the heat deposition and of the cooling rate, a promising microstructure without sigma secondary phase precipitation and enhanced mechanical properties can be achieved. Later, Lervåg et al. [111] showed the precipitation of chromium nitrides, CrN, mainly in the HAZ rather than in the base material (BM) in the same superduplex stainless steel UNS S32507, as seen in Figure 5. Such precipitation is found in the ferrite phase after the rapid cooling from high temperatures, due to the supersaturation of nitrogen in the ferrite. Thus, chromium nitrides were not found in the built walls after WAAM, due to the lower content of ferrite phase. The formed layers of the material were obtained through the Cold Metal Transfer (CMT) process-based WAAM technique under different heat inputs in the range of 0.40–0.87 kJ/mm and the absence of intermetallic phases at low heat input was reported. The lower heat input is due to the reciprocating motions of the filler wire and the short-circuiting mode of the material, which produce reduced sputter in comparison to the conventional arc mode [101].

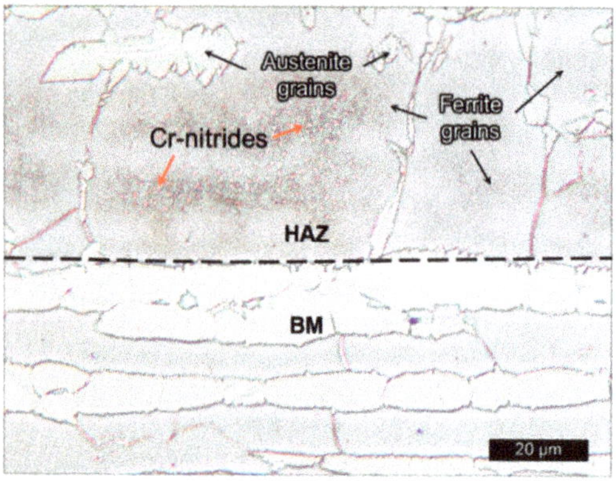

Figure 5. Chromium nitrides in HAZ in a SDSS 2507 superduplex steel (SDSS) as support plate (BM) and SDSS wire after WAAM process [111].

Hejripour et al. [112] demonstrated that a controlled cooling rate promotes austenite formation in ferrite matrix and prevents the formation of intermetallic phases during the process of building two different parts of 2209 DSS using the WAAM process. However, due to the complex thermal cycles and cooling rates, the ferrite content after the process showed a lower quantity in comparison to the same material processed by the conventional technique. Due to the required high cooling rate in the first layer (shorter distance between substrate and layer), a finer microstructure was obtained. The sigma phase and other deleterious intermetallic phases were not noticed in the microstructure and the ferrite content in the WAAM parts was lower than in the base metal produced by conventional methods.

Posch et al. [114] have shown the successful application of the CMT-WAAM (Cold Metal Transfer-WAAM) process to build blade-shaped geometry by a robot layer-by-layer using 2209 duplex stainless steel filler wire. No porosity, and non-harmful phases were detected in the transition microstructure of two scams, and a δ-ferrite content of 26 vol%–29 vol% was found by EBSD analyses. Therefore, the microstructure of the additively manufactured DSS is comparable to conventional GMAW DSS weld metals. The EBSD phase mapped the ferrite (δ)/austenite (γ) microstructure of the material after the process without any further heat treatment.

The flux-cored WAAM (FCWAAM) process is another technique to produce DSS weld metal, consisting of flux-cored wire instead of solid wires. Hence, there is more adjustability in the wire composition in order to better control the obtained microstructure of the fabricated material. Zhang et al. [115] have also found an excessive austenite content (around 74% average ratio) in a DSS component, using commercial 2009 DSS welding wire as filling material in the Flux-cored WAAM process. Although the mechanical properties of the obtained component were desirable, there was a difference in properties in the bottom and top layers in the horizontal direction. Therefore, efforts must be made to promote a balanced ferrite (δ)/austenite (γ) microstructure. For instance, authors have made heat treatments afterwards at 1250 °C and 1300 °C in order to achieve an approximately equal ratio of both austenite and ferrite phases, obtaining promising pitting corrosion resistance which is comparable to a hot-rolled 2205 DSS plate. Another alternative that allowed researches to modify the austenite ratio was developing specific wires for the WAAM process of DSS by altering the composition of the flux in the flux-cored wire. For example, Cr content was higher and Ni content was lower in the new FCWAAM DSS components in comparison to the commercial welding wire.

Later, Zhang et al. [116] showed, in another investigation, the satisfying pitting corrosion resistance of heat-treated WAAM DSS components, comparable to as-built WAAM DSS and hot-rolled 2205 DSS. The recommended post-AM heat treatment temperature for WAAM DSS was 1300 °C. In addition, the corrosion behavior of the as-built components was worse than the hot-rolled 2205 DSS, because of the different austenite ratio and chromium nitrides (Cr_2N) and a chemically different composition of austenite, the as-known secondary austenite (γ2) were detected. The lack of chromium nitrides after WAAM DSS were heat treated at 1250 °C and 1300° showed desirable pitting corrosion resistance.

Tavares et al. [117] reported results on the corrosion behavior of superduplex stainless steel 2705. Similar to a 2205 duplex stainless steel, a balanced microstructure was crucial for preventing pitting decrease. For instance, PREN over 40 was verified by Rajesh et al. [118] for a SDSS 2594 after WAAM process, showing the similar behavior of three different parts of the material, the base metal, middle part, and top layer, and their better pitting corrosion resistance in comparison to the same composition wrought alloy. The obtained microstructure showed no traces of intermetallic phases, which was reflected in the excellent pitting corrosion properties. Moreover, the similar corrosion behavior of three different parts in the wall layers showed a homogeneous microstructure after the process. However, some transients in current, indicating the occurrence of repassivating pitting events, were observed. Such transient pitting coincides with the middle layer, which is expected to have the higher load of defects. Defects such as LOF, porosity, or residual stresses may influence the corrosion resistance of the material; thus, the optimization of the layer deposition in addition to the control of the intermetallic phase formation is of crucial importance.

Overall, it could be summarized that WAAM processes are based on a deposition of subsequent layers and, thus, continuous heating occurs during the process, requiring extra high cooling rates because of the dissipation of the heat along and perpendicularly to the deposition direction [118,119]. On the contrary, when slower cooling rates are performed, there is enough time for the transformation of ferrite to austenite, leading to an excessive austenite ratio in the final layers. Obtaining a balanced austenite/ferrite microstructure is still a challenge in comparison to other processes. There are alternative techniques such as CMT-WAAM processes or the flux-cored wire used of FC-WAAM and, due to the lower heat input or flux-cored wires, the excessive austenite ratio could be modified. Moreover, post-heat treatments are mostly required.

5.1.2. Selective Laser Melting Processes

Selective laser melting (SLM) is another of the powder bed fusion techniques that are widely used as AM processing for duplex stainless steels. After SLM processing, the as-built microstructure of DSS is almost ferritic. Consequently, solution annealing is required afterwards for achieving a balanced ferrite/austenite microstructure [120,121]. On the other hand, due to temperature changes, resulting from the fast-thermal heating and cooling cycles, a high number of residual stresses appear in the as-built microstructure during and after SLM processing. Such residual stresses can be removed during the post-annealing treatment or by reducing their presence by decreasing the temperature gradient at the initial point of the process by preheating the building platform [122].

Several authors have shown microstructural features that manifest the importance of the heat-annealing treatment after applying SLM. Papula et al. [122] determined that, to obtain a ferrite-austenite microstructure, 2205 DSS had to be solution-annealed from

950 °C to 1100 °C after the SLM process. In addition, the as-built SLM-processed material presented an almost fully ferritic microstructure (99%) and, after the corresponding heat treatment, 40–46% of austenitic content was reached. The as-built SLM-processed material was mainly formed of ferrite phase, whereas heat-treated samples for different times showed an austenite phase increment up to 46%.

Shang et al. [123] observed, using EBSD, that secondary phases, such as sigma, were precipitated after SLM processing on 2707 HDSS. However, after the same processing and additional solution annealing treatments, at different temperatures, it gave place to a balanced ferrite-austenite microstructure. Characterization after solution annealing showed that the temperature must be carefully selected, as the sigma phase was still detected and mostly austenite phase was present in the microstructure. The corrosion behavior was evaluated and a particular ratio of α to γ showed better corrosion resistance. Therefore, phase arrangement is responsible for the higher pitting resistance compared to the other samples wherein which the precipitation of intermetallic sigma phase has occurred. Thus, the content of ferrite phase/vol.% was varied. σ phase is mainly found at α/α, α/γ, and γ/γ, and it may have led to depleted Cr areas at such phase boundaries, which further resulted in decreased pitting resistance. On the other hand, passive films were also investigated and showed better stability within the increase in grain boundaries due to the grain refinement. Thus, an improved passive film is formed due to the improved chromium distribution [124].

Davidson et al. [125] reported a high content of ferrite phase in a 2705 SDSS after the SLM process with some chromium nitrides. However, after a heat treatment, the ratio changed again. The high cooling rate after the process mainly restricts the austenite phase nucleation and an almost fully ferrite microstructure is observed in the as-built material. After a suitable heat treatment, a balanced austenite-ferrite microstructure is obtained, as was also seen by Hengsback et al. [126] for 1803 DSS material after the SLM process.

Table 4 shows a summary of different duplex stainless steel grades processed by WAAM and SLM processes, the post-AM heat treatments applied, and the pitting corrosion potential.

Table 4. Different DSS types processed by AM and post-AM heat treatments revealing critical issues related to the corrosion behaviour.

Material	Main Process	Solution Annealing [°C]	Qualitative Corrosion Aspects	Reference
2205 DSS	WAAM	1250–1350	Balanced ratio ferrite-austenite; improved pitting corrosion resistance after 1300 °C heat treatment	[116]
2594 SDSS	WAAM	-	Stable micropits in the middle part without decreasing corrosion resistance	[118]
2205 DSS	SLM	950–1100	Improved pitting resistance after heat treatments	[122]
2707 HDSS	SLM	1050–1200	Balanced ratio ferrite-austenite; higher pitting corrosion resistance after solution annealing	[123]

In conclusion, additive manufacturing processes are a promising alternative method to fabricate duplex stainless steels, especially when complex and irregular shapes are required in addition to the density of the materials, in comparison to the conventional techniques. However, high cooling rates and control during the processes are of extreme importance in order to avoid the formation of deleterious secondary phases that can affect mechanical and corrosion resistance properties.

On the other hand, it is crucial to maintain the balanced ferrite/austenite microstructure. Hence, most of the reviewed investigations have tried to modify the ratio of austenite by combining WAAM with other process, for instance, using cold wire feed [119] or using heat treatments after the AM process [127,128]. The high cooling rates and, most of the time, additional treatments are crucial; for instance, post-AM heat treatments are required to balance the austenite/ferrite ratio and maintain a desirable pitting corrosion resistance of duplex stainless steels.

AM processes such as SLM or EBM present some undesirable effects or limitations. The latest research is looking further into WAAM processes and their relationship to corrosion resistance [129], as well as new and recently-developed μ-plasma arc additive manufacturing process, which are quite promising [130].

Finally, AM is still an emergent manufacturing concept with many parameters in consideration to obtain the "perfect" component. However, attending the latest 2022 trends, there is a new concept in discussion related to the corrosion proofing of AM alloy parts. For instance, the low-temperature infusion of interstitial solute, exemplified for a high-entropy alloy (Alloy 22) to simulate thermochemical treatment, has been discussed by Illing et al. [131]. The post-processing of these parts is carried out through a newly developed process of LTNC-SRP (low-temperature nitrocarburization by solid-reagent pyrolysis). Future research should focus on the definition of proper corrosion tests to characterize the strong relationship between complex AM microstructures and corrosion resistance with accuracy. It should take into consideration the role of each defect, since each defect could create local electrochemical conditions leading to local corrosion cells. In respect, there should be a normal standard for these types of materials as they are processed by AM. In order to assess and develop apropiate testing for qualification purposes, it is critical to understand the long-term performance of AM materials in different corosion environments and correlate them to the short-term laboratory studies, which is also discussed in reference [7].

6. Conclusions

Additive manufacturing is an attractive alternative manufacturing technique for metallic components. AM allows for the fabrication of dense materials with complex shapes and comparable properties to the same conventionally manufactured materials. In this review, different AM processes, the most commonly occurring defects, and the corrosion resistance of light metals, duplex stainless steels, and other metallic materials obtained by different AM processes have been analyzed.

It has been reported by many researchers that light metals such as Ti and Al alloys may present some problems during the AM process, and are mainly related to the processing temperature that drives the formation of different intermetallic phases or oxides that could act as defect nucleation site with their impact on corrosion behavior. It is necessary to point out that more studies on the corrosion behavior of AM, in particular SLM-produced materials, are needed to understand the corrosion resistance of the AM-produced parts related to AM biomedical Ti alloys.

The results of adding Pt group metals to other metallic systems, with the improvement in their corrosion resistance, gives some interesting expectations to enhance the corrosion properties of other composites, even in switching to the AM technique.

Phase precipitation is one of the main problems related to the AM techniques for duplex stainless steels. Moreover, it is the maintenance of the dual phase microstructure that is of crucial importance in order to provide comparable mechanical and corrosion resistance properties to the conventionally fabricated materials. It has been found in the literature that AM could be a successful and powerful process with regards to pitting corrosion resistance properties, whenever austenite/ferrite ratio is maintained. With most of the AM processes, the as-built DSS components showed an almost fully ferrite microstructure and, thus, post-AM heat treatments are required to balance the austenite ratio.

Corrosion testing of AM components must take into account the high degree of anisotropy of all kinds of defects, which may appear in such objects, depending on the kind of process. When planning corrosion tests, the correct grinding and polishing procedures (allowing no modification of morphology or size of the defects, especially for surface porosity), the surface after post-processing (if applicable), the bulk material—tested surface perpendicular to the build direction (depth is relevant), and bulk material—tested surface parallel to the build direction must be considered. Not all of these test variables will be relevant for the corrosion susceptibility under service conditions; however, they may yield valuable information on the kind and distribution of defects on and in the AM material.

Finally, it must be pointed out that a real effort should be undergone to further investigate the corrosion behavior of metal additive manufacturing alloys, since there is not much literature that has studied it. For the last seven years, from 2015 to 2022, there have been more than 20,000 published studies that consider additively manufactured metallic systems. The majority of these studies are related to mechanical properties, mainly fatigue behavior and dimensional issues of the components. The existence of alternative methods to analyze corrosion behavior as LRP could be the beginning of understanding the interrelation between manufacture variables, as well as the impacts of microstructure and corrosion resistance. Carrying out the right actions to improve corrosion behavior and preventing its decrease is crucial to avoid undesirable situations.

Author Contributions: Conceptualization, M.V.B.-M., N.L.-I. and P.L.; methodology, J.G.-L., A.B.-T. and M.V.B.-M.; investigation, A.B.-T., J.G.-L. and M.V.B.-M.; resources, A.B.-T. and J.G.-L.; data curation, A.B.-T., J.G.-L., M.V.B.-M. and N.L.-I.; writing—original draft preparation, A.B.-T. and J.G.-L.; writing—review and editing, M.V.B.-M., N.L.-I., P.L. and A.B.-T.; supervision, M.V.B.-M., N.L.-I., P.L. and A.B.-T. All authors have read and agreed to the published version of the manuscript.

Funding: This research received no external funding.

Data Availability Statement: Data sharing not applicable.

Conflicts of Interest: The authors declare no conflict of interest.

Glossary

Abbreviation	Item
AM	Additive Manufacturing
BM	Base Metal
CMT	Cold Metal Transfer
CNC	Computer Numerical Control
CT	Computed Tomography
DED	Direct Energy Deposition
DLD	Direct Laser Deposition
DSS	Duplex Stainless Steels
EBM	Electron Beam Melting
EBSD	Electron Backscatter Diffraction
FC	Flux-Cored
FGM	Functionally Graded Materials
GMAW	Gas Metal Arc Welding
GP	Gas Porosity
HAZ	Heat-Affected Zone
HDSS	Hyperduplex Stainless Steels
HIP	Hot Isostatic Pressing
LMD	Laser Metal Deposition
LOF	Lack-of-Fusion

LTNC-SRP	Low temperature nitrocarburization by solidreagent pyrolisis
MAM	Metal additive manufacturing
MPB	Melt Pool Border
MPZ	Melt Pool Zone
NAB	Nickel-Aluminium-Bronze alloys
PBF	Powder Bed Fusion
PBFAM	Powder bed fusion additive manufacturing
SDSS	Superduplex Stainless Steels
SLM	Selective Laser Melting
SLS	Selective Laser Sintering
VSCE	Electrode potential measured versus Saturated Calomel Electrode
WAAM	Wire Arc Additive Manufacturing

References

1. Culmone, C.; Smit, G.; Breedveld, P. Additive manufacturing of medical instruments: A state-of-the-art review. *Addit. Manuf.* **2019**, *27*, 461–473. [CrossRef]
2. Zhang, Q.; Guan, Y. Application of metal additive manufacturing in oral dentistry. *Curr. Opin. Biomed. Eng.* **2023**, *25*, 100441. [CrossRef]
3. Seiti, M.; Ginestra, P. Additive Manufacturing for orthopedic applications: Case study on market impact. *Procedia Comput. Sci.* **2023**, *217*, 737–745. [CrossRef]
4. Artaza, T.; Suárez, A.; Veiga, F.; Braceras, I.; Tabernero, I.; Larrañaga, O.; Lamikiz, A. Wire arc additive manufacturing Ti6Al4V aeronautical parts using plasma arc welding: Analysis of heat-treatment processes in different atmospheres. *J. Mater. Res. Technol.* **2020**, *9*, 15454–15466. [CrossRef]
5. Madhavadas, V.; Srivastava, D.; Chadha, U.; Raj, S.A.; Sultan, M.T.H.; Shahar, F.S.; Shah, A.U.M. A review on metal additive manufacturing for intricately shaped aerospace components. *CIRP J. Manuf. Sci. Technol.* **2022**, *39*, 18–36. [CrossRef]
6. Monteiro, H.; Carmona-Aparicio, G.; Lei, I.; Despeisse, M. Energy and material efficiency strategies enabled by metal additive manufacturing—A review for the aeronautic and aerospace sectors. *Energy Rep.* **2022**, *8*, 298–305. [CrossRef]
7. Schindelholz, E.J.; Melia, M.A.; Rodelas, J.M. Corrosion of Additively Manufactured Stainless Steels—Process, Structure, Performance: A Review. *Corrosion* **2021**, *77*, 484–503. [CrossRef]
8. Sander, G.; Tan, J.; Balan, P.; Gharbi, O.; Feenstra, D.; Singer, L.; Thomas, S.; Kelly, R.; Scully, J.; Birbilis, N. Corrosion of Additively Manufactured Alloys: A Review. *Corrosion* **2018**, *74*, 1318–1350. [CrossRef]
9. Praveena, B.A.; Lokesh, N.; Buradi, A.; Santhosh, N.; Praveena, B.L.; Vignesh, R. A comprehensive review of emerging additive manufacturing (3D printing technology): Methods, materials, applications, challenges, trends and future potential. *Mater. Today Proc.* **2021**, *52*, 1309–1313. [CrossRef]
10. Orzolek, S.M.; Semple, J.K.; Fisher, C.R. Influence of processing on the microstructure of nickel aluminum bronze (NAB). *Addit. Manuf.* **2022**, *56*, 10285. [CrossRef]
11. *ISO/ASTM 52900:2015*; Additive Manufacturing-General Principles-Terminology. International Organization for Standarization: Geneva, Switzerland, 2015. Available online: www.iso.orgwww.astm.org (accessed on 19 January 2023).
12. Kumar, S.A.; Prasad, R. Basic principles of additive manufacturing: Different additive manufacturing technologies. In *Additive Manufacturing*; Woodhead Publishing: Cambridge, UK, 2021, pp. 17–35. [CrossRef]
13. Frazier, W.E. Metal Additive Manufacturing: A Review. *J. Mater. Eng. Perform.* **2014**, *23*, 1917–1928. [CrossRef]
14. Faludi, J.; Bayley, C.; Bhogal, S.; Iribarne, M. Comparing environmental impacts of additive manufacturing vs traditional machining via life-cycle assessment. *Rapid Prototyp. J.* **2015**, *21*, 14–33. [CrossRef]
15. Javaid, M.; Ariz, A.; Tasneem, I.; Bharti, D.; Vaish, A.; Haleem, A. Is additive manufacturing of patient-specific implant is beneficial for orthopedics. *Apollo Med.* **2021**, *18*, 33. [CrossRef]
16. Pereira, T.; Kennedy, J.V.; Potgieter, J. A comparison of traditional manufacturing vs additive manufacturing, the best method for the job. *Procedia Manuf.* **2019**, *30*, 11–18. [CrossRef]
17. Fathi, P.; Mohammadi, M.; Duan, X.; Nasiri, A.M. A comparative study on corrosion and microstructure of direct metal laser sintered AlSi10Mg_200C and die cast A360.1 aluminum. *J. Mater. Process. Technol.* **2019**, *259*, 1–14. [CrossRef]
18. Rafi, H.K.; Karthik, N.V.; Gong, H.; Starr, T.L.; Stucker, B.E. Microstructures and Mechanical Properties of Ti6Al4V Parts Fabricated by Selective Laser Melting and Electron Beam Melting. *J. Mater. Eng. Perform.* **2013**, *22*, 3872–3883. [CrossRef]
19. Zhai, Y.; Galarraga, H.; Lados, D.A. Microstructure, static properties, and fatigue crack growth mechanisms in Ti-6Al-4V fabricated by additive manufacturing: LENS and EBM. *Eng. Fail. Anal.* **2016**, *69*, 3–14. [CrossRef]
20. Revilla, R.I.; Verkens, D.; Rubben, T.; De Graeve, I. Corrosion and Corrosion Protection of Additively Manufactured Aluminium Alloys—A Critical Review. *Materials* **2020**, *13*, 4804. [CrossRef]
21. Dawes, J.; Bowerman, R.; Trepleton, R. Introduction to the Additive Manufacturing Powder Metallurgy Supply Chain. *Johns. Matthey Technol. Rev.* **2015**, *59*, 243–256. [CrossRef]

22. Martin, J.H.; Yahata, B.D.; Hundley, J.M.; Mayer, J.A.; Schaedler, T.A.; Pollock, T.M. 3D printing of high-strength aluminium alloys. *Nature* **2017**, *549*, 365–369. [CrossRef]
23. Sander, G.; Thomas, S.; Cruz, V.; Jurg, M.; Birbilis, N.; Gao, X.; Brameld, M.; Hutchinson, C.R. On The Corrosion and Metastable Pitting Characteristics of 316L Stainless Steel Produced by Selective Laser Melting. *J. Electrochem. Soc.* **2017**, *164*, C250–C257. [CrossRef]
24. Sun, G.; Shen, X.; Wang, Z.; Zhan, M.; Yao, S.; Zhou, R.; Ni, Z. Laser metal deposition as repair technology for 316L stainless steel: Influence of feeding powder compositions on microstructure and mechanical properties. *Opt. Laser Technol.* **2018**, *109*, 71–83. [CrossRef]
25. Trelewicz, J.R.; Halada, G.P.; Donaldson, O.K.; Manogharan, G. Microstructure and Corrosion Resistance of Laser Additively Manufactured 316L Stainless Steel. *JOM* **2016**, *68*, 850–859. [CrossRef]
26. Zhang, X.; Lv, Y.; Tan, S.; Dong, Z.; Zhou, X. Microstructure and corrosion behaviour of wire arc additive manufactured AA2024 alloy thin wall structure. *Corros. Sci.* **2021**, *186*, 109453. [CrossRef]
27. McAndrew, A.R.; Rosales, M.A.; Colegrove, P.A.; Hönnige, J.R.; Ho, A.; Fayolle, R.; Eyitayo, K.; Stan, I.; Sukrongpang, P.; Crochemore, A.; et al. Interpass rolling of Ti-6Al-4V wire + arc additively manufactured features for microstructural refinement. *Addit. Manuf.* **2018**, *21*, 340–349. [CrossRef]
28. Bhavar, V.; Kattire, P.; Patil, V.; Khot, S.; Gujar, K.; Singh, R. A review on powder bed fusion technology of metal additive manufacturing. In *Additive Manufacturing Handbook*; CRC Press: Boca Raton, FL, USA, 2018; pp. 251–253. [CrossRef]
29. Kong, D.; Dong, C.; Ni, X.; Li, X. Corrosion of metallic materials fabricated by selective laser melting. *npj Mater. Degrad.* **2019**, *3*, 24. [CrossRef]
30. Otero, E.; Pardo, A.; Utrilla, V.; Saenz, E.; Álvarez, J. Corrosion behaviour of aisi 304l and 316l stainless steels prepared by powder metallurgy in the presence of sulphuric and phosphoric acid. *Corros. Sci.* **1998**, *40*, 1421–1434. [CrossRef]
31. Maximenko, A.L.; Olevsky, E.A. Pore filling during selective laser melting—Assisted additive manufacturing of composites. *Scr. Mater.* **2018**, *149*, 75–78. [CrossRef]
32. Housholder, R.F. Molding Process. U.S. Patent US06099333, 3 December 1979.
33. Beaman, J.J.; Deckard, C.R. Selective Laser Sintering with Assisted Powder Handling. U.S. Patent US4938816A, 5 September 1989.
34. Al-Shebeeb, O.A. An Investigation of the Metal Additive Manufacturing Issues and Perspective for Solutions Approach. In *Concepts, Applications and Emerging Opportunities in Industrial Engineering*; Books on Demand: Norderstedt, Germany, 2021. [CrossRef]
35. Ma, F.; Zhang, H.; Hon, K.; Gong, Q. An optimization approach of selective laser sintering considering energy consumption and material cost. *J. Clean. Prod.* **2018**, *199*, 529–537. [CrossRef]
36. Vayre, B.; Vignat, F.; Villeneuve, F. Metallic additive manufacturing: State-of-the-art review and prospects. *Mech. Ind.* **2012**, *13*, 89–96. [CrossRef]
37. Richardson, B.S.; Lind, R.F.; Lloyd, P.D.; Noakes, M.W.; Love, L.J.; Post, B.K. The design of an additive manufactured dual arm manipulator system. *Addit. Manuf.* **2018**, *24*, 467–478. [CrossRef]
38. Murr, L. Metallurgy of additive manufacturing: Examples from electron beam melting. *Addit. Manuf.* **2015**, *5*, 40–53. [CrossRef]
39. Zafer, Y.E.; Goel, S.; Ganvir, A.; Jansson, A.; Joshi, S. Encapsulation of Electron Beam Melting Produced Alloy 718 to Reduce Surface Connected Defects by Hot Isostatic Pressing. *Materials* **2020**, *13*, 1226. [CrossRef] [PubMed]
40. Najmon, J.C.; Raeisi, S.; Tovar, A. Review of additive manufacturing technologies and applications in the aerospace industry. In *Additive Manufacturing for the Aerospace Industry*; Elsevier: Amsterdam, The Netherlands, 2019; pp. 7–31. [CrossRef]
41. Jin, W.; Zhang, C.; Jin, S.; Tian, Y.; Wellmann, D.; Liu, W. Wire Arc Additive Manufacturing of Stainless Steels: A Review. *Appl. Sci.* **2020**, *10*, 1563. [CrossRef]
42. Gao, W.; Zhang, Y.; Ramanujan, D.; Ramani, K.; Chen, Y.; Williams, C.B.; Wang, C.C.L.; Shin, Y.C.; Zhang, S.; Zavattieri, P.D. The status, challenges, and future of additive manufacturing in engineering. *Comput. Aided Des.* **2015**, *69*, 65–89. [CrossRef]
43. Coemert, S.; Traeger, M.F.; Graf, E.C.; Lueth, T.C. Suitability Evaluation of various Manufacturing Technologies for the Development of Surgical Snake-like Manipulators from Metals Based on Flexure Hinges. *Procedia CIRP* **2017**, *65*, 1–6. [CrossRef]
44. Bartlett, J.L.; Li, X. An overview of residual stresses in metal powder bed fusion. *Addit. Manuf.* **2019**, *27*, 131–149. [CrossRef]
45. Jabnoun, N. Control processes for total quality management and quality assurance. *Work. Study* **2002**, *51*, 182–190. [CrossRef]
46. Webster, S.; Lin, H.; Iii, F.M.C.; Ehmann, K.; Cao, J. Physical mechanisms in hybrid additive manufacturing: A process design framework. *J. Mater. Process. Technol.* **2021**, *291*, 117048. [CrossRef]
47. Wu, B.; Pan, Z.; Ding, D.; Cuiuri, D.; Li, H.; Xu, J.; Norrish, J. A review of the wire arc additive manufacturing of metals: Properties, defects and quality improvement. *J. Manuf. Process.* **2018**, *35*, 127–139. [CrossRef]
48. Kim, F.H.; Moylan, S.P. *Literature Review of Metal Additive Manufacturing Defects*; National Institute of Standards and Technology: Gaithersburg, MD, USA, 2018; pp. 1–17. [CrossRef]
49. Busachi, A.; Erkoyuncu, J.; Colegrove, P.; Martina, F.; Ding, J. Designing a WAAM Based Manufacturing System for Defence Applications. *Procedia CIRP* **2015**, *37*, 48–53. [CrossRef]

50. Zhang, B.; Li, Y.; Bai, Q. Defect Formation Mechanisms in Selective Laser Melting: A Review. *Chin. J. Mech. Eng.* **2017**, *30*, 515–527. [CrossRef]
51. Xu, Z.; Wang, Q.; Wang, X.; Tan, C.; Guo, M.; Gao, P. High cycle fatigue performance of AlSi10mg alloy produced by selective laser melting. *Mech. Mater.* **2020**, *148*, 103499. [CrossRef]
52. Siddique, S.; Imran, M.; Rauer, M.; Kaloudis, M.; Wycisk, E.; Emmelmann, C.; Walther, F. Computed tomography for characterization of fatigue performance of selective laser melted parts. *Mater. Des.* **2015**, *83*, 661–669. Available online: https://www.sciencedirect.com/science/article/pii/S026412751500413X (accessed on 10 February 2022). [CrossRef]
53. Rollett, A. Finding Keyholes in Metal Additive Manufacturing | 2020-03-11 | Industrial Heating. Available online: https://www.industrialheating.com/articles/95529-finding-keyholes-in-metal-additive-manufacturing (accessed on 10 February 2022).
54. Khairallah, S.A.; Martin, A.A.; Lee, J.R.I.; Guss, G.; Calta, N.P.; Hammons, J.A.; Nielsen, M.H.; Chaput, K.; Schwalbach, E.; Shah, M.N.; et al. Controlling interdependent meso-nanosecond dynamics and defect generation in metal 3D printing. *Science* **2020**, *368*, 660–665. [CrossRef] [PubMed]
55. Collins, P.C.; Bond, L.J.; Taheri, H.; Bigelow, T.A.; Shoaib, M.R.B.M.; Koester, L.W. Powder-based additive manufacturing—A review of types of defects, generation mechanisms, detection, property evaluation and metrology. *Int. J. Addit. Subtractive Mater. Manuf.* **2017**, *1*, 172. [CrossRef]
56. Snow, Z.; Diehl, B.; Reutzel, E.W.; Nassar, A. Toward in-situ flaw detection in laser powder bed fusion additive manufacturing through layerwise imagery and machine learning. *J. Manuf. Syst.* **2021**, *59*, 12–26. [CrossRef]
57. Nassar, A.R.; Gundermann, M.A.; Reutzel, E.W.; Guerrier, P.; Krane, M.H.; Weldon, M.J. Formation processes for large ejecta and interactions with melt pool formation in powder bed fusion additive manufacturing. *Sci. Rep.* **2019**, *9*, 5038. [CrossRef] [PubMed]
58. Monzón Vivas, A. Thickness and Orientation Dependent Surface Roughness and Internal Defect Characterization of SLM Ti-6Al-4V. Available online: https://riunet.upv.es/handle/10251/163329 (accessed on 10 February 2022).
59. Blinn, B.; Klein, M.; Gläßner, C.; Smaga, M.; Aurich, J.C.; Beck, T. An Investigation of the Microstructure and Fatigue Behavior of Additively Manufactured AISI 316L Stainless Steel with Regard to the Influence of Heat Treatment. *Metals* **2018**, *8*, 220. [CrossRef]
60. Roehling, T.T.; Wu, S.S.; Khairallah, S.A.; Roehling, J.D.; Soezeri, S.S.; Crumb, M.F.; Matthews, M.J. Modulating laser intensity profile ellipticity for microstructural control during metal additive manufacturing. *Acta Mater.* **2017**, *128*, 197–206. Available online: https://www.sciencedirect.com/science/article/pii/S1359645417301167 (accessed on 10 February 2022). [CrossRef]
61. Gu, J.; Ding, J.; Williams, S.W.; Gu, H.; Ma, P.; Zhai, Y. The effect of inter-layer cold working and post-deposition heat treatment on porosity in additively manufactured aluminum alloys. *J. Mater. Process. Technol.* **2016**, *230*, 26–34. [CrossRef]
62. Derekar, K.; Lawrence, J.; Melton, J.; Addison, A.; Zhang, X.; Xu, L. Influence of interpass temperature on wire arc additive manufacturing (WAAM) of aluminium alloy components. In Proceedings of the IIW 2018—International Conference on Advanced Welding and Smart Fabrication Technologies, Bali, Indonesia, 15–20 July 2018; EDP Sciences: Les Ulis, France, 2019; Volume 269, p. 05001. [CrossRef]
63. Hauser, T.; Reisch, R.T.; Breese, P.P.; Lutz, B.S.; Pantano, M.; Nalam, Y.; Bela, K.; Kamps, T.; Volpp, J.; Kaplan, A.F. Porosity in wire arc additive manufacturing of aluminium alloys. *Addit. Manuf.* **2021**, *41*, 101993. [CrossRef]
64. Ibrahim, H.; Jahadakbar, A.; Dehghan, A.; Moghaddam, N.S.; Amerinatanzi, A.; Elahinia, M. In Vitro Corrosion Assessment of Additively Manufactured Porous NiTi Structures for Bone Fixation Applications. *Metals* **2018**, *8*, 164. [CrossRef]
65. Qiu, X. Microstructure and corrosion properties of Al2CrFeCo CuNiTi high entropy alloys prepared by additive manufacturing. *J. Alloy. Compd.* **2021**, *887*, 161422. [CrossRef]
66. Javadi, Y.; MacLeod, C.N.; Pierce, S.G.; Gachagan, A.; Lines, D.; Mineo, C.; Ding, J.; Williams, S.; Vasilev, M.; Mohseni, E.; et al. Ultrasonic phased array inspection of a Wire + Arc Additive Manufactured (WAAM) sample with intentionally embedded defects. *Addit. Manuf.* **2019**, *29*, 100806. Available online: https://www.sciencedirect.com/science/article/pii/S221486041930524X (accessed on 11 February 2022). [CrossRef]
67. Ye, D.; Hong, G.S.; Zhang, Y.; Zhu, K.; Fuh, J.Y.H. Defect detection in selective laser melting technology by acoustic signals with deep belief networks. *Int. J. Adv. Manuf. Technol.* **2018**, *96*, 2791–2801. [CrossRef]
68. Romano, S.; Abel, A.; Gumpinger, J.; Brandão, A.; Beretta, S. Quality control of AlSi10Mg produced by SLM: Metallography versus CT scans for critical defect size assessment. *Addit. Manuf.* **2019**, *28*, 394–405. [CrossRef]
69. Nourian-Avval, A.; Fatemi, A. Characterization and Analysis of Porosities in High Pressure Die Cast Aluminum by Using Metallography, X-Ray Radiography, and Micro-Computed Tomography. *Materials* **2020**, *13*, 3068. [CrossRef]
70. Ng, G.K.L.; Jarfors, A.E.W.; Bi, G.; Zheng, H.Y. Porosity formation and gas bubble retention in laser metal deposition. *Appl. Phys. A* **2009**, *97*, 641–649. [CrossRef]
71. Kang, Y.-J.; Yang, S.; Kim, Y.-K.; AlMangour, B.; Lee, K.-A. Effect of post-treatment on the microstructure and high-temperature oxidation behaviour of additively manufactured inconel 718 alloy. *Corros. Sci.* **2019**, *158*, 108082. [CrossRef]
72. Chandrasekaran, S.; Hari, S.; Amirthalingam, M. Wire arc additive manufacturing of functionally graded material for marine risers. *Mater. Sci. Eng. A* **2020**, *792*, 139530. [CrossRef]
73. Bobbio, L.D.; Otis, R.A.; Borgonia, J.P.; Dillon, R.P.; Shapiro, A.A.; Liu, Z.-K.; Beese, A.M. Additive manufacturing of a functionally graded material from Ti-6Al-4V to Invar: Experimental characterization and thermodynamic calculations. *Acta Mater.* **2017**, *127*, 133–142. Available online: https://www.sciencedirect.com/science/article/pii/S1359645416310187 (accessed on 11 February 2022). [CrossRef]

74. Shang, Y.; Yuan, Y.; Li, D.; Li, Y.; Chen, J. Effects of scanning speed on in vitro biocompatibility of 316L stainless steel parts elaborated by selective laser melting. *Int. J. Adv. Manuf. Technol.* **2017**, *92*, 4379–4385. [CrossRef]
75. Han, Q.; Gu, Y.; Soe, S.; Lacan, F.; Setchi, R. Effect of hot cracking on the mechanical properties of Hastelloy X superalloy fabricated by laser powder bed fusion additive manufacturing. *Opt. Laser Technol.* **2019**, *124*, 105984. [CrossRef]
76. Qiu, Z.; Wu, B.; Zhu, H.; Wang, Z.; Hellier, A.; Ma, Y.; Li, H.; Muransky, O.; Wexler, D. Microstructure and mechanical properties of wire arc additively manufactured Hastelloy C276 alloy. *Mater. Des.* **2020**, *195*, 109007. Available online: https://www.sciencedirect.com/science/article/pii/S0264127520305414 (accessed on 11 February 2022). [CrossRef]
77. Kim, J.H.; Anderson, J.; Hoden, B. Electronics Chassis with Oscillating Heat Pipe (OHP). U.S. Patent No. 11,112,840, 7 September 2021. Available online: https://patents.google.com/patent/US11112840B2/en (accessed on 14 February 2022).
78. US11097487Patents | PatentGuru. Available online: https://www.patentguru.com/search?q=US11097487 (accessed on 14 February 2022).
79. US11097350Patents | PatentGuru. Available online: https://www.patentguru.com/search?q=US11097350 (accessed on 14 February 2022).
80. Monteiro, W.A. *Light Metal Alloys Applications*; Books on Demand GmbH: Norderstedt, Germany, 2014. [CrossRef]
81. Aboulkhair, N.T.; Simonelli, M.; Parry, L.; Ashcroft, I.; Tuck, C.; Hague, R. 3D printing of Aluminium alloys: Additive Manufacturing of Aluminium alloys using selective laser melting. *Prog. Mater. Sci.* **2019**, *106*, 100578. Available online: https://www.sciencedirect.com/science/article/pii/S007964251930060X (accessed on 14 February 2022). [CrossRef]
82. Maamoun, A.H.; Xue, Y.F.; Elbestawi, M.A.; Veldhuis, S.C. The Effect of Selective Laser Melting Process Parameters on the Microstructure and Mechanical Properties of Al6061 and AlSi10Mg Alloys. *Materials* **2018**, *12*, 12. [CrossRef]
83. Leon, A.; Aghion, E. Effect of surface roughness on corrosion fatigue performance of AlSi10Mg alloy produced by Selective Laser Melting (SLM). *Mater. Charact.* **2017**, *131*, 188–194. [CrossRef]
84. Nezhadfar, P.; Thompson, S.; Saharan, A.; Phan, N.; Shamsaei, N. Structural integrity of additively manufactured aluminum alloys: Effects of build orientation on microstructure, porosity, and fatigue behavior. *Addit. Manuf.* **2021**, *47*, 102292. [CrossRef]
85. Amanov, A. Effect of local treatment temperature of ultrasonic nanocrystalline surface modification on tribological behavior and corrosion resistance of stainless steel 316L produced by selective laser melting. *Surf. Coat. Technol.* **2020**, *398*, 126080. [CrossRef]
86. Laleh, M.; Haghdadi, N.; Hughes, A.E.; Primig, S.; Tan, M.Y. Enhancing the repassivation ability and localised corrosion resistance of an additively manufactured duplex stainless steel by post-processing heat treatment. *Corros. Sci.* **2022**, *198*, 110106. [CrossRef]
87. Fathi, P.; Mohammadi, M.; Duan, X.; Nasiri, A.M. Effects of Surface Finishing Procedures on Corrosion Behavior of DMLS-AlSi10Mg_200C Alloy Versus Die-Cast A360.1 Aluminum. *JOM* **2019**, *71*, 1748–1759. [CrossRef]
88. Zeng, F.-L.; Wei, Z.-L.; Li, J.-F.; Li, C.-X.; Tan, X.; Zhang, Z.; Zheng, Z.-Q. Corrosion mechanism associated with Mg2Si and Si particles in Al–Mg–Si alloys. *Trans. Nonferrous Met. Soc. China* **2011**, *21*, 2559–2567. [CrossRef]
89. Revilla, R.I.; De Graeve, I. Influence of Si Content on the Microstructure and Corrosion Behavior of Additive Manufactured Al-Si Alloys. *J. Electrochem. Soc.* **2018**, *165*, C926. [CrossRef]
90. Gu, X.-H.; Zhang, J.-X.; Fan, X.-L.; Zhang, L.-C. Corrosion Behavior of Selective Laser Melted AlSi10Mg Alloy in NaCl Solution and Its Dependence on Heat Treatment. *Acta Met. Sin. Engl. Lett.* **2019**, *33*, 327–337. [CrossRef]
91. Gu, X.; Zhang, J.; Fan, X.; Dai, N.; Xiao, Y.; Zhang, L.-C. Abnormal corrosion behavior of selective laser melted AlSi10Mg alloy induced by heat treatment at 300 °C. *J. Alloy. Compd.* **2019**, *803*, 314–324. [CrossRef]
92. Berlanga-Labari, C.; Biezma-Moraleda, M.V.; Rivero, P.J. Corrosion of Cast Aluminum Alloys: A Review. *Metals* **2020**, *10*, 1384. [CrossRef]
93. Liu, D.; Atkinson, H.V.; Kapranos, P.; Jirattiticharoean, W.; Jones, H. Microstructural evolution and tensile mechanical properties of thixoformed high performance aluminium alloys. *Mater. Sci. Eng. A* **2003**, *361*, 213–224. [CrossRef]
94. Dutta, B.; Froes, F.H.S. The Additive Manufacturing (AM) of titanium alloys. *Met. Powder Rep.* **2017**, *72*, 96–106. [CrossRef]
95. Laquai, R.; Müller, B.R.; Kasperovich, G.; Haubrich, J.; Requena, G.; Bruno, G. X-ray refraction distinguishes unprocessed powder from empty pores in selective laser melting Ti-6Al-4V. *Mater. Res. Lett.* **2018**, *6*, 130–135. [CrossRef]
96. Zhang, L.-C.; Liu, Y.; Li, S.; Hao, Y. Additive Manufacturing of Titanium Alloys by Electron Beam Melting: A Review. *Adv. Eng. Mater.* **2018**, *20*, 1700842. [CrossRef]
97. Bermingham, M.; McDonald, S.; Nogita, K.; John, D.S.; Dargusch, M. Effects of boron on microstructure in cast titanium alloys. *Scr. Mater.* **2008**, *59*, 538–541. [CrossRef]
98. Chiu, T.-M.; Mahmoudi, M.; Dai, W.; Elwany, A.; Liang, H.; Castaneda, H. Corrosion assessment of Ti-6Al-4V fabricated using laser powder-bed fusion additive manufacturing. *Electrochim. Acta* **2018**, *279*, 143–151. [CrossRef]
99. Mwamba, I.A.; Cornish, L.A.; Van der Lingen, E. Effect of platinum group metal addition on microstructure and corrosion behaviour of Ti–47·5 at-%Al. *Corros. Eng. Sci. Technol.* **2014**, *49*, 180–188. [CrossRef]
100. Lo, K.H.; Kwok, C.T.; Chan, W.K.; Kuan, H.C.; Lai, K.K.; Wang, K.Y. Duplex Stainless Steels. In *Encyclopedia of Iron, Steel, and Their Alloys*; CRC Press: Boca Raton, FL, USA, 2016; pp. 1150–1160. [CrossRef]
101. Alvarez-Armas, I.; Degallaix-Moreuil, S. *Duplex Stainless Steels*; John Wiley and Sons: Hoboken, NJ, USA, 2009. [CrossRef]
102. Nilsson, J.O.; Chai, G. The physical metallurgy of duplex stainless steels. *Proc. Duplex Stainl. Steel* **1997**, *97*, 73–82.
103. Llorca-Isern, N.; López-Luque, H.; López-Jiménez, I.; Biezma, M.V. Identification of sigma and chi phases in duplex stainless steels. *Mater. Charact.* **2016**, *112*, 20–29. [CrossRef]

104. Biserova-Tahchieva, A.; Chatterjee, D.; van Helvoort, A.T.; Llorca-Isern, N.; Cabrera, J.M. Effect of the nanostructuring by high-pressure torsion process on the secondary phase precipitation in UNS S32750 Superduplex stainless steel. *Mater. Charact.* **2022**, *183*, 111639. [CrossRef]
105. Zhao, Y.; Wang, Y.; Tang, S.; Zhang, W.; Liu, Z. Edge cracking prevention in 2507 super duplex stainless steel by twin-roll strip casting and its microstructure and properties. *J. Mater. Process. Technol.* **2019**, *266*, 246–254. [CrossRef]
106. Tavares, S.; Pardal, J.; Almeida, B.; Mendes, M.; Freire, J.; Vidal, A. Failure of superduplex stainless steel flange due to inadequate microstructure and fabrication process. *Eng. Fail. Anal.* **2018**, *84*, 1–10. [CrossRef]
107. Sun, Z.J.; Tan, X.P.; Tor, S.B.; Yeong, W.Y. Selective laser melting of stainless steel 316L with low porosity and high build rates. *Mater. Des.* **2016**, *104*, 197–204. [CrossRef]
108. Wang, G.; Liu, Q.; Rao, H.; Liu, H.; Qiu, C. Influence of porosity and microstructure on mechanical and corrosion properties of a selectively laser melted stainless steel. *J. Alloy. Compd.* **2020**, *831*, 154815. [CrossRef]
109. Eriksson, M.; Lervåg, M.; Sørensen, C.; Robertstad, A.; Brønstad, B.M.; Nyhus, B.; Aune, R.; Ren, X.; Akselsen, O. Additive manufacture of superduplex stainless steel using WAAM. *MATEC Web Conf.* **2018**, *188*, 03014. [CrossRef]
110. A Hosseini, V.; Högström, M.; Hurtig, K.; Bermejo, M.A.V.; Stridh, L.-E.; Karlsson, L. Wire-arc additive manufacturing of a duplex stainless steel: Thermal cycle analysis and microstructure characterization. *Weld. World* **2019**, *63*, 975–987. [CrossRef]
111. Lervåg, M.; Sørensen, C.; Robertstad, A.; Brønstad, B.M.; Nyhus, B.; Eriksson, M.; Aune, R.; Ren, X.; Akselsen, O.M.; Bunaziv, I. Additive Manufacturing with Superduplex Stainless Steel Wire by CMT Process. *Metals* **2020**, *10*, 272. [CrossRef]
112. Hejripour, F.; Binesh, F.; Hebel, M.; Aidun, D.K. Thermal modeling and characterization of wire arc additive manufactured duplex stainless steel. *J. Mater. Process. Technol.* **2019**, *272*, 58–71. [CrossRef]
113. Knezović, N.; Garašić, I.; Jurić, I. Influence of the Interlayer Temperature on Structure and Properties of Wire and Arc Additive Manufactured Duplex Stainless Steel Product. *Materials* **2020**, *13*, 5795. [CrossRef]
114. Posch, G.; Chladil, K.; Chladil, H. Material properties of CMT—Metal additive manufactured duplex stainless steel blade-like geometries. *Weld. World* **2017**, *61*, 873–882. [CrossRef]
115. Zhang, Y.; Cheng, F.; Wu, S. The microstructure and mechanical properties of duplex stainless steel components fabricated via flux-cored wire arc-additive manufacturing. *J. Manuf. Process.* **2021**, *69*, 204–214. [CrossRef]
116. Zhang, Y.; Cheng, F.; Wu, S. Improvement of pitting corrosion resistance of wire arc additive manufactured duplex stainless steel through post-manufacturing heat-treatment. *Mater. Charact.* **2020**, *171*, 110743. [CrossRef]
117. Tavares, S.; Pardal, J.; Lima, L.; Bastos, I.; Nascimento, A.; de Souza, J. Characterization of microstructure, chemical composition, corrosion resistance and toughness of a multipass weld joint of superduplex stainless steel UNS S32750. *Mater. Charact.* **2007**, *58*, 610–616. [CrossRef]
118. Kannan, A.R.; Shanmugam, N.S.; Rajkumar, V.; Vishnukumar, M. Insight into the microstructural features and corrosion properties of wire arc additive manufactured super duplex stainless steel (ER2594). *Mater. Lett.* **2020**, *270*, 127680. [CrossRef]
119. Stützer, J.; Totzauer, T.; Wittig, B.; Zinke, M.; Jüttner, S. GMAW Cold Wire Technology for Adjusting the Ferrite–Austenite Ratio of Wire and Arc Additive Manufactured Duplex Stainless Steel Components. *Metals* **2019**, *9*, 564. [CrossRef]
120. Saeidi, K.; Kevetkova, L.; Lofaj, F.; Shen, Z. Novel ferritic stainless steel formed by laser melting from duplex stainless steel powder with advanced mechanical properties and high ductility. *Mater. Sci. Eng. A* **2016**, *665*, 59–65. [CrossRef]
121. Wittig, B.; Zinke, M.; Jüttner, S. Influence of arc energy and filler metal composition on the microstructure in wire arc additive manufacturing of duplex stainless steels. *Weld. World* **2021**, *65*, 47–56. [CrossRef]
122. Papula, S.; Song, M.; Pateras, A.; Chen, X. B.; Brandt, M.; Easton, M.; Yagodzinskyy, Y.; Virkkunen, I.; Hänninen, H. Selective Laser Melting of Duplex Stainless Steel 2205: Effect of Post-Processing Heat Treatment on Microstructure, Mechanical Properties, and Corrosion Resistance. *Materials* **2019**, *12*, 2468. [CrossRef] [PubMed]
123. Shang, F.; Chen, X.; Wang, Z.; Ji, Z.; Ming, F.; Ren, S.; Qu, X. The Microstructure, Mechanical Properties, and Corrosion Resistance of UNS S32707 Hyper-Duplex Stainless Steel Processed by Selective Laser Melting. *Metals* **2019**, *9*, 1012. [CrossRef]
124. Zheng, Z.; Gao, Y.; Gui, Y.; Zhu, M. Corrosion behaviour of nanocrystalline 304 stainless steel prepared by equal channel angular pressing. *Corros. Sci.* **2012**, *54*, 60–67. [CrossRef]
125. Davidson, K.; Singamneni, S. Selective Laser Melting of Duplex Stainless Steel Powders: An Investigation. *Mater. Manuf. Process.* **2016**, *31*, 1543–1555. [CrossRef]
126. Hengsbach, F.; Koppa, P.; Duschik, K.; Holzweissig, M.J.; Burns, M.; Nellesen, J.; Tillmann, W.; Tröster, T.; Hoyer, K.-P.; Schaper, M. Duplex stainless steel fabricated by selective laser melting—Microstructural and mechanical properties. *Mater. Des.* **2017**, *133*, 136–142. [CrossRef]
127. Kim, S.-T.; Jang, S.-H.; Lee, I.-S.; Park, Y.-S. Effects of solution heat-treatment and nitrogen in shielding gas on the resistance to pitting corrosion of hyper duplex stainless steel welds. *Corros. Sci.* **2011**, *53*, 1939–1947. [CrossRef]
128. Zhang, Z.; Zhao, H.; Zhang, H.; Hu, J.; Jin, J. Microstructure evolution and pitting corrosion behavior of UNS S32750 super duplex stainless steel welds after short-time heat treatment. *Corros. Sci.* **2017**, *121*, 22–31. [CrossRef]
129. Nemani, A.V.; Ghaffari, M.; Salahi, S.; Nasiri, A. On the microstructural characteristics and corrosion performance of as-printed and heat-treated PH 13–8Mo martensitic stainless steel fabricated by wire arc additive manufacturing. *Mater. Today Commun.* **2023**, *34*, 105477. [CrossRef]

130. Kumar, P.; Jain, N.K.; Jaiswal, S.; Gupta, S. Development of Ti–Ta–Nb–Mo–Zr high entropy alloy by μ-plasma arc additive manufacturing process for knee implant applications and its biocompatibility evaluation. *J. Mater. Res. Technol.* **2023**, *22*, 541–555. [CrossRef]
131. Illing, C.; Bestic, M.; Ernst, F. Additive Manufacturing: Corrosion Proofing by Infusion of Interstitial Solute—Exemplified for Alloy 22. *Metals* **2023**, *13*, 127. [CrossRef]

materials

MDPI

Article

Effect of Cu Addition on the Corrosion and Antifouling Properties of PEO Coated Zinc-Aluminized Steel

Luca Pezzato [1,*], Alessio Giorgio Settimi [1], Daniel Fanchin [1], Emanuela Moschin [2], Isabella Moro [2] and Manuele Dabalà [1]

[1] Department of Industrial Engineering, University of Padova, Via Marzolo 9, 35131 Padova, Italy
[2] Department of Biology, University of Padova, Via U. Bassi, 58/B, 35121 Padova, Italy
* Correspondence: luca.pezzato@unipd.it; Tel.: +39-0498275498

Abstract: In the present work, Plasma Electrolytic Oxidation (PEO) coatings were produced on zinc-aluminized carbon steels (Galvalume commercial treatment). In addition, copper particles of various sizes were introduced into the coating in order to produce samples with antifouling properties. The particles were successfully embedded into the coating. A higher number of embedded particles was observed when these are in sub-micrometric size and obtained in pulsed current. The presence of particles produces significant antifouling properties on the sample's surfaces during the first 20 days of immersion. The presence of the particles reduces the corrosion resistance in comparison to the samples PEO coated without the particles; however, the corrosion resistance remain higher than the one of the untreated sample.

Keywords: Galvalume; Plasma Electrolytic Oxidation; antifouling; anti-corrosion coatings

Citation: Pezzato, L.; Settimi, A.G.; Fanchin, D.; Moschin, E.; Moro, I.; Dabalà, M. Effect of Cu Addition on the Corrosion and Antifouling Properties of PEO Coated Zinc-Aluminized Steel. *Materials* **2022**, *15*, 7895. https://doi.org/0.3390/ma15227895.

Academic Editors: Amir Mostafaei and Liyuan Sheng

Received: 19 September 2022
Accepted: 4 November 2022
Published: 8 November 2022

1. Introduction

Steel is one of the most used materials in marine applications, especially for large ships, where carbon steels are often employed. However, this material is subjected to corrosion problems and the substitution with stainless steel is not always possible due to economic or technological problems. Surface treatments on steels and in detail the production of oxide ceramic coatings for corrosion protection is the largely diffused way to improve their corrosion resistance. Between the various surface treatments that aims to produce protective oxide coatings on the surface of metals, PEO seems one of the most promising [1,2].

PEO of metals is a complex electrochemical process, developed from traditional anodizing, that works with higher voltages and current densities [3] in comparison with traditional processes [4]. Due to the high voltage (that has to be above the dielectric breakdown potential of the oxide layer), anodic micro-discharges are formed over the processed surface and produce the growth of an oxide ceramic coating.

PEO coating on aluminum [5], titanium [6] and magnesium alloys [7] has been widely studied and many results are reported in literature, in particular regarding the improvement of the corrosion and wear behavior of the samples [8]. The characteristics of the coatings depends on several process parameters, in particular by the current mode [9] and the current literature showed that coatings produced with pulsed current mode are characterized by improved properties in comparison with the ones produced in direct current [10].

Compared to the PEO coating on light alloys, there are relatively fewer studies regarding these coatings produced on steels [11–14]. Moreover, the quality of PEO coatings obtained on steels is often reduced if compared with the ones obtained on light alloys. One of the possibilities studied in literature to produce PEO coatings on ferrous metals is to perform, before the PEO, a pre-treatment with valve metals. In particular, some works can be found regarding the realization of PEO coatings on aluminized steels [15–17]. The main problem regarding this approach is the fact the aluminization of steels is not a so common treatment. Zinc–Aluminum coating (ZA) is instead quite common in steel production, and provides excellent corrosion protection and forming properties compared to traditional zinc layers. Moreover, ZA coating is more common and easy to obtain compared to aluminum coatings.

In the literature are reported some studies regarding the production and characterization of PEO coatings on pure zinc [18,19]. The quality of the obtained coatings was not good and their corrosion protection was quite low. In particular, it was found that despite the thickness of the obtained coatings, the ZnO layers are only slightly protective, due to their semiconducting properties and the presence of cracks and porosities. Better results were found for ZA alloys in different studies published by Pezzato et al. [20], Bian et al. [21] and Guangyin Li et al. [22]. In all the cases was found that the PEO method can produce continuous and dense coating on the ZA27 alloy using silicate, aluminate, and aluminate/borate electrolytes.

The idea of the present work is to work with a very high aluminum content in the ZA layer using a commercial Galvalume coated carbon steel (with 55% Al and 45% Zn in the coatings). This is due to the fact that an increase in the aluminum should increase the quality of the obtained PEO coating. Moreover, Galvalume coated carbon steel was never employed in literature as substrate for Plasma Electrolytic Oxidation coatings.

One of the other characteristics of PEO coating, thanks to the formation of discharge channels, and consequently of a porous surface, is the fact that it also offers a wide range of possibilities to further functionalize the coated surface thanks to the addition of specific additives into the electrolyte [23]. In particular, considering possible marine applications, the possibility of incorporating copper particles into the coatings obtained on aluminum alloys, in order to give to the samples antifouling properties, was already studied by the authors of [24,25]. However, to the best of our knowledge, no works in the literature report the production of antifouling PEO coatings on steels, even if this is of high technological importance, considering that steels, and in particular, carbon steels, are more often used in comparison to aluminum in the production of large ships that work in marine environment.

In this work the possibility to incorporate copper particles into PEO coatings produced on carbon steel, pre-treated with Galvalume treatment, in order to study a possible application in marine environment, was studied. The main goal of the work was to produce on carbon steel a copper-containing PEO coating characterized by good corrosion resistance and antifouling properties that can be employed in marine applications, such as ships. In particular, the idea is to produce an antifouling ceramic coating that can be employed as possible pre-treatment for commercial painting, in order to increase the adhesion of the painting and to assure anticorrosion and antifouling properties in the presence of scratches or defects in the paintings. The microstructure of the coatings and the presence of the particles was studied through SEM and XRD analysis. Corrosion properties were evaluated trough electrochemical tests and trough in situ immersion tests in circulating seawater. After the immersion in circulating sea water, the colonization of the samples after different immersion times in order to estimate the antifouling properties of the copper-containing coatings and to compare them with the ones of the untreated sample and with the ones of the sample PEO coated without copper, was also evaluated. The presence of the copper particles produces a significant antifouling effect during the first 20 days of immersion and slightly reduces the corrosion properties of the coatings that, however, remain higher in comparison to the ones of the untreated sample.

2. Materials and Methods

Commercial Galvalume coated carbon steel samples (55% Al, 45% Zn) were used as substrate for PEO coatings. The samples were previously degreased by ultrasonic bath in acetone for 10 min. The electrolyte used was an aqueous solution of 25 g/L of Na_2SiO_3 and 2.5 g/L of NaOH with the addition of 15 g/L of metallic copper particles (all chemicals from Sigma Aldrich, St. Louis, MO, USA). Two different particle sizes were tested; in particular, the particles defined as "large" are characterized by an average size of about

1 μm and the "small" particles by an average size of about 0.45 μm. A SEM micrograph of the different particles can be found in Figure 1 (Figure 1A "large" particles, Figure 1B "small" particles). The dispersion of both "large" and "small" particles in the electrolyte was good and was maintained thanks to continuous magnetic stirring of the electrolyte.

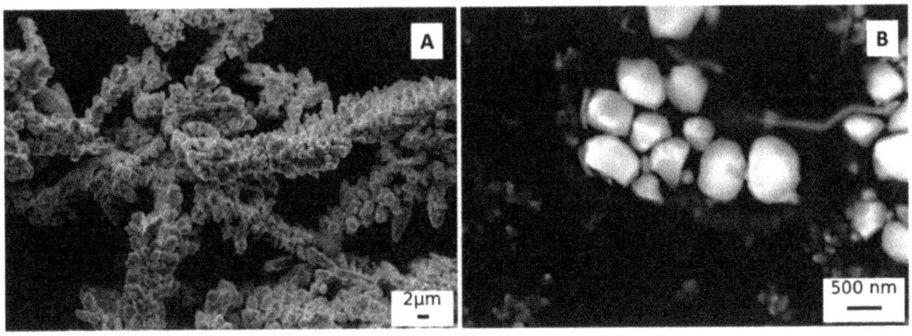

Figure 1. SEM image of the "large" copper particles (**A**) and of the "small" copper particles (**B**).

The DC generator used for the PEO coating is a TDK Lambda power supply (TDK-Lambda, Achern, Germany) capable of delivering 2400 W (315 V, 8 A). During the treatment the sample worked as an anode while the cathode was made up of a carbon steel cage. The treatments were performed in galvanostatic mode. Two different operating modes were tested. (a) Direct Current (DC) mode: Current density 0.5 A/cm^2 in direct current. (b) Unipolar Pulsed Current mode (UPC): 1 A/cm^2 in pulsed current (with a frequency of 20 Hz and a duty cycle of 50%). Higher current density was employed in the samples obtained in UPC mode due to compensation of the off-time present in pulsed mode, in order to obtain coatings with similar thickness.

The samples produced in direct current modes were treated for 90 s while those obtained in pulsed current for 120 s. The differences in the current density and in the treatment time between the two different modes were chosen on the base of preliminarily tests and of previous works in order to obtain coatings with similar thickness.

After treatment, the samples were washed with deionized water and ethanol and dried with compressed air. The surfaces of the various samples obtained were observed using a Zeiss Stemi 2000-C stereo microscope (Carl Zeiss, Jena, Germany). The cross-sections of the treated samples were obtained by cutting the sample and mounting in epoxy resin, then polished using standard metallographic technique (grinding with SiC abrasive papers from 500 to 1200 grit and polishing with clothes with 6- and 1-micron diamond suspensions). Both the surfaces and the cross sections were analyzed with a SEM Cambridge Stereoscan LEO 440 scanning electron microscope (Leica Microsystem S.r.l., Milan, Italy), equipped with Philips PV9800 EDS microanalysis (Leica Microsystem S.r.l., Milan, Italy), in order to evaluate the morphological characteristics, the thickness of the coating and the elemental composition, as well as clear presence, or not, of the particles. Moreover, EDS elemental maps and line scan analysis were recorded in order to identify elemental distribution into the coatings. The composition of the phases constituting the coating was evaluated by X-ray diffraction analysis (XRD) using a Siemens D500 diffractometer (Siemens, Munich, Germany) using Cu-Kα radiation (step size 0.05°, counting time 5 s).

Corrosion properties of the various samples were evaluated with potentiodynamic polarization (PDP) and electrochemical impedance spectroscopy (EIS) tests at room temperature. PDP tests were performed in a solution containing 0.1 M Na_2SO_4 and 0.05 M NaCl, in order to simulate a moderate aggressive environment, with an AMEL 2549 potentiostat (Amel Electrochemistry S.r.l., Milan, Italy), using a saturated calomel electrode as the reference electrode (SCE) and a platinum electrode as the counter electrode with a scan rate of 0.5 mV s^{-1}. The EIS measurements were carried out in the previous described solution and electrochemical cell at the value of the open circuit potential and in a frequency range between 10^5 Hz and 10^{-2} Hz with a perturbation amplitude of 10 mV. The impedance measurements were recorded with a Materials Instrument Spectrometer (Amel Electrochemistry S.r.l., Milan, Italy) coupled with the 2549 Potentiostat and the Z-View software (version 3.3) was used for the fitting of impedance spectra. Both PDP and EIS tests were performed using and exposed area of 1 cm^2.

To test the antifouling properties, four sets of untreated and treated samples were immersed in triplicate in tanks with circulating seawater at the Hydrobiological Station Umberto D'Ancona in Chioggia (University of Padua).

After the collection, samples were immediately preserved by adding an amount of filtered seawater with 4% formalin, neutralized with hexamethylenetetramine. The biological coverage of the sample surfaces was evaluated after 10, 20, 30 and 40 days of immersion with visual observation with a Zeiss Stemi 2000-C stereomicroscope (Carl Zeiss, Jena, Germany). For the determination of the microphytobenthos community structure, the surface of each sample was scraped and the microorganisms were observed with an inverted Leitz Diavert microscope (Leitz, Stuttgart, Germany), equipped with phase contrast. The scraped biofilms were distributed in a settling chamber and the microalgae analyzed from 60 randomly selected fields of view with a magnification of 400×. Microalgae were identified using standard keys, in particular for diatoms [26–28] were used as reference.

Moreover, to identify the microphytobenthos taxa present in the biofilms a series of samples was analyzed through SEM Cambridge Stereoscan LEO 440, after proper gold sputtering.

3. Results and Discussion

3.1. Production and Characterization of PEO Coatings

Metallic copper particles were inserted into PEO coatings, produced on aluminum-galvanized steel, in order to functionalize the surfaces and confer antifouling properties. The insertion of the particles into the coatings took place by simply adding them to the electrolyte used for the process. In order to verify the presence of particles inside the coating and qualitatively assess the number of particles incorporated, the various samples were observed under optical and electronic microscope. The results of the observation of the surfaces under the stereo microscope can be seen in Figure 2, in order to understand the global microstructure of the coating at low magnification. The presence of copper particles, which are then successfully incorporated into PEO coatings, is evident both in the samples obtained with large particles (Figure 2B,E) and in those obtained with small particles (Figure 2C,F). An evident correlation can also be observed between the particle size and the amount of Cu incorporated, with small particles that are more easily incorporated within the coating (compare regarding this Figure 2C with Figures 2B and 2F with Figure 2E). This fact can be linked with the dimension of the pores; the particles that have a smaller size than the pores are in fact more easily incorporated into the coating. This, according to Lu et al. [23], showed that pores on the coating surface can be considered as uptake paths for particles.

The method of application of the current also influences the quantity of particles that are incorporated into the coating; in fact, in pulsed current mode (Figure 2E,F), a greater quantity of particles incorporated into the coating can be observed in comparison to DC mode (Figure 2B,C). This fact can be noted both in the case of large particles (Figure 2E in comparison with Figure 2B) and in the case of small particles (Figure 2F in comparison with Figure 2C) and is also in accordance with literature [29].

In order to study the coatings obtained in more detail, SEM observations of the sample were also carried out and the results are shown in Figure 3 for the surfaces and in Figure 4 for the cross sections. Considering the observation of the samples' surfaces, first of all, the typical porous morphology of PEO coatings in Figure 3A (average pore size 10 microns) and 3D (average pore size 30 microns) can be observed, and it can be also noted that the UP mode produces a denser PEO layer. The consideration performed on the quantity of incorporated particles can be clearly confirmed and a mixed mechanism of inert and reactive incorporation can be observed. Part of the particles are, in fact, inertly incorporated and maintain their shape, whereas another part of the particles is melted and re-solidified due to the extremely high local temperatures that can be reached during PEO process. The melting point of copper is, in fact, remarkably lower than the temperature of arc plasma that, as evidenced by Lee et al. [30], ranges between 1800 and 2370 °C. The melted particles resulted as grey-white zones, without clear shape, on the surface of the coatings (Figure 3B,C,E,F).

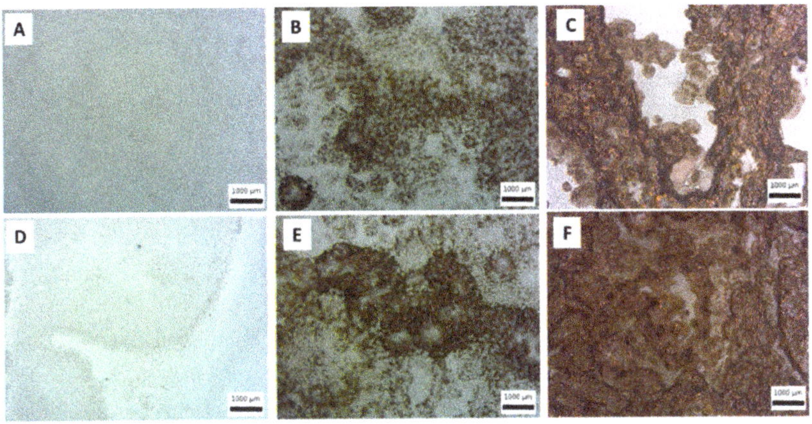

Figure 2. Stereo microscope images of the surfaces of PEO coatings obtained working at 0.5 A/cm^2 in DC mode without particles (**A**), with large particles (**B**) and with small particles (**C**) and working at 1 A/cm^2 in UPC without particles (**D**), with large particles (**E**) and with small particles (**F**).

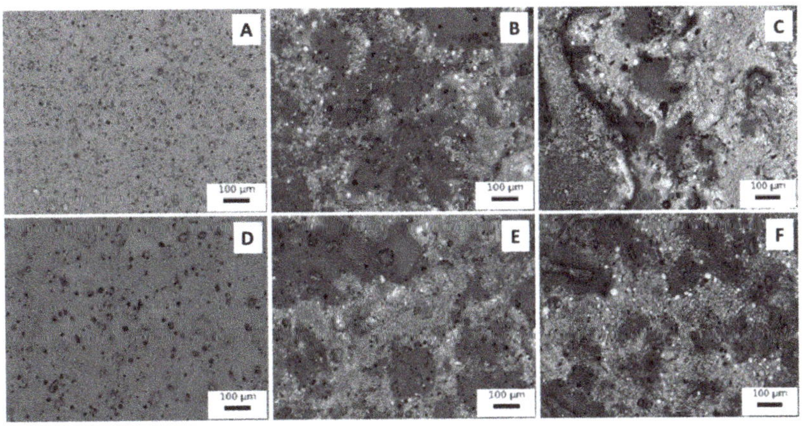

Figure 3. SEM images of the surfaces of PEO coatings obtained working at 0.5 A/cm^2 in DC mode without particles (**A**), with large particles (**B**) and with small particles (**C**), and working at 1 A/cm^2 in UPC without particles (**D**), with large particles (**E**) and with small particles (**F**).

Considering the cross section (Figure 4), in all samples, obtained both with and without particles, the presence of both the Galvalume layer and the PEO coating can be observed.

In all samples, the PEO coating has a thickness of about 20 microns, except in the sample obtained with small particles at 0.5 A/cm^2 in DC mode (Figure 4C). In this case, the PEO coating appears not completely formed, due to the too violent discharge phenomena created by the direct current with the presence of small particles. In the samples produced with particles in the electrolyte, the presence of the particles themselves (white areas) within the PEO coating is observed. These appear as large agglomerates in the case of the use of large particles (Figure 4B,E) while they are small in the case of the use of small particles (Figure 4C,F). The size of the particles significantly affects their distribution mode within the coating. In fact, the large particles (Figure 4B,E) are mainly concentrated on the surface of the coating, while the small particles, in particular working in pulsed current (Figure 4F), are distributed along the entire thickness of the coating. This appears to be in accordance with the literature [31] where it is highlighted how particles of sub micrometric size are able to enter the discharge channels that are formed during the formation of the PEO coating and, consequently, remain trapped inside the ceramic layer. Large particles, on the other hand, tend to simply "decorate" the surface of the coating itself.

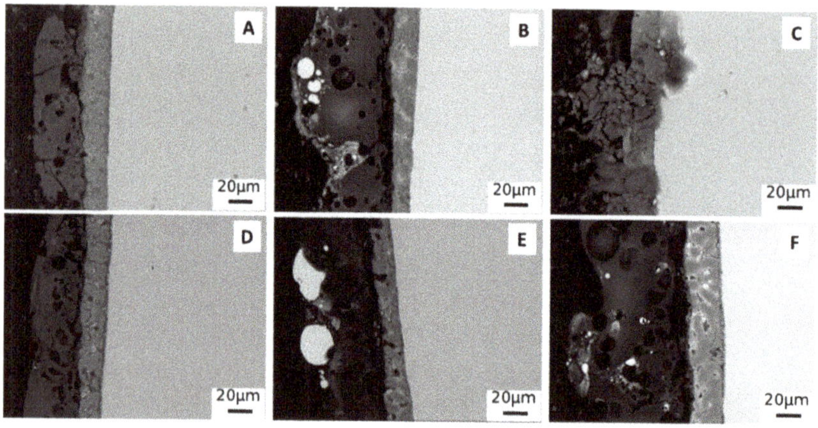

Figure 4. SEM images of the cross section of PEO coatings obtained working at 0.5 A/cm^2 in DC mode without particles (**A**), with large particles (**B**) and with small particles (**C**), and working at 1 A/cm^2 in UPC without particles (**D**), with large particles (**E**) and with small particles (**F**).

To confirm these observations and to evaluate the chemical distribution of the elements into the coatings, EDS elemental maps were performed on the most promising samples so the ones obtained with the particles in UPC mode, and the results are reported in Figure 5. The uniform distribution of Si, Al and O and the presence in the Galvalume Sub-layer of Al and Zn can be clearly noted in the PEO coated samples. Considering the copper particles, the previously reported observations can be confirmed: large particles "decorate" only the surface of the coating (Figure 5A), whereas small particles (Figure 5B) are also present inside the coatings, trapped in the discharge channels.

In order to also evaluate the composition and the thickness of the ZA layer and to study the interface between the ZA layer and the PEO coating, EDS line analysis was performed and the results obtained in the sample produced at 1 A/cm^2 in UPC with small particles are reported in Figure 6. First of all, it can be noted that the ZA layer is composed almost equally by Zinc and Aluminum and the thickness of the layer is around 20 microns. The resulting PEO layer is instead composed mainly of Al, Si and O with the Zinc not contributing significantly in the formation of the coating. Considering the interface between the ZA layer and the PEO coating, a decrease in the aluminum in comparison with the PEO layer can be observed. Therefore, at the interface, there is a concentration in the silicon content with the aluminum that increase instead in the external part of the coating. This is probably related to the fact that silicon oxidation occurs before that of aluminum, causing an increase in the silicon oxide content at the interface between the ZA layer and the PEO layer.

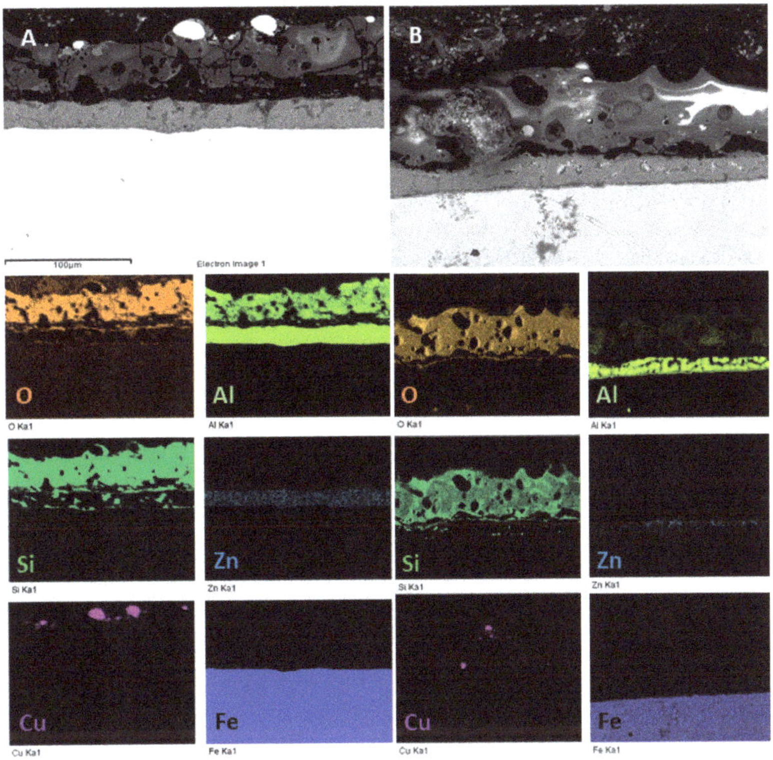

Figure 5. EDS elemental maps of the samples obtained working at 1 A/cm^2 in UPC with large particles (**A**) and with small particles (**B**).

To identify the various phases, present within the coating, XRD analysis was also performed, and the results are reported in Figure 7. From the spectra, it can be seen that the coating consists mainly of aluminum oxide (Al_2O_3) and aluminosilicates (Al_2SiO_5 and $NaAlSiO_4$). There are also clearly present peaks of copper, relative to the particles in the samples obtained with the particles, and those of Al and Zn, relative to the underlying Galvalume coating. The composition of the obtained coating resulted in accordance with the ones of the electrolyte and of the substrate and is also confirmed by results previously obtained in the literature [24]. XRD spectra of the sample obtained at 0.5 A/cm^2 in DC with small particles are not reported due to failure of the coating. From the comparison between the different samples, no significant differences can be observed in terms of phase composition.

3.2. Corrosion Properties

In order to preliminarily evaluate the effect of the presence of copper particles on the corrosion resistance of the obtained coatings, potentiodynamic polarization tests were carried out on the various samples. The results of the samples obtained with the particles were compared with those of the samples obtained without particles and with those of the untreated sample (steel with Galvalume coating). The results can be found plotted in Figure 8 and the values of corrosion potentials and corrosion current densities, graphically extrapolated, are reported in Table 1. Generally, it can be observed that the addition of the particles leads to a decrease in corrosion resistance (and so to an increase in the corrosion current) compared to samples obtained by PEO treatment without particles, probably due to galvanic contact phenomena between the copper and the underlying metal (with the formation of Cu-Al and Cu-Zn galvanic couples) and to variations in the discharge mechanisms due to the presence of copper in the electrolyte. Samples obtained with small particles show a greater decline in corrosion resistance than those obtained with large particles, probably due to the greater number of particles incorporated.

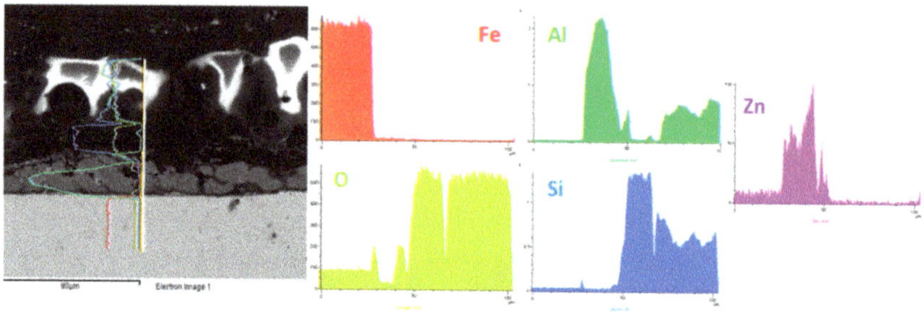

Figure 6. EDS elemental line analysis of the sample obtained working at 1 A/cm^2 in UPC with small particles.

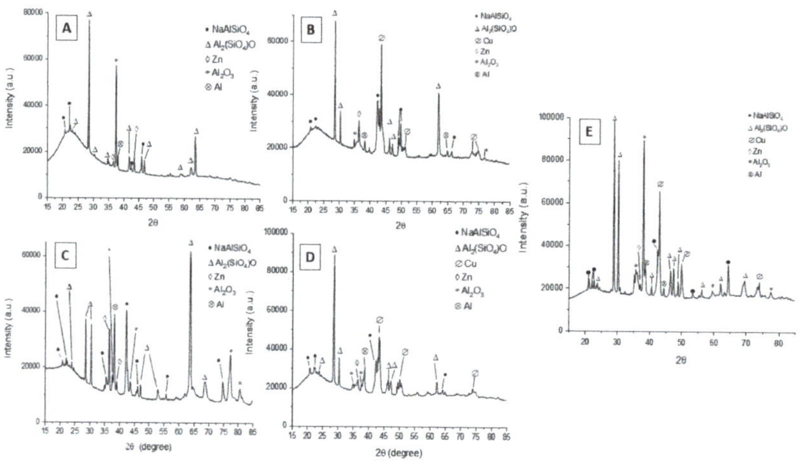

Figure 7. X-ray diffraction patterns of the samples obtained working at 0.5 A/cm^2 in DC mode without particles (**A**) and with large particles (**B**) and working at 1 A/cm^2 in UPC without particles (**C**), with large particles (**D**) and with small particles (**E**). Sample obtained at 0.5 A/cm^2 in DC with small particles not reported due to failure of the coating.

In order to quantitatively evaluate the corrosion performance of the samples and to study the effect of particles addition on the corrosion resistance, EIS tests were also performed in the same electrolyte employed in PDP tests. The results in term of Nyquist plot are reported in Figure 9, where dots represent the experimental data. The data from EIS tests were also fitted using the circuit reported in Figure 10 and the results of the fitting, that are graphically represented in Figure 9 as lines, can be found reported in Table 2.

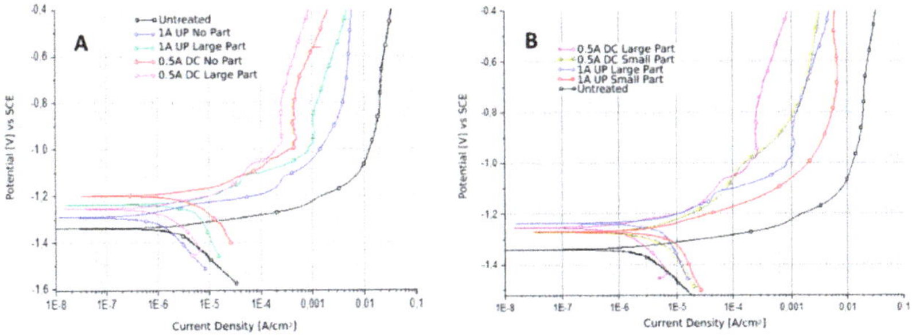

Figure 8. Results of potentiodynamic polarization tests (PDP) performed on the samples produced without particles and with "large" particles (**A**) and of the samples produced with "small" and "large" particles (**B**). Test electrolyte: 0.1 M Na_2SO_4 and 0.05 M NaCl.

Table 1. Corrosion current densities and corrosion potentials graphically extrapolated from the potentiodynamic polarization plots.

Sample	E_{corr} (V)	I_{corr} (A/cm^2)
Untreated	−1.34	2.04×10^{-6}
0.5 DC No part	−1.19	2.28×10^{-6}
0.5 DC large part	−1.25	7.83×10^{-7}
0.5 DC small part	−1.27	2.07×10^{-6}
1A UP No part	−1.28	5.17×10^{-7}
1A UP large part	−1.23	2.46×10^{-6}
1A UP small part	−1.26	3.22×10^{-6}

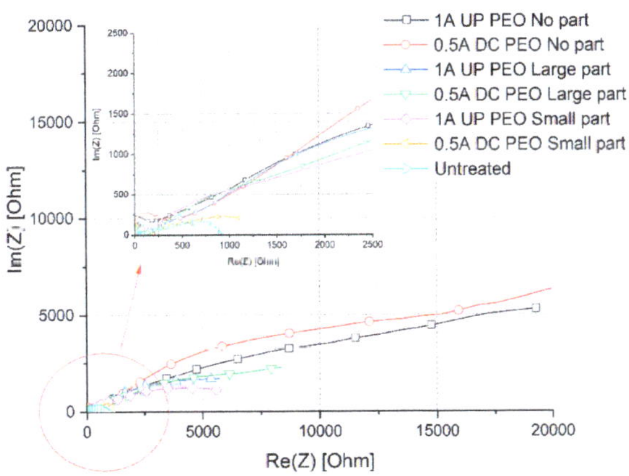

Figure 9. Results of the EIS tests in term of Nyquist plot. In the high left, the zone at the high frequencies can be observed in more detail. Test electrolyte: 0.1 M Na_2SO_4 and 0.05 M NaCl.

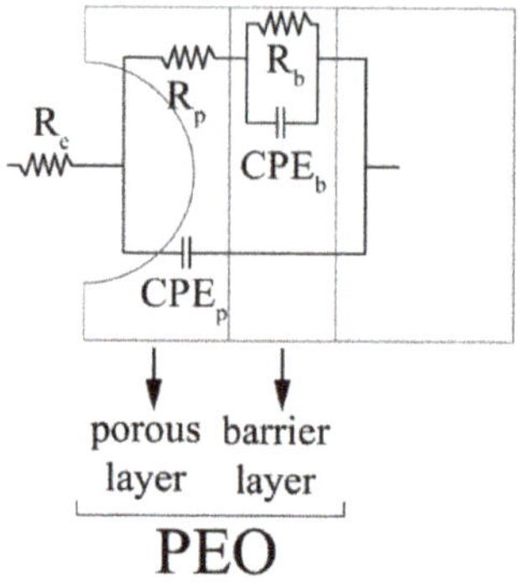

Figure 10. Equivalent circuit employed to fit the experimental data coming from EIS test.

Table 2. Results of the fitting of the experimental data coming from EIS tests.

Sample	R_e (Ωcm^2)	R_P (Ωcm^2)	R_B (Ωcm^2)	Q_P (F $cm^{-2}Hz^{1-n}$)	n_P	Q_B (F $cm^{-2}Hz^{1-n}$)	n_B	X^2
Untreated	20	-	950	-	-	3.8×10^{-5}	0.8	0.0001
0.5 DC No part	21	442	32,129	2.97×10^{-5}	0.51	1.22×10^{-7}	0.7	0.0003
0.5 DC large part	20	432	12,351	1.57×10^{-5}	0.84	1.3×10^{-5}	0.62	0.0008
0.5 DC small part	17	83	1148	8.2×10^{-4}	0.75	1.7×10^{-7}	0.75	0.0004
1A UP No part	22	200	32,172	5.51×10^{-5}	0.65	5.93×10^{-6}	0.8	0.0006
1A UP large part	20	250	13,145	3.1×10^{-4}	0.62	1.49×10^{-6}	0.71	0.0004
1A UP small part	16	448	9126	3.06×10^{-6}	0.84	1.8×10^{-5}	0.68	0.0001

The equivalent circuit was chosen on the basis of the literature on PEO coatings [32] that suggests to employ a double circuit (Figure 10) to fit data coming from PEO treated samples. In particular, this permits consideration of the presence of an inner and an external layer, called, respectively, barrier layer and porous layer. A good fitting quality was obtained, as confirmed by the low values of chi-squared in Table 2 and by the good correspondence between dots and lines in Figure 9. The untreated sample was instead fitted using a simple Randles circuit in order to consider the presence of the natural oxide layer. Considering the physical meaning of the different elements of the equivalent circuits in Figure 10, R_e represents the resistance of the electrolyte, R_p and CPE_p represents the porous layer of PEO coating, and R_b and CPE_b the barrier layer. CPEi (Constant Pahse Elements) were used in the equivalent circuits instead of capacitances due to the fact that the measured capacitance is not ideal. The impedance of a CPE has the form:

$$1/Z = Y = Q^\circ \text{ (j omega)}^n \quad (1)$$

where Q° has the numerical value of the admittance (1/Z) at omega = 1 rad/s. When this equation describes a capacitor, Y = C (the capacitance) and the exponent n = 1. For a constant phase element, the exponent n < 1. The "double-layer capacitor" on real cells often behaves like a CPE instead of like a capacitor. The values of Q_i and n_i in Table 2 refer to the CPE_i of the two layers that constitute the PEO coating. The value of X^2 represents the statistical error during the fitting of the experimental data.

In the Nyquist plot reported in Figure 9, the real part of the impedance at low frequencies can be considered as qualitative indication of the corrosion properties of the sample. Considering this, it can be clearly noted that the EIS results confirm the indication given by the PDP tests. First of all, it can be clearly noted that all the PEO treated samples are characterized by improved corrosion properties in comparison with the untreated one. The sample with the lower corrosion properties was, in fact, the one 0.5 DC with small particles, in which the PEO coating was not completely formed, whereas the best corrosion properties are obtained on the samples without the copper particles. All the samples obtained with particles (expect for the one 0.5DC small particles) are characterized by similar corrosion resistance, lower than the one of the samples without particles, due to the galvanic couple caused by the presence of copper. The qualitative indication obtained from the observation of the Nyquist plot is also confirmed by the results of the fitting of the experimental data reported in Table 2. Considering this data, first of all, it can be observed in all the samples, as typical for PEO coatings, that the value of R_B is higher than the one of R_P, due to the increased corrosion properties of the barrier layer. The value of R_B for the samples treated without particles (1A UP and 0.5 DC) was around 30,000 ohm cm^2, whereas the one of the samples treated with the particles was around 10,000 ohm cm^2, confirming the reduction in the corrosion properties due to the galvanic couple between the copper and the aluminum substrate. Excluding the sample 0.5DC small particles, characterized by a value of R_B around 1000 ohm cm^2 due to incomplete coating formation, the sample with the lower value of R_B was the one 1A UP with small particles (9126 ohm cm^2). This can be linked with the SEM observation where the sample 1A UP with small particles was the one with the higher amount of incorporated particles and so with higher problems of galvanic coupling. In all the PEO treated samples, the polarization resistance was higher than the one of the untreated sample, characterized by the presence of only the natural oxide layer (950 ohm cm^2).

3.3. Antifouling Properties

The antifouling properties were evaluated by immersing the samples in circulating seawater for 40 days and analyzing the surfaces after each 10 days. In Figures 11–13, the visual and the stereo microscope observations of the immersed samples after different immersion times are reported. In particular, in Figure 11, the untreated sample is reported, in Figure 12, the PEO treated sample (1A UP without particles), and in Figure 13, the sample PEO treated with the presence of copper particles (1A UP with small particles). The immersion tests were performed on the sample 1A UP with small particles; from the previously reported observations, this was in fact the one with the higher number of incorporated particles, and so the more promising in terms of antifouling properties. In fact, as reported in the literature [24], the quantity of the antifouling agent plays a key role in the colonization. For comparison reasons, the immersed sample without particles was the corresponding one, 1A UP without particles. Considering the untreated sample (so the one with only the ZA layer, Figure 11), already after 10 days of immersion, a remarkable colonization can be clearly observed and after 40 days almost all the surface resulted covered. Moreover, remarkable corrosion phenomena can be noted with the presence of evident cavities on the surface already after 10 days. The PEO coatings without particles (Figure 12) do not reduce the fouling phenomena that instead were increased, as can be observed by the presence of bigger green zones in the surface after 40 days of immersion. This fact was in accordance with previous work of the authors [25] and is linked with the surface morphology of the PEO layer. As was also observed in the SEM analysis, the surface of PEO coatings is, in fact, porous and in these pores, the algae can grow faster. Instead, no remarkable corrosion phenomena can be observed on the surface, even after 40 days of immersion, thanks to the protection given by the PEO layer. Considering the sample PEO treated with the copper particles (Figure 13), a remarkable antifouling effect of the copper particles during the first 10 and 20 days of immersion can be observed. In fact, after 10 and 20 days, no colonization is observable. After 30 and 40 days, the colonization seems to start also in the sample PEO treated with the copper particles, probably due to the fact that the complete coverage of the surface of the samples by the biofilm significantly reduces the antifouling effect of the copper particles present in the PEO underlayer. Furthermore, the corrosion properties of the samples PEO/Cu were good with no evidence of corrosion phenomena even after 40 days. Considering both these facts, the PEO coating with copper particles can be considered as promising pre-treatment for antifouling paints on steels. In fact, the PEO layer will protect from corrosion in presence of scratches of defects in the varnish and the presence of the copper particles will assure the antifouling properties, at least during the first days of immersion before that proper maintenance can be performed. Moreover, the porous morphology of the PEO surface make it particularly suitable for increasing the mechanical bond between the metal and the organic coating.

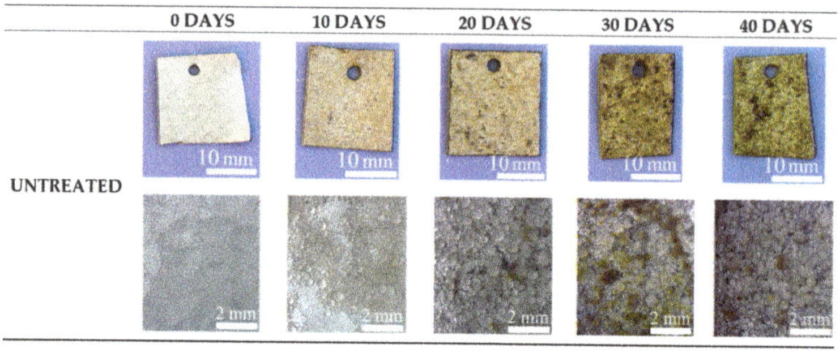

Figure 11. Visual observation (on the upper part) and stereo-microscope observation (on the bottom part) of the untreated samples after 0, 10, 20, 30 and 40 days of immersion in circulating seawater at the Hydrobiological Station Umberto D'Ancona.

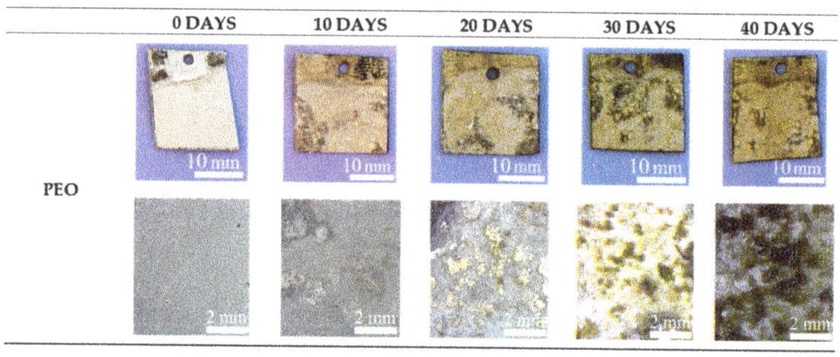

Figure 12. Visual observation (on the upper part) and stereo-microscope observation (on the bottom part) of the PEO treated samples (1A UP No part) samples after 0, 10, 20, 30 and 40 days of immersion in circulating seawater at the Umberto D'Ancona Hydrobiological Station.

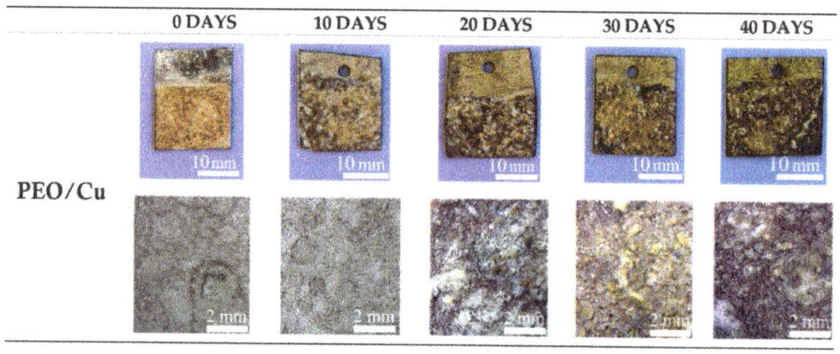

Figure 13. Visual observation (on the upper part) and stereo-microscope observation (on the bottom part) of the PEO treated samples with copper particles (1A UP small part) samples after 0, 10, 20, 30 and 40 days of immersion in circulating seawater at the Umberto D'Ancona hydrobiological station.

3.4. Biological Colonization

The observation at the light microscope of the microphytobenthos community of the samples showed a predominance of diatoms, represented by pennate forms, while the centric ones were absent. This was due to their capability to attach and glide on a substrate thanks to the presence of raphe or pores, structures involved in the secretion of extracellular polymeric substances (EPS) [33,34]. In the later phases of the colonization (30 and 40 days), filamentous cyanobacteria were also present, mainly in the PEO treated samples (PEO and PEO/Cu).

After 10 days, PEO and PEO/Cu treatments showed a lower number of taxa than the control one, suggesting a selective effect of PEO on the adhesion of some taxa (Table 3). All the samples showed erect (attached to surfaces by stalks), adnate (forms closely appressed to the substratum) and motile species, mainly belonging to the genera *Amphora* (Figure 14A), *Cylindrotheca, Haslea, Navicula* (Figure 14B,C), *Licmophora* (Figure 13D), *Nitzschia* and *Tabularia*. In the untreated sample, the cell number for each taxon was almost homogeneous, in the PEO and PEO/Cu treated samples, some genera (e.g., for PEO, *Psammodyction panduriforme*, for PEO/Cu *Thalassionema nitzschioides*) were present with a high number of cells.

Table 3. Number of taxa registered on the different samples during the experimental period.

	10 Days	20 Days	30 Days	40 Days
Untreated	21	19	23	21
PEO	11	17	16	20
PEO/Cu	14	16	22	22

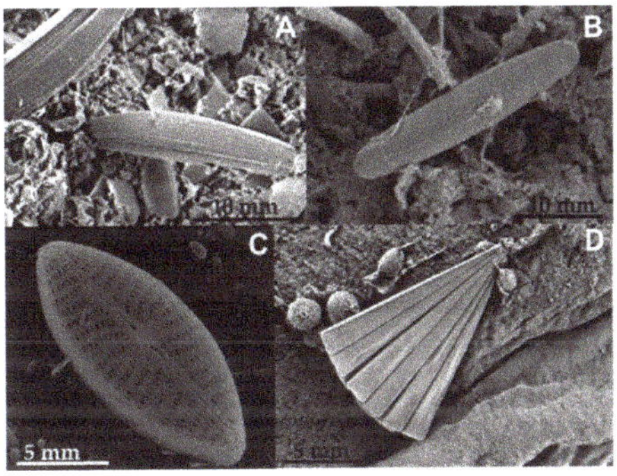

Figure 14. Scanning electron micrographs showing some diatoms found in the samples. (**A**) *Amphora* sp., bar = 10 μm; (**B**) *Navicula* sp., bar = 10 μm; (**C**) *Navicula* sp., bar = 5 μm, (**D**) *Licmophora* sp., bar = 5 μm.

In the following immersion times (20 and 30 days), while PEO maintained a scarce number of taxa, PEO/Cu showed an increase in the biodiversity, until reaching the same level as the untreated sample. This could be probably due to the roughness of the surfaces, favoring a better adhesion of the benthic microorganisms.

After 40 days of immersion, a high number of taxa was registered in all the samples (control and PEO treatments). The decrease in the antifouling capability in the samples treated with PEO and PEO/Cu showing a number of taxa similar to the control one could be probably due to the complete coverage of the substrate by microalgal biofilm that cancels the effect of the treatments for the colonization by the upper foulers.

4. Conclusions

The following conclusion can be summarized:

Copper particles were successfully incorporated into PEO coatings obtained on steels after Galvalume treatment.

The number of incorporated particles increases by reducing the particle size and switching from a direct current to a pulsed current mode.

The presence of particles reduces the corrosion resistance of PEO coatings, which still remains higher than that of steel after Galvalume treatment.

The presence of the copper particles produces a significant antifouling effect during the first 20 days of immersion, whereas for increased immersion times, the presence of copper particles does not show effect in reducing the colonization phenomena.

PEO treatment with copper particles can be considered as promising pre-treatment for painting steel due to the fact that assure corrosion and antifouling properties even in presence of defect or scratches in the varnish, at least during the first 20 days of immersion before that proper maintenance can be performed.

Author Contributions: Conceptualization, L.P., M.D. and I.M.; methodology, L.P. and E.M.; software, D.F. and A.G.S.; validation, D.F. and A.G.S.; formal analysis, A.G.S.; investigation, D.F., L.P. and E.M.; resources, I.M. and M.D.; data curation, L.P. and A.G.S.; writing—original draft preparation, L.P. and I.M.; writing—review and editing, L.P. and I.M.; visualization, A.G.S. and E.M.; supervision, M.D. and I.M.; project administration, M.D.; funding acquisition, M.D. and I.M. All authors have read and agreed to the published version of the manuscript.

Funding: Project supported by the BIRD 2020 program of the University of Padova (Project BIRD202558/20).

Institutional Review Board Statement: Not applicable.

Informed Consent Statement: Not applicable.

Data Availability Statement: The raw/processed data required to reproduce these findings cannot be shared at this time as the data also form part of an ongoing study.

Acknowledgments: The authors want to thank the hydrobiological station Umberto D'Ancona of the University of Padova in Chioggia for the antifouling tests.

Conflicts of Interest: The authors declare no conflict of interest.

References

1. Wang, Y.L.; Wang, M.; Zhou, M.; Li, B.J.; Amoako, G.; Jiang, Z.H. Microstructure characterisation of alumina coating on steel by PEO. *Surf. Eng.* **2013**, *29*, 271–275. [CrossRef]
2. Attarzadeh, N.; Molaei, M.; Babaei, K.; Fattah-alhosseini, A. New Promising Ceramic Coatings for Corrosion and Wear Protection of Steels: A Review. *Surf. Interfaces* **2021**, *23*, 100997. [CrossRef]
3. Yerokhin, A.L.; Snizhko, L.O.; Gurevina, N.L.; Leyland, A.; Pilkington, A.; Matthews, A. Spatial characteristics of discharge phenomena in plasma electrolytic oxidation of aluminium alloy. *Surf. Coat. Technol.* **2004**, *177–178*, 779–783. [CrossRef]
4. Hryniewicz, T. Plasma electrolytic oxidation of metals and alloys. *Metals* **2018**, *8*, 1058. [CrossRef]
5. Egorkin, V.S.; Gnedenkov, S.V.; Sinebryukhov, S.L.; Vyaliy, I.E.; Gnedenkov, A.S.; Chizhikov, R.G. Increasing thickness and protective properties of PEO-coatings on aluminum alloy. *Surf. Coat. Technol.* **2018**, *334*, 29–42. [CrossRef]
6. Kostelac, L.; Pezzato, L.; Settimi, A.G.; Franceschi, M.; Gennari, C.; Brunelli, K.; Rampazzo, C.; Dabalà, M. Investigation of hydroxyapatite (HAP) containing coating on grade 2 titanium alloy prepared by plasma electrolytic oxidation (PEO) at low voltage. *Surf. Interfaces* **2022**, *30*, 101888. [CrossRef]
7. Pezzato, L.; Lorenzetti, L.; Tonelli, L.; Bragaggia, G.; Dabalà, M.; Martini, C.; Brunelli, K. Effect of SiC and borosilicate glass particles on the corrosion and tribological behavior of AZ91D magnesium alloy after PEO process. *Surf. Coat. Technol.* **2021**, *428*, 127901. [CrossRef]
8. Pezzato, L.; Vranescu, D.; Sinico, M.; Gennari, C.; Settimi, A.; Pranovi, P.; Brunelli, K.; Dabalà, M. Tribocorrosion Properties of PEO Coatings Produced on AZ91 Magnesium Alloy with Silicate- or Phosphate-Based Electrolytes. *Coatings* **2018**, *8*, 202. [CrossRef]
9. Hussein, R.O.; Zhang, P.; Nie, X.; Xia, Y.; Northwood, D.O. The effect of current mode and discharge type on the corrosion resistance of plasma electrolytic oxidation (PEO) coated magnesium alloy AJ62. *Surf. Coat. Technol.* **2011**, *206*, 1990–1997. [CrossRef]
10. Gao, Y.; Yerokhin, A.; Matthews, A. Effect of current mode on PEO treatment of magnesium in Ca- and P-containing electrolyte and resulting coatings. *Appl. Surf. Sci.* **2014**, *316*, 558–567. [CrossRef]
11. Wang, Y.; Jiang, Z.; Yao, Z.; Tang, H. Microstructure and corrosion resistance of ceramic coating on carbon steel prepared by plasma electrolytic oxidation. *Surf. Coat. Technol.* **2010**, *204*, 1685–1688. [CrossRef]
12. Wang, Y.; Jiang, Z.; Yao, Z. Preparation and properties of ceramic coating on Q235 carbon steel by plasma electrolytic oxidation. *Curr. Appl. Phys.* **2009**, *9*, 1067–1071. [CrossRef]
13. Wang, Y.; Jiang, Z.; Yao, Z. Microstructure, bonding strength and thermal shock resistance of ceramic coatings on steels prepared by plasma electrolytic oxidation. *Appl. Surf. Sci.* **2009**, *256*, 650–656. [CrossRef]

14. Pezzato, L.; Brunelli, K.; Dolcet, P.; Dabalà, M. Plasma electrolytic oxidation coating produced on 39NiCrMo3 steel. *Surf. Coat. Technol.* **2016**, *307*, 73–80. [CrossRef]
15. Saikiran, A.; Hariprasad, S.; Arun, S.; Rama Krishna, L.; Rameshbabu, N. Effect of electrolyte composition on morphology and corrosion resistance of plasma electrolytic oxidation coatings on aluminized steel. *Surf. Coat. Technol.* **2019**, *372*, 239–251.
16. Wu, Z.; Xia, Y.; Li, G.; Xu, F. Structure and mechanical properties of ceramic coatings fabricated by plasma electrolytic oxidation on aluminized steel. *Appl. Surf. Sci.* **2007**, *253*, 8398–8403. [CrossRef]
17. Gu, W.C.; Lv, G.H.; Chen, H.; Chen, G.L.; Feng, W.R.; Zhang, G.L.; Yang, S.Z. Preparation of ceramic coatings on inner surface of steel tubes using a combined technique of hot-dipping and plasma electrolytic oxidation. *J. Alloys Compd.* **2007**, *430*, 308–312. [CrossRef]
18. Stojadinović, S.; Tadić, N.; Vasilić, R. Formation and characterization of ZnO films on zinc substrate by plasma electrolytic oxidation. *Surf. Coat. Technol.* **2016**, *307*, 650–657. [CrossRef]
19. Rocca, E.; Veys-Renaux, D.; Guessoum, K. Electrochemical behavior of zinc in KOH media at high voltage: Micro-arc oxidation of zinc. *J. Electroanal. Chem.* **2015**, *754*, 125–132. [CrossRef]
20. Pezzato, L.; Settimi, A.G.; Cerchier, P.; Gennari, C.; Dabalà, M.; Brunelli, K. Microstructural and corrosion properties of PEO coated zinc-aluminized (ZA) steel. *Coatings* **2020**, *10*, 448. [CrossRef]
21. Bian, G.; Wang, L.; Wu, J.; Zheng, J.; Sun, H.; DaCosta, H. Effects of electrolytes on the growth behavior, microstructure and tribological properties of plasma electrolytic oxidation coatings on a ZA27 alloy. *Surf. Coat. Technol.* **2015**, *277*, 251–257. [CrossRef]
22. Li, G.; Mao, Y.; Li, Z.; Wang, L.; DaCosta, H. Tribological and Corrosion Properties of Coatings Produced by Plasma Electrolytic Oxidation on the ZA27 Alloy. *J. Mater. Eng. Perform.* **2018**, *27*, 2298–2305. [CrossRef]
23. Lu, X.; Mohedano, M.; Blawert, C.; Matykina, E.; Arrabal, R.; Kainer, K.U.; Zheludkevich, M.L. Plasma electrolytic oxidation coatings with particle additions—A review. *Surf. Coat. Technol.* **2016**, *307*, 1165–1182. [CrossRef]
24. Cerchier, P.; Pezzato, L.; Gennari, C.; Moschin, E.; Moro, I.; Dabalà, M. PEO coating containing copper: A promising anticorrosive and antifouling coating for seawater application of AA 7075. *Surf. Coat. Technol.* **2020**, *393*, 125774. [CrossRef]
25. Cerchier, P.; Pezzato, L.; Moschin, E.; Coelho, L.B.; Olivier, M.G.M.; Moro, I.; Magrini, M. Antifouling properties of different Plasma Electrolytic Oxidation coatings on 7075 aluminium alloy. *Int. Biodeterior. Biodegrad.* **2018**, *133*, 70–78. [CrossRef]
26. Peragallo, H.; Peragallo, M. *Diatomees Marine de France et des Districts Maritimes Voisins*; Micrographe Editeur: Grez sur Loing, France, 1898.
27. Husted, F. Die Kiesealgen Von Deutschland, Österreichs Und der Schweiz Mit Berusichtigung der Übrigen Länder Europas Sowie der Angrenzender Mehresgebiete. In *Rabenhorst's Kriptogamen-Flora von Deutschland, Österreichs und der Schweiz*; M.B.H, Verlag.: Leipzig, Germany, 2011.
28. Van der Werff, A.; Hulls, H. *Diatomeeën Flora van Nederland*; Koeltz, O., Ed.; Science Publishers: Koeningstein, Germany, 1957–1974.
29. Lu, X.; Blawert, C.; Mohedano, M.; Scharnagl, N.; Zheludkevich, M.L.; Kainer, K.U. Influence of electrical parameters on particle uptake during plasma electrolytic oxidation processing of AM50 Mg alloy. *Surf. Coat. Technol.* **2016**, *289*, 179–185. [CrossRef]
30. Lee, K.M.; Lee, B.U.; Yoon, S.I.; Lee, E.S.; Yoo, B.; Shin, D.H. Evaluation of plasma temperature during plasma oxidation processing of AZ91 Mg alloy through analysis of the melting behavior of incorporated particles. *Electrochim. Acta* **2012**, *67*, 6–11. [CrossRef]
31. O'Hara, M.; Troughton, S.C.; Francis, R.; Clyne, T.W. The incorporation of particles suspended in the electrolyte into plasma electrolytic oxidation coatings on Ti and Al substrates. *Surf. Coat. Technol.* **2020**, *385*, 125354. [CrossRef]
32. Pezzato, L.; Coelho, L.B.; Bertolini, R.; Settimi, A.G.; Brunelli, K.; Olivier, M.; Dabalà, M. Corrosion and mechanical properties of plasma electrolytic oxidation coated AZ80 magnesium alloy. *Mater. Corros.* **2019**, *70*, 2103–2112. [CrossRef]
33. Gomathi Sankar, G.; Sathya, S.; Sriyutha Murthy, P.; Das, A.; Pandiyan, R.; Venugopalan, V.P.; Doble, M. Polydimethyl siloxane nanocomposites: Their antifouling efficacy invitro and in marine conditions. *Int. Biodeterior. Biodegrad.* **2015**, *104*, 307–314. [CrossRef]
34. Molino, P.J.; Wetherbee, R. The biology of biofouling diatoms and their role in the development of microbial slimes. *Biofouling* **2008**, *24*, 365–379. [CrossRef] [PubMed]

materials

MDPI

Article

Development of Micro Laser Powder Bed Fusion for Additive Manufacturing of Inconel 718

Saeed Khademzadeh, Claudio Gennari, Andrea Zanovello, Mattia Franceschi, Alberto Campagnolo and Katya Brunelli *

Industrial Engineering Department, University of Padova, 35121 Padova, Italy;
saeed.khademzadeh@unipd.it (S.K.); claudio.gennari@unipd.it (C.G.);
andrea.zanovello.1@studenti.unipd.it (A.Z.); mattia.franceschi@phd.unipd.it (M.F.);
alberto.campagnolo@unipd.it (A.C.)
* Correspondence: katya.brunelli@unipd.it; Tel.: +39-049-8275500

Abstract: The development of laser powder bed fusion (LPBF) additive manufacturing techniques for microfabrication raises the need for the employment of new process configurations and parameters. In this study, micro-LPBF of Ni-based superalloy Inconel 718 using a spot laser of 30 μm was examined. The response surface method with a central composite design was employed to determine the optimum process parameter. A wide range of heat treatment cycles was applied to additively manufacture Inconel samples. The mechanical behavior of heat-treated Inconel 718 parts fabricated via micro-LPBF was investigated and correlated to the microstructural characteristics. The result showed that using optimum input energy density led to a homogenous distribution of nanosized (<10 nm) circular γ' and plate-like γ'' particles in the γ matrix. Uniaxial tensile tests on heat-treated samples showed that ageing temperature is the most determinant factor in the mechanical strength of additively manufactured Inconel 718.

Keywords: Ni-based superalloys; micro laser powder bed fusion; heat treatment; mechanical properties

Citation: Khademzadeh, S.; Gennari, C.; Zanovello, A.; Franceschi, M.; Campagnolo, A.; Brunelli, K. Development of Micro Laser Powder Bed Fusion for Additive Manufacturing of Inconel 718. *Materials* **2022**, *15*, 5231. https://doi.org/10.3390/ma15155231

Academic Editor: Thomas Niendorf

Received: 24 June 2022
Accepted: 26 July 2022
Published: 28 July 2022

1. Introduction

Inconel 718 (Inc718) is the most widely used nickel-based superalloy [1] due to its high strength, creep, and fatigue resistance at high temperatures (up to 700 °C). Excellent resistance to oxidation and hot corrosion makes Inc718 an alloy of choice for many applications such as the aerospace industry [2]. In particular, Inc718 is used for extreme operating conditions such as gas turbines, turbochargers, nuclear reactors, liquid propellant rockets, combustors, heat exchangers, turbine blades, high-pressure vessels, and cryogenic storage tanks [3]. Due to rapid advances in technology, fabricated components are required to have increasingly complex part geometries (e.g., cooling channels in turbine blades). However, high hardness and low thermal conductivity make Inc718 a difficult to cut material because of tool over-wear and poor workpiece surface integrity after conventional machining processes [4]. Additive Manufacturing (AM) is one of the main emerging advanced manufacturing methods and has demonstrated vast application opportunities. AM technologies are capable of making complex functional parts that cannot be obtained using conventional techniques [5,6]. One of the main laser-based AM processes is laser powder bed fusion (LPBF) which is based on the melting of certain areas of the metal powder layers using a high-intensity laser, following the indications of computer-aided designed models (CAD data) [7]. The LPBF appears to be one of the most promising metal AM techniques [8] for the fabrication of geometrically complex components with high dimensional accuracy and good surface integrity [9]. A vast number of research activities have been carried out on LPBF of engineering metallic materials such as titanium and nickel alloys [10]. Therefore, LPBF as a tool-free production method was found to be an excellent candidate for the fabrication of complex geometries from Nickel-based superalloys. Inc718 parts require heat treatments to improve mechanical properties by precipitation hardening, regardless of the process used for fabrication. Generally, these treatments consist of solubilization followed by age hardening to induce the precipitation of reinforcement phases (γ' and γ'' phases) [11].

Selecting the correct heat treatment parameters is necessary to achieve proper changes in the microstructure and mechanical properties. Chlebus et al. [12] found that solution annealing of Inc718 produced by laser powder bed fusion requires a higher temperature than that typically involved (~1100 °C instead of ~1000 °C). Huang et al. [13] studied the influence of solution time, temperature, and cooling rate as well as age hardening on the microstructure and mechanical properties of LPBF Inc718. In order to increase the productivity of the LPBF process, a laser spot greater than 50 µm is generally used for the LPBF of Inc718 [14,15]. For precision applications, on the other hand, a reduced spot size allows for producing parts with a higher resolution, recreating their features with high dimensional accuracy [16]. Recently, there has been an ever-increasing demand for micro-fabrication technologies to meet the push toward miniaturization that is taking place in various sectors [17]. Compared with other techniques commonly used for micro-AM, micro-LPBF is interesting due to several factors such as faster cycle time, material versatility, and process simplicity [18]. In the present study, several heat treatments were performed on Inc718 samples produced by micro-LPBF using a combination of optimized process parameters. Aiming to develop LPBF for micro-fabrication, a 30 µm laser spot was employed which led to a change in input energy densities and microstructural characteristics.

2. Materials and Methods

2.1. Starting Powder

Gas-atomized Inc718 powder produced by Böhler was used as starting material. Powder particles ranged from 15 to 45 µm. The chemical composition of the initial powder is presented in Table 1. A scanning electron microscope (SEM) showed a spherical overall shape for powder particles and the presence of irregular satellites (Figure 1). The flowability of the powder is a crucial factor in powder bed additive manufacturing technologies such as LPBF. The rheology of the powders was measured by FT4 Powder Rheometer (Freeman Technology, Tewkesbury, UK). This experiment revealed that the basic flowability energy (BFE) of the Inconel 718 powder was 940 mJ that provided a flawless layer of powder after recoating.

Table 1. Chemical compositions of Inc718 powder.

El	Ni	Cr	Fe	Nb	Mo	Ti	Al	Co	Si	Mn	Cu	C	P	N	B
wt%	53.70	17.93	18.17	5.20	2.96	0.95	0.48	0.33	0.08	0.08	0.05	0.025	0.009	0.004	0.0025

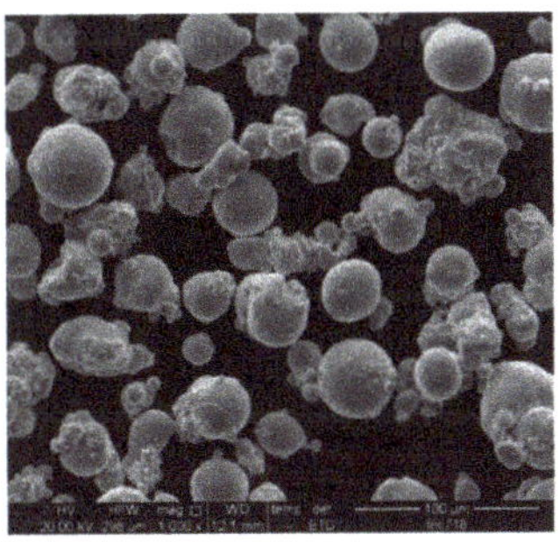

Figure 1. SEM micrograph of Inc718 powder produced by gas atomization.

2.2. LPBF Process Set Up for Microfabrication

LPBF machine of SISMA (MYSINT100TM) was used for the fabrication of Inc718 samples. The machine is equipped with a 200 W fiber laser. A galvo scanner system consisting of quartz F-Theta Lens provides the planar movement of the laser beam. To enhance the resolution of the process, laser spot diameter was set at 30 μm which enables microfabrication via LPBF. The machine uses an x-lip rubber-like blade to spread the powder particles on the building platform. A constant argon flow over the building platform was used to remove unwanted melting products such as burnt or oxidized particles. The whole process was carried out in an argon atmosphere with an oxygen level below 500 ppm to minimize unwanted chemical reactions. The layer thickness was kept constant at 20 μm in all experiments. Support generation and slicing of computer-aided designed (CAD) models were performed using the software Magics^{TM} (Materialise, SISMA, Leuven, Belgium).

2.3. Attaining Process Parameters

In this study, a comprehensive design of experiment (DOE) was employed to evaluate the density and surface topography of LPBF parts corresponding to the interaction among the main LPBF process parameters such as laser power, scanning speed, hatch distance, and scanning strategy. Dense cubic (10 mm × 10 mm × 10 mm) samples were built for DOE. Input ranges of process parameters for three numeric factors and one categorical factor are summarized in Table 2. The Design–Expert (Stat-Ease Inc., Minneapolis, MN, USA) software was employed to implement a two-step DOE using response surface methodology (RSM) and subsequent full-factorial technique. A central composite design (CCD) was used which is the most widely used RSM design for the experiment. CCD efficiently locates the sampling points to increase the accuracy of the response surfaces after experimental measurement of parameters for sampling points [19]. Each numeric factor was set to five levels: high and low levels (factorial points), plus and minus alpha (axial points), and the central point. For each categorical factor, the central composite design was duplicated for every combination of the categorical factor levels. Bidirectional (Figure 2a) and alternating bidirectional (Figure 2b) scanning strategies were employed for the fabrication of CCD samples.

Table 2. Input ranges of process parameter used in CCD.

Parameter	Unit	Numeric Factors				
		$-\alpha$	Low	Central	High	$+\alpha$
Laser Power	W	42	50	75	100	108
Scanning speed	mm/s	400	450	600	750	800
Hatch distance	mm	0.04	0.05	0.07	0.09	0.1
		Categorical factor				
Scanning strategy		Level 1 Bidirectional			Level 2 Alternating Bidirectional	

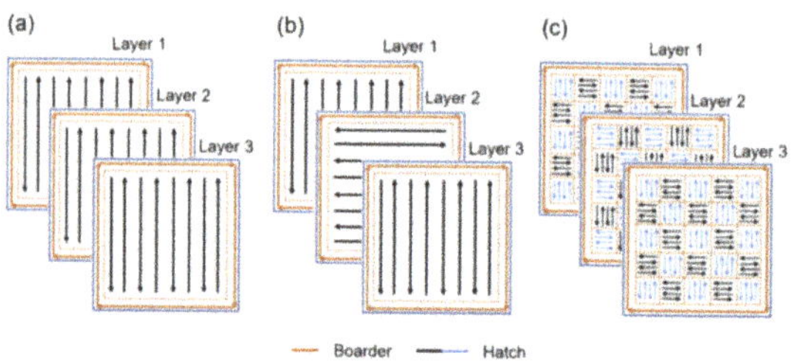

Figure 2. Different LPBF scanning strategies used for processing of Inc718; (**a**) bidirectional, (**b**) alternating bidirectional, (**c**) alternating chessboard.

2.4. Heat Treatment Cycles

Based on existing standard heat treatments for wrought and casting Inc718 (see Table 3), a systematic heat treatment study was performed through different heat treatment cycles performed on Inc718 cubic samples produced by optimized LPBF process parameters. These consist of a Solution Annealing (SA) followed by single Age Hardening (AH) or Double Age Hardening (DAH) at two levels of ageing temperature (620 °C, 720 °C) for 4 h and 8 h. The furnace cooling rate from the solution annealing temperature to the ageing temperature was set at 100 °C/h. The same cooling rate was applied for double ageing. Heat treatment cycles are summarized in Table 4 and schematically presented in Figure 3. All heat treatments were carried out in a CWF laboratory chamber furnace (Carbolite®, Derbyshire, UK) at a heating rate of 10 °C/min, in argon atmosphere.

Table 3. Industrial standard heat treatment cycles for wrought and casting Inc718.

	Step 1	Step 2	Step 3
Wrought Inc718 [20]	SA: 980 °C (1 h), AC	DAH: 720 °C (8 h), FC at 55 °C/h to 620 °C (8 h), AC	
Casting Inc718 [21]	H: 1080 °C (1.5 h), AC	SA: 980 °C (1 h), AC	DAH: 720 °C (8 h), FC at 55 °C/h to 620 °C (8 h), AC

SA: solution annealing; **H**: homogenization; **DAH**: double age hardening; **AC**: air cooling; **FC**: furnace cooling.

Table 4. Heat treatment cycles performed on Inc718 micro-LPBF samples.

Solution Annealing	Cooling 1	Ageing 1	Ageing 2	Cooling 2	Type
1080 °C, 1.5 h	AC	–	–	–	SA
1080 °C, 1.5 h	AC	620 °C, 4 h	–	WQ	1AH
1080 °C, 1.5 h	AC	620 °C, 8 h	–	WQ	2AH
1080 °C, 1.5 h	AC	720 °C, 4 h	–	WQ	3AH
1080 °C, 1.5 h	AC	720 °C, 8 h	–	WQ	4AH
1080 °C, 1.5 h	AC	720 °C, 4 h	620 °C, 4 h	WQ	1DAH

SA: solution annealed; AH: age hardening; DAH: double age hardening: AC: air cooling: WQ: water quenched.

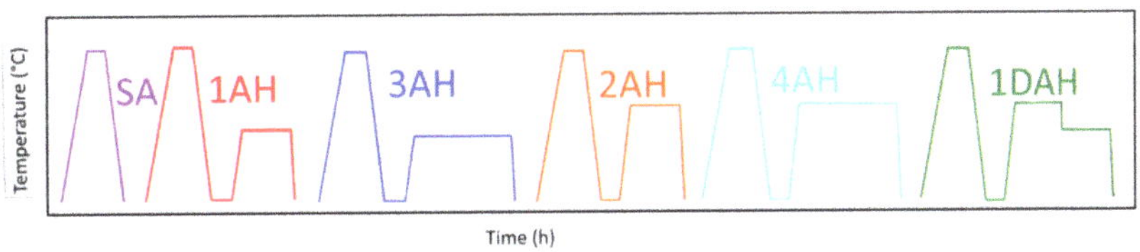

Figure 3. Schematic representation of 6 heat treatment cycles performed in this research.

2.5. *Characterization of LPBF Products*

2.5.1. Density Measurement

Archimedes' principle was employed to determine the density of LPBF processed Inc718. A precision balance with an accuracy of 0.1 mg was used to measure the mass of produced samples in the air (m_{air}) and while submerged in de-ionized water (m_w). Both measurements were performed five times and the arithmetic mean values were applied in Equation (1) to calculate the density:

$$\rho_s = \frac{m_{air}}{m_{air} - m_w}(\rho_w - \rho_{air}) + \rho_{air} \tag{1}$$

where ρ_s is the density of the as-built sample, ρ_w is the density of de-ionized water and ρ_{air} is the density of the air. Density values were reported relative to the theoretical bulk density of Inc718 (ρ_{Bulk} = 8400 kg/m^3).

2.5.2. Surface Roughness Measurement

The surface roughness of the printed Inc718 samples was measured using 3D optical profilometry. In this work, a Sensofar SNeox optical 3D profiler was used in focus variation mode with a 20× objective lens (NA 0.45, field of view 877 × 660 µm, spatial sampling equal to 0.65 µm and optical resolution 0.31 µm). For each sample, a surface with an area equal to 3.68 × 3.29 mm was scanned and extracted to compute the surface parameter *Sa* which is the extension of *Ra* (arithmetical mean height of a line) to a surface.

2.5.3. Microstructural Analysis

The cross-sections of cubic specimens in the building direction were prepared by wet grinding using abrasive papers (400–4000 grit) and by subsequent polishing with 1 μm diamond suspension. To reveal the microstructural features, as-built specimens were electrolytically etched using a mixture of 70 mL H_3PO_4 and 30 mL H_2O at 5 V for 10 s. The heat-treated samples were instead etched with Kalling's etchant (100 mL HCl, 100 mL ethanol, 5 g $CuCl_2$). The microstructure was examined using a Leica DMRE optical microscope (OM, Leica Microsystems S.r.l., Milan, Italy) and a Leica/Cambridge Leo Stereoscan S-440 scanning electron microscope (SEM, Leica Microsystems S.r.l., Milan, Italy), and Zeiss Sigma HD field-emission SEM (FESEM) (Carl Zeiss S.p.A., Milan, Italy). Moreover, a JEOL 200CX (Jeol Ltd., Tokyo, Japan) transmission electron microscope (TEM) was employed to investigate the precipitation hardening after heat treatment cycles. For this purpose, thin foils were obtained from the samples and mechanically thinned down to approximately 50 μm in thickness. Then, 3 mm diameter disks were punched out from the thin foils and further electropolished down to electron transparency with a specific solution (1:2:9 perchloric acid, butoxyethanol, methanol) at −10 °C and 22 V, with a twin-jet polisher STRUERS TENUPOL-3 (Struers S.A.S., Milan, Italy). Phase identification was carried out through a Siemens D500 (Siemens, Munich, Germany), diffractometer equipped with Cu radiation tube with Ni filter on the tube side and monochromator on the detector side (0.05° step and 5 s counting time).

2.5.4. Mechanical Tests

Vickers microhardness measurements were performed on a Leitz™ DURIMET (Leica Microsystem S.r.l., Milan, Italy) testing machine using a load of 0.5 kg (i.e., a load of 4.905° N) with 30 s of dwell time. Ten measurements were performed in different areas of the polished surface of the sample. The uncertainty for microhardness measurement was calculated as the standard deviation on the measurement. Vertically oriented tensile test samples were directly manufactured according to ASTM E8 [22] as shown in Figure 4a using an alternating chessboard strategy with 3 mm × 3 mm rectangles and 90° of rotation between successive layers (Figure 2c). An MTS Minibionix servo-hydraulic testing machine (MTS System Corporation, Eden Prairie, MN, USA) having a load capacity of 15 kN equipped with an MTS TestStar IIm controller was adopted to perform tensile static tests on as-built, and heat-treated samples at room temperature. For each condition, three samples were tested for mechanical properties evaluation. The samples were loaded under displacement controlled at a rate equivalent to a strain rate of 10^{-4} s^{-1}. The uniaxial MTS extensometer Model No. 632.29 F-30 having a gauge length of 5 mm was used to measure the strain as shown in Figure 4b.

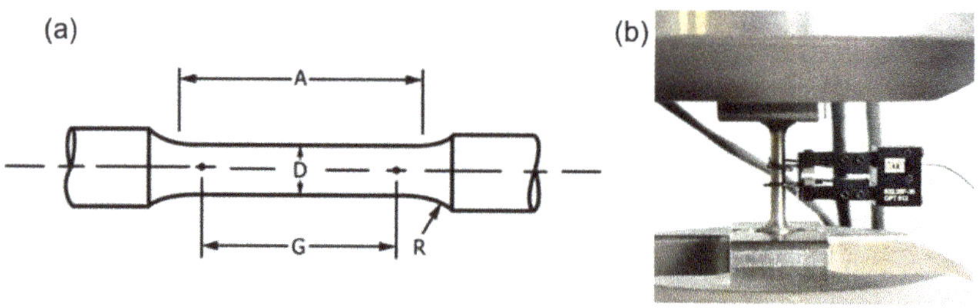

Figure 4. LPBF of Inc718; (**a**) tensile samples geometry, according to ASTM E8 (A = 20 mm, D = 4 mm, G = 16 mm, R = 4 mm), (**b**) tensile test set up.

3. Results and Discussion

3.1. Optimization of LPBF Process Parameters for Inc718

The response surface method with a central composite design was employed to determine the optimum process parameter in LPBF of Inc718 using a spot laser of 30 μm. The relative density measurement of specimens produced by process parameters presented in Table 2 revealed a strong correlation between laser power and relative density (see Figure 5a). As can be seen in Figure 5b, a weak correlation was noted between scanning speed and relative density at this step. This can be explained by the fact that the maximum value of relative density in the first step was obtained to be lower than 98%. In the first DOE using RSM, relative density was considered as a single output that ranged from 90.11% to 97.8%. The model fit summary table corresponding to CCD is presented in Figure 5e. The 2FI model with a sequential p-value of 0.0204 was selected for further statistical analysis. Analysis of variance (ANOVA) for the selected model (2FI) showed a p-value of 0.0018 which confirms the significance of the selected model. The Model F-value was calculated to be 7.63 which implies that the model is significant. There is only a 0.18% chance that an F-value this large can occur due to noise. Details of the ANOVA are presented in Figure 5f for the selected model, input parameters, and model parameters; p-values less than 0.05 indicate model terms are significant. In this case, A, B, AC, and BD are significant model terms. In this model, A, B, C, and D factors represent laser power, scanning speed, hatch distance, and scanning strategy, respectively. Values greater than 0.1 indicate the model terms are not significant. The optimization module of the design expert software was used to predict the optimum process parameters aiming to achieve full density in LPBF Inc718 specimens. The optimization criteria were set on the maximum density and minimum surface roughness. In particular, five star and three star importance were considered for relative density and surface roughness, respectively. In order to develop the LPBF process for microfabrication, laser power of 95 W was considered a limit in the optimization process. Solution results of the optimization process showed that a set of process parameters consisting of laser power of 85 W, scanning speed of 737 mm/s, and hatch distance of 0.08 mm with an alternating bidirectional scanning strategy may lead to the full density LPBF product. This set of process parameters was then placed as the central point for the subsequent full factorial design of the experiment. Three levels of laser power (80 W, 85 W, and 95 W), two levels of scanning speed (650 mm/s and 750 mm/s), and two levels of hatch distance (0.06 and 0.08 mm) were considered in a full factorial design. Overall, relative density steadily increased up to the highest point above 99.9% by the raise in laser power, while it decreased by increasing scanning speed (Figure 6a,b). It is well-understood that excessive input energy density leads to evaporation-induced porosity [23]. As can be seen in Figure 6a,d, low scanning speed (650 mm/s) intensified evaporation and increased the porosity and balling effect. Comparing Figure 6a,b shows that the effect of scanning speed is more pronounced in samples produced by a lower hatch distance of 0.06 mm. Adjacent single tracks are highlighted by dashed red lines in Figure 6e,I which correspond to hatch distances of 0.06 mm and 0.08 mm, respectively.

It can be concluded that small hatch distances lead to distorted single tracks due to significant overlap between adjacent tracks. This phenomenon led to increased surface roughness in samples produced with 0.06 mm hatch distance (20–37 μm) in comparison with those with 0.08 mm hatch distance (10–18 μm).

From Figure 6a,b, it can be concluded that the sample produced by 95 W of laser power, 750 mm/s of scanning speed, and 0.08 mm of hatch distance presented the optimal condition due to its highest relative density (>99.9%) and lowest surface roughness. This set of process parameters was then utilized for the fabrication of all samples that underwent the various heat treatment cycles.

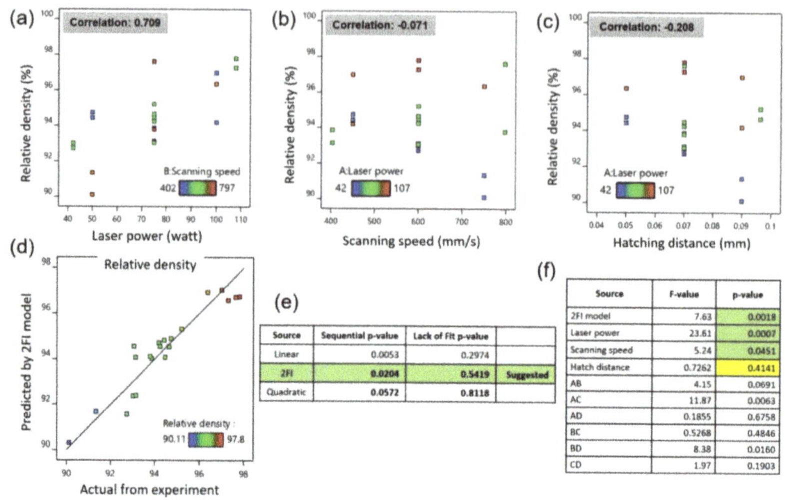

Figure 5. Optimization of micro-LPBF process parameters using response surface method for processing of Inc718; correlation between relative density and (**a**) laser power, (**b**) scanning speed and (**c**) hatch distance; (**d–f**) fitting a model corresponding to central composite design.

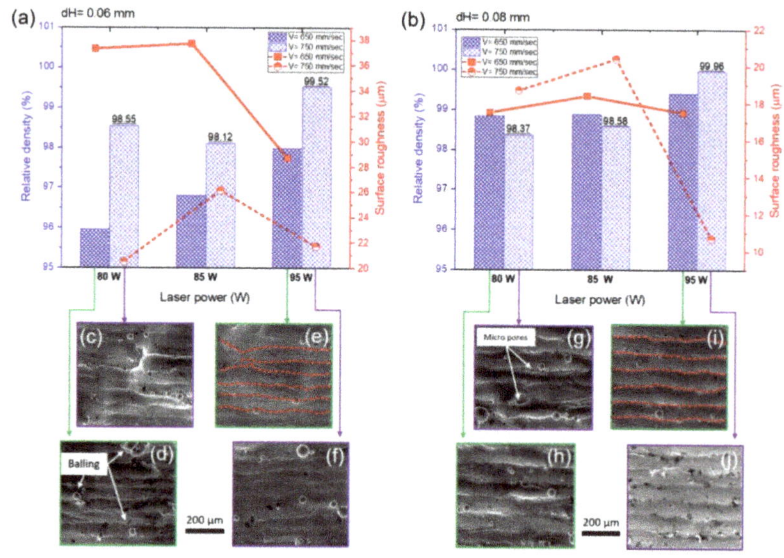

Figure 6. (**a**,**b**) Relative density and surface roughness of Inc718 samples produced by micro-LPBF; full factorial design with laser power of 80 W, 85 W, and 95 W, scanning speed of 650 mm/s, and 750 mm/s, hatch distance of 0.06 mm, and 0.08 mm; (**c**–**j**) surface topography of 12 samples from full factorial design showing surface defects such as porosity and balling.

3.2. Microhardness

The microhardness measurement was performed on the samples produced by optimized parameters, to evaluate which heat treatment cycles induced the higher hardness values. The results are reported in Figure 7. The hardness of as-built samples resulted in 324 HV, slightly higher than reported values in the literature. For example, Tucho et al. [24] reported 288 ± 7 HV and 304 ± 9 HV for vertically and horizontally printed samples, respectively. Xing Li et al. [25] reported a hardness value of 300 HV for as-built Inc718. This difference can be attributed to a finer grain microstructure induced by small melt pool dimension and rapid solidification using the spot laser of 30 μm. As can be seen in Figure 7, hardness dropped from 324 HV to 232 HV after solution annealing. This loss of hardness can be attributed to the high temperature of solution annealing treatment (1080 °C) that induced a complete recrystallization. A single-step solution annealing at 620 °C for 4 h (1AH) and 8 h (2AH) did not increase the hardness significantly (261 HV and 308 HV, respectively). This can be easily attributed to low temperature and ageing treatment which led to the insufficient precipitation of secondary phases. Increasing the ageing temperature from 620 °C to 720 °C showed a substantial increase in hardness, and the values of 430 and 498 HV were reached after 4 h (3AH) and 8 h (4AH) of treatment, respectively. With double age hardening (1DAH), the hardness reached values of approximately 450 HV.

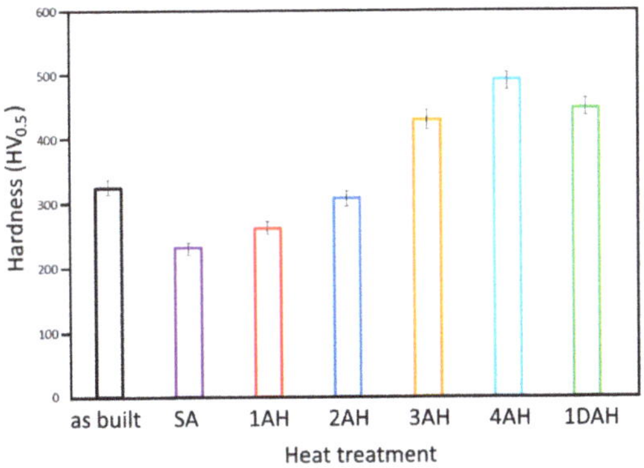

Figure 7. The average values of Vickers hardness in micro-LPBF Inc718 samples before and after different heat treatment cycles.

In conclusion, a single treatment for 8 h at 720 °C (4AH) resulted in the highest value (approximately 500 HV) of hardness. Double ageing treatment at 720 °C and 620 °C for 4 h (1DAH) showed the second-highest value of hardness (450 HV).

After this preliminary test, the results of samples 4AH and 1DAH were considered the most promising in terms of hardness value and time of treatment; therefore, the subsequent characterizations focused mainly on the comparison of these two samples with as-built and solution annealed samples.

3.3. Microstructure

3.3.1. Micro-LPBF As-Built Inc718

Observation of the as-built sample in the building direction using optical microscope showed a typical LPBF microstructure as presented in Figure 8. LPBF microstructure is characterized by melt pools and specific features such as columnar dendrites, columnar grains, and cellular sub-structures whose origin can be traced to the high thermal gradients of the process.

SEM analysis of as-built samples evidenced the presence of cellular and dendritic sub-structures (Figure 9a) parallel to the building direction. Similar microstructures are reported in the literature for Inc718 produced by LPBF [14,15]. As shown in Figure 9b, a fine microstructure was formed because of the high cooling rate induced by high input energy density (grain boundaries are indicated by yellow dash lines). As shown by the yellow arrows in Figure 9b, dendrite growth direction varies in different melt pools which indicates various temperature gradients in adjacent melt pools. It should be mentioned that the overall heat flow direction was parallel to the building direction [7]. However, the heat flow directions were not necessarily parallel to the building direction. It is well-understood that process parameters such as scanning strategies and thermal fields can significantly affect the heat flow directions and consequently dendrite growth direction [26]. Primary dendrite arm spacing was approximately 100–200 nm which suggests a very fine cellular-dendrite microstructure in comparison with the previous reports [27]. The grain size distribution was inhomogeneous with an average value of 10 μm. Each grain contained several columnar/cellular sub-grains.

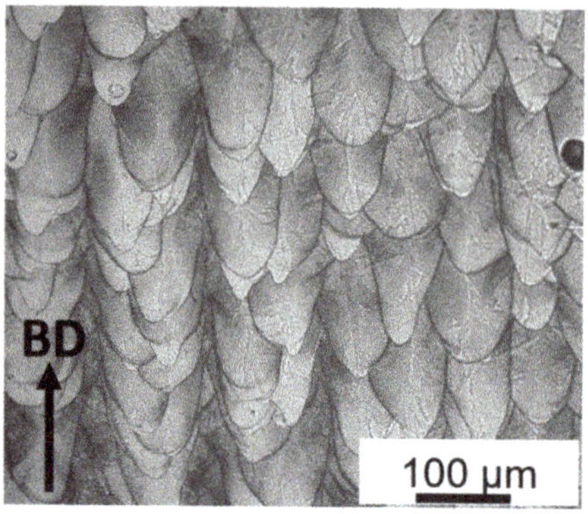

Figure 8. OM image of as-built micro-LPBF Inc718.

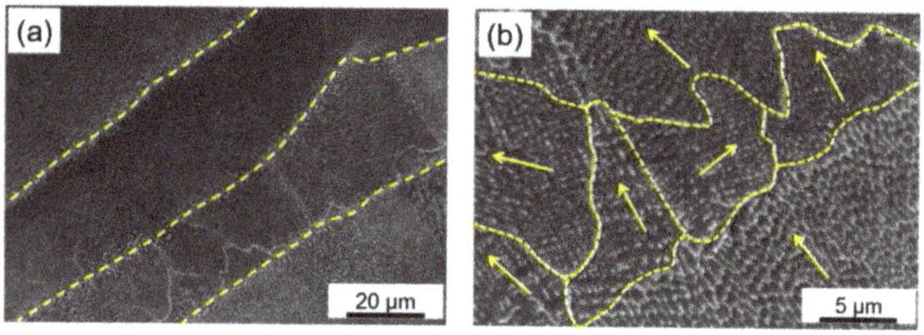

Figure 9. SEM micrographs of as-built sample parallel to the building direction; (**a**) SEM image in low magnification shows columnar/cellular structure in successive layers; (**b**) grains and columnar/cellular sub-grains (arrows in (**b**) indicate dendrite growth directions).

TEM analysis revealed more details of the columnar and cellular substructure of the as-built sample as presented in Figure 10a,b, respectively. Cellular substructures suggest equiaxed sub-grains whereas a columnar substructure consists of elongated sub-grains. In addition, the cellular microstructure was confirmed to be γ phase from the SAD pattern identified in Figure 10b. The average thickness of columnar sub-grains and average equiaxed sub-grains size was measured to be the same at 200 nm. A network of dislocations entangled at the low angle grain boundaries of cellular substructure can be seen in Figure 9b. The high-density dislocation microstructure of the as-built sample is due to the plastic deformation caused by residual stresses, induced in turn by the high thermal gradient and rapid solidification typically present in the LPBF process [13,28]. The same dimension for columnar and cellular sub-grains of 500 nm was previously reported by Tucho et al. [24]. Xing Li et al. reported 700 nm for the average dimension of cellular substructure in LPBF of Inc718 by using a laser spot diameter of 70 μm [25]. Moreover, the grain size distribution of 14.9 μm and a fine sub-structure of 452 nm were reported by Huang et al. [13] in LPBF of Inc718 using a self-developed LPBF machine equipped with a 500 W fiber laser. The finer columnar/cellular sub-grain structure in the current study with respect to the literature can be attributed to the smaller spot laser used in the micro-LPBF process (30 μm). A smaller spot laser diameter led to the formation of a smaller melt pool. The cooling rate of a small melt pool is very high which resulted in a steep temperature gradient. On the other hand, a steep temperature gradient is known as the main driving force for recrystallization and consequent fine microstructure of LPBF built parts [12,13].

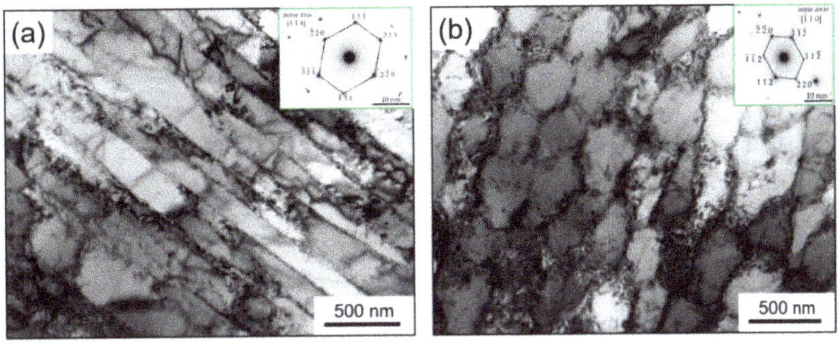

Figure 10. TEM bright field images showing (**a**) columnar and (**b**) equiaxed cellular substructure in as-built Inc718 sample produced by micro-LPBF. Inserts show the corresponding SAD patterns.

3.3.2. Solution Annealing of Inc718 Produced by LPBF

In this work, solution annealing at 1080 °C for 1.5 h was performed as the initial step in all heat treatment procedures, as presented in Table 4. OM, SEM, and TEM observations of the SA sample evidenced a notable change in the microstructure in comparison with the as-built samples: the melt pools disappeared, and recrystallization and grain growth occurred with the formation of grains ranging between 30 and 80 μm (Figure 11a). The presence of annealing twins was also detected (Figure 11a–c), which are typical in metals with an FCC structure subjected to annealing treatment [29].

Figure 11. (**a**) OM image; (**b**) SEM image, and (**c**) TEM image of solution annealed micro-LPBF Inc718.

3.3.3. Ageing Hardening of Inc718 Produced by LPBF

Considering the results of the hardness test, samples 4AH (720 °C for 8 h) and 1DAH (720 °C for 4 h + 620 °C for 4 h) were selected for further microstructural analysis. The SEM images of low magnification evidenced the grain size of the two samples. The 4AH sample was characterized by a slightly larger grain size in comparison with the 1DAH one due to a longer holding at one higher temperature (Figure 12). To study the microstructure more in detail, observations were also performed with a Field Emission SEM.

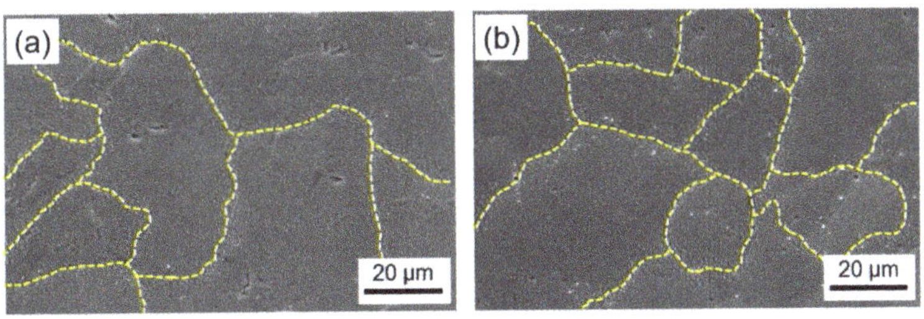

Figure 12. SEM image of (**a**) 4AH and (**b**) 1DAH samples of micro-LPBF Inc718.

The microstructure of the two samples was characterized by the presence of very fine precipitates homogeneously distributed in the matrix. In the 4AH sample, it was possible to discern the spherical precipitates that suggested the presence of γ' phase, and long disc-shaped ones, attributed to γ'' phase (Figure 13a). The size of both phases varies between 10 and 25 nm. In the 1DAH sample, the microstructure was smaller compared with the 4AH one (Figure 13b) and only at higher magnification it was possible to distinguish the presence of γ' from γ'' (Figure 13c). The size of the phases resulted in approximately 10 nm. To deeply analyze the microstructure of 1DAH samples, TEM observations were carried out.

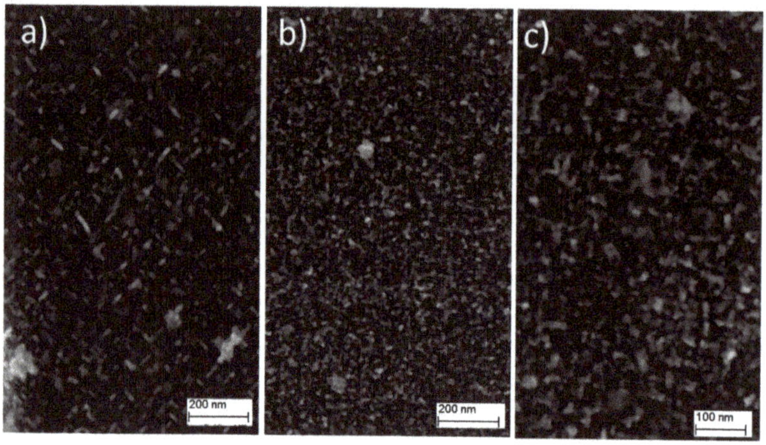

Figure 13. FESEM image of (**a**) 4AH and (**b**,**c**) 1DAH samples of micro-LPBF Inc718.

Figure 14 shows the bright-field TEM image that revealed the presence of very fine (<10 nm) circular γ' and plate-like γ'' particles distributed homogenously in the γ matrix. A typical SAD pattern with zone axis <001> is shown in Figure 14. It can be seen that all the reflections of the three types of orientation relationship of γ'' and reflection of the [010]γ'' are stretched along the [010]γ'' direction, while [100]γ'' are stretched along the [001]γ'' direction, no stretching of the [001]γ'' variant can be noted.

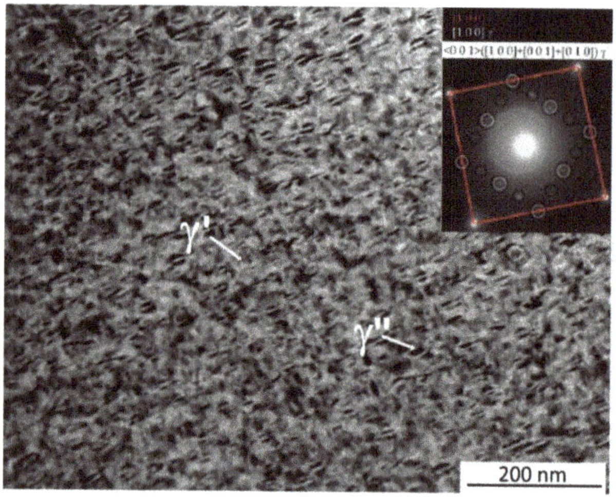

Figure 14. TEM bright field images of 1DAH sample with corresponding SAD patterns of γ and γ'' phases along the $[1\bar{1}2]$ and $[1\bar{1}1]$ zone axes, respectively.

3.4. X-Ray Diffraction

The XRD patterns from the cross-sections of the as-built and heat-treated (SA, 4AH and 1DAH) samples in the building direction are shown in Figure 15. The main peaks observed for the as-built and SA samples were from the matrix, γ phase, and no peaks related to the presence of δ and the Laves phase were observed. The XRD patterns of 4AH and 1DAH samples are similar to ones of as-built and SA samples: the only difference is an enlargement of the peaks. The peaks of γ' and γ'' phases cannot be distinguished since the lattice constants of γ, γ', and γ'' phases are very close and each of the diffraction peaks was an overlap of peaks associated with the individual phases. The enlargement of the peaks in 4AH and 1DAH samples was most likely due to the presence of γ' and γ'' phases. To separate the overlapping peaks and to calculate the volume fractions of γ, γ', and γ'' phases, the Gaussian fitting method in the Origin software was used in the literature [24,30,31]. In this work, the Gaussian fitting method present in the HighScore Plus software was employed.

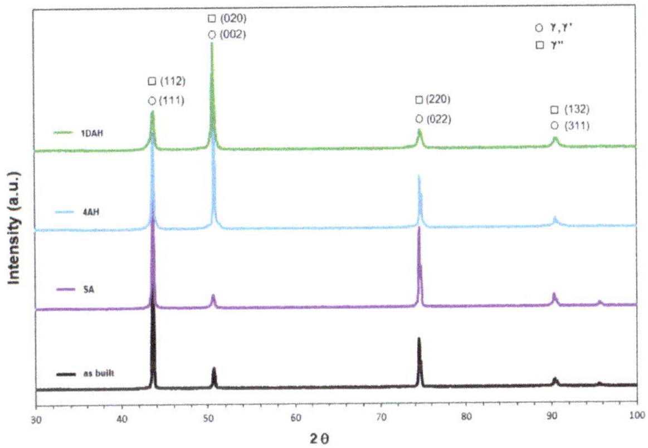

Figure 15. XRD patterns of the micro-LPBF Inc 718 samples before and after heat treatment.

The fitting results on the strongest peak 002 in 4AH and 1DAH samples are shown in Figure 16. Based on the integral area of the separated peaks, the volume fractions of phases in heat-treated specimens were: 31% of γ' and 16% of γ'' in the 4AH sample, 16% of γ' and 20% of γ'' in the 1DAH sample. These results evidenced that the prolonged heat treatment induced a higher precipitation of γ' phase than γ'', whereas the double treatment promoted the precipitation of γ''.

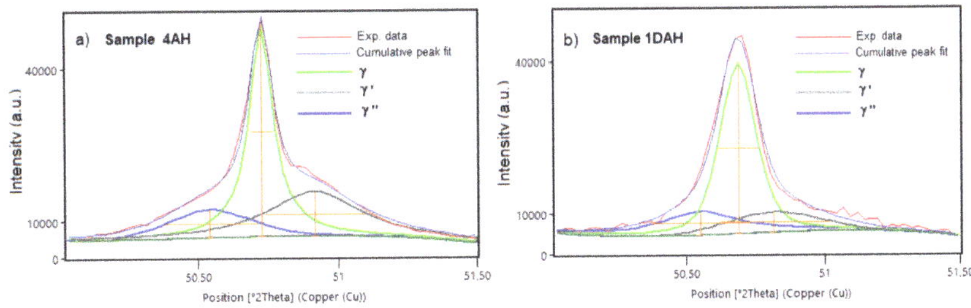

Figure 16. The fitting results of the peaks 200 for (**a**) 4AH sample and (**b**) 1DAH sample.

3.5. Mechanical Tests

3.5.1. Uniaxial Tensile Test

Mechanical performance of as-built and heat-treated micro-LPBF Inc718 samples was investigated through a uniaxial tensile test at room temperature. The stress-strain curves

- Precipitation of very fine spherical γ' phase and long disc-shaped γ'' phase was detected in Inc718 samples subjected to solution annealing and subsequent age hardening. The FESEM and TEM analysis revealed that double age hardening resulted in smaller precipitates (<10 nm) compared with a single-step age hardening;
- Age hardening heat treatment significantly increased the yield stress (28% after single age hardening) and ultimate tensile strength (39% after single age hardening) due to the formation and even distribution of the strengthening γ' and γ'' phases;
- Single age hardening (720 °C for 8 h) induced the precipitation of a higher amount of γ' precipitates and higher values of mechanical strength than double age hardening (720 °C and 620 °C for 4 h).

Author Contributions: S.K.: Conceptualization, Methodology, Original draft, Supervision of Manufacturing phase; C.G.: SEM and TEM analysis; A.Z.: Data collection, Methodology; M.F.: XRD analysis; A.C.: Mechanical tests; K.B.: Writing-Reviewing and Editing, Supervision of characterization phase. All authors have read and agreed to the published version of the manuscript.

Funding: This research received no external funding.

Data Availability Statement: The raw/processed data required to reproduce these findings is available upon the request.

Conflicts of Interest: The authors declare no conflict of interest.

References

1. Akca, E.; Gürsel, A. A Review on Superalloys and IN718 Nickel-Based INCONEL Superalloy. *Period. Eng. Nat. Sci.* **2015**, *3*. [CrossRef]
2. Hosseini, E.; Popovich, V.A. A review of mechanical properties of additively manufactured Inconel 718. *Addit. Manuf.* **2019**, *30*, 100877. [CrossRef]
3. Xu, Z.; Cao, L.; Zhu, Q.; Guo, C.; Li, X.; Hu, X.; Yu, Z. Creep property of Inconel 718 superalloy produced by selective laser melting compared to forging. *Mater. Sci. Eng. A* **2020**, *794*, 139947. [CrossRef]
4. Roy, S.; Kumar, R.; Anurag; Panda, A.; Das, R.K. A brief review on machining of inconel 718. *Mater. Today Proc.* **2018**, *5*, 18664–18673. [CrossRef]
5. Holmström, J.; Partanen, J.; Tuomi, J.; Walter, M. Rapid manufacturing in the spare parts supply chain: Alternative approaches to capacity deployment. *J. Manuf. Technol. Manag.* **2010**, *21*, 687–697. [CrossRef]
6. Zhao, Y.; Du, C.; Wang, P.; Meng, W.; Li, C. The Mechanism of In-Situ Laser Polishing and Its Effect on the Surface Quality of Nickel-Based Alloy Fabricated by Selective Laser Melting. *Metals* **2022**, *12*, 778. [CrossRef]
7. Wang, X.; Gong, X.; Chou, K. Review on powder-bed laser additive manufacturing of Inconel 718 parts. *Proc. Inst. Mech. Eng. Part B J. Eng. Manuf.* **2017**, *231*, 1890–1903. [CrossRef]
8. Uriondo, A.; Esperon-Miguez, M.; Perinpanayagam, S. The present and future of additive manufacturing in the aerospace sector: A review of important aspects. *Proc. Inst. Mech. Eng. Part G J. Aerosp. Eng.* **2015**, *229*, 2132–2147. [CrossRef]
9. Yadroitsev, I.; Smurov, I. Selective laser melting technology: From the single laser melted track stability to 3D parts of complex shape. *Phys. Procedia* **2010**, *5*, 551–560. [CrossRef]
10. Yap, C.Y.; Chua, C.K.; Dong, Z.L.; Liu, Z.H.; Zhang, D.Q.; Loh, L.E.; Sing, S.L. Review of selective laser melting: Materials and applications. *Appl. Phys. Rev.* **2015**, *2*, 041101. [CrossRef]
11. Jia, Q.; Gu, D. Selective laser melting additive manufacturing of Inconel 718 superalloy parts: Densification, microstructure and properties. *J. Alloys Compd.* **2014**, *585*, 713–721. [CrossRef]
12. Chlebus, E.; Gruber, K.; Kuźnicka, B.; Kurzac, J.; Kurzynowski, T. Effect of heat treatment on the microstructure and mechanical properties of Inconel 718 processed by selective laser melting. *Mater. Sci. Eng. A* **2015**, *639*, 647–655. [CrossRef]
13. Huang, W.; Yang, J.; Yang, H.; Jing, G.; Wang, Z.; Zeng, X. Heat treatment of Inconel 718 produced by selective laser melting: Microstructure and mechanical properties. *Mater. Sci. Eng. A* **2019**, *750*, 98–107. [CrossRef]
14. Liu, P.; Hu, J.; Sun, S.; Feng, K.; Zhang, Y.; Cao, M. Microstructural evolution and phase transformation of Inconel 718 alloys fabricated by selective laser melting under different heat treatment. *J. Manuf. Process.* **2019**, *39*, 226–232. [CrossRef]
15. Wang, Z.; Guan, K.; Gao, M.; Li, X.; Chen, X.; Zeng, X. The microstructure and mechanical properties of deposited-IN718 by selective laser melting. *J. Alloys Compd.* **2012**, *513*, 518–523. [CrossRef]
16. Nagarajan, B.; Hu, Z.; Song, X.; Zhai, W.; Wei, J. Development of Micro Selective Laser Melting: The State of the Art and Future Perspectives. *Engineering* **2019**, *5*, 702–720. [CrossRef]
17. Qin, Y.; Brockett, A.; Ma, Y.; Razali, A.; Zhao, J.; Harrison, C.; Pan, W.; Dai, X.; Loziak, D. Micro-manufacturing: Research, technology outcomes and development issues. *Int. J. Adv. Manuf. Technol.* **2010**, *47*, 821–837. [CrossRef]

18. Khademzadeh, S. Precision Additive Manufacturing of NiTi shape memory parts using Micro Laser Powder Bed Fusion. *Prog. Addit. Manuf.* **2021**, *7*, 419–432. [CrossRef]
19. Antony, J. *Design of Experiments for Engineers and Scientists*; Elsevier: Amsterdam, The Netherlands, 2003.
20. Rao, G.A.; Kumar, M.; Srinivas, M.; Sarma, D.S. Effect of standard heat treatment on the microstructure and mechanical properties of hot isostatically pressed superalloy inconel 718. *Mater. Sci. Eng. A* **2003**, *355*, 114–125. [CrossRef]
21. Zhao, X.; Chen, J.; Lin, X.; Huang, W. Study on microstructure and mechanical properties of laser rapid forming Inconel 718. *Mater. Sci. Eng. A* **2008**, *478*, 119–124. [CrossRef]
22. *ASTM E8/E8M*; Standard Test Methods for Tension Testing of Metallic Materials. American Society for Testing and Materials (ASTM): West Conshohocken, PA, USA, 2021.
23. Khademzadeh, S.; Carmignato, S.; Parvin, N.; Zanini, F.; Bariani, P.F. Micro porosity analysis in additive manufactured NiTi parts using micro computed tomography and electron microscopy. *Mater. Des.* **2016**, *90*, 745–752. [CrossRef]
24. Tucho, W.M.; Cuvillier, P.; Sjolyst-Kverneland, A.; Hansen, V. Microstructure and hardness studies of Inconel 718 manufactured by selective laser melting before and after solution heat treatment. *Mater. Sci. Eng. A* **2017**, *689*, 220–232. [CrossRef]
25. Li, X.; Shi, J.J.; Wang, C.H.; Cao, G.H.; Russell, A.M.; Zhou, Z.J.; Li, C.P.; Chen, G.F. Effect of heat treatment on microstructure evolution of Inconel 718 alloy fabricated by selective laser melting. *J. Alloys Compd.* **2018**, *764*, 639–649. [CrossRef]
26. Liu, J.; Li, G.; Sun, Q.; Li, H.; Sun, J.; Wang, X. Understanding the effect of scanning strategies on the microstructure and crystallographic texture of Ti-6Al-4V alloy manufactured by laser powder bed fusion. *J. Mater. Process. Technol.* **2022**, *299*, 117366. [CrossRef]
27. Choi, J.P.; Shin, G.H.; Yang, S.; Yang, D.Y.; Lee, J.S.; Brochu, M.; Yu, J.H. Densification and microstructural investigation of Inconel 718 parts fabricated by selective laser melting. *Powder Technol.* **2017**, *310*, 60–66. [CrossRef]
28. Zhang, D.; Niu, W.; Cao, X.; Liu, Z. Effect of standard heat treatment on the microstructure and mechanical properties of selective laser melting manufactured Inconel 718 superalloy. *Mater. Sci. Eng. A* **2015**, *644*, 32–40. [CrossRef]
29. Mahajan, S.; Pande, C.S.; Imam, M.A.; Rath, B.B. Formation of annealing twins in f.c.c. crystals. *Acta Mater.* **1997**, *45*, 2633–2638. [CrossRef]
30. Ling, L.S.B.; Yin, Z.; Hu, Z.; Liang, J.H.; Wang, Z.Y.; Wang, J.; Sun, B. De Effects of the γ''-Ni3Nb phase on mechanical properties of Inconel 718 superalloys with different heat treatments. *Materials* **2020**, *13*, 151. [CrossRef] [PubMed]
31. Yan, S.; Wang, Y.; Wang, Q.; Zhang, C.; Chen, D.; Cui, G. Enhancing mechanical properties of the spark plasma sintered inconel 718 alloy by controlling the nano-scale precipitations. *Materials* **2019**, *12*, 3336. [CrossRef] [PubMed]
32. Deng, D.; Peng, R.L.; Brodin, H.; Moverare, J. Microstructure and mechanical properties of Inconel 718 produced by selective laser melting: Sample orientation dependence and effects of post heat treatments. *Mater. Sci. Eng. A* **2018**, *713*, 294–306. [CrossRef]
33. Popovich, V.A.; Borisov, E.V.; Popovich, A.A.; Sufiiarov, V.S.; Masaylo, D.V.; Alzina, L. Impact of heat treatment on mechanical behaviour of Inconel 718 processed with tailored microstructure by selective laser melting. *Mater. Des.* **2017**, *131*, 12–22. [CrossRef]
34. Kruth, J.-P.; Deckers, J.; Yasa, E.; Wauthlé, R. Assessing and comparing influencing factors of residual stresses in selective laser melting using a novel analysis method. *Proc. Inst. Mech. Eng. Part B J. Eng. Manuf.* **2012**, *226*, 980–991. [CrossRef]

materials

MDPI

Article

The Effect of Deposited Dust on SCC and Crevice Corrosion of AISI 304L Stainless Steel in Saline Environment

Chun-Ping Yeh *, Kun-Chao Tsai and Jiunn-Yuan Huang

Institute of Nuclear Energy Research (INER), 1000 Wenhua Rd., Longtan District, Taoyuan City 32546, Taiwan; tsaijohn@iner.gov.tw (K.-C.T.); jyhuang@iner.gov.tw (J.-Y.H.)
* Correspondence: cpyeh@iner.gov.tw; Tel.: +886-3-471-1400

Abstract: Crevice corrosion has become an important issue of the safety of AISI 304L austenitic stainless steel canister when exposed to the chloride environments located in coastal areas. Moreover, dust deposited on the canister surface may enhance the corrosion effect of 304L stainless steel. In this work, white emery was adopted to simulate the dust accumulated on the as-machined specimen surface. To investigate the effect of deposited white emery, chloride concentration, and relative humidity on the crevice corrosion behavior, an experiment was conducted on 304L stainless steel specimens at 45 °C with 45%, 55%, and 70% relative humidity (RH) for 7000 h. The surface features and crack morphology of the tested 304L stainless steel specimens were examined by SEM equipped with energy-dispersive spectrometry (EDS) and electron back scatter diffraction (EBSD). From the experimental results, a threshold RH for the stress corrosion cracking (SCC) initiation of AISI 304L austenitic stainless steel with different concentrations of chloride was proposed.

Keywords: chloride concentration; relative humidity; crevice corrosion; stress corrosion cracking; dust; stainless steel

check for updates

Citation: Yeh, C.-P.; Tsai, K.-C.; Huang, J.-Y. The Effect of Deposited Dust on SCC and Crevice Corrosion of AISI 304L Stainless Steel in Saline Environment. *Materials* **2021**, *14*, 6834. https://doi.org/10.3390/ma14226834

Academic Editors: Luca Pezzato and Claudio Gennari

Received: 12 October 2021
Accepted: 10 November 2021
Published: 12 November 2021

1. Introduction

The facilities of interim storage for storing spent nuclear fuel are located in coastal areas in Taiwan. The dry cask storage systems are expected to operate for approximately 40–60 years [1]. As a result, it is necessary to maintain the integrity of the stainless steel canisters for guaranteeing the safety of spent nuclear fuel storage. Canisters are stored in passively ventilated overpacks and accumulate dust containing chloride salts on their surface over a long period of time. When dust containing chloride salts is deposited on canister surfaces, it creates an environment where water accumulates and corrodes the region, which can lead to localized crevice corrosion.

Stainless steel is susceptible to suffering from localized corrosion, particularly crevice corrosion occurring on surfaces [2]. It has been suggested that the crevice corrosion behavior of stainless steel is directly related to the passive film evolution process inside the crevice [3]. With the increase in chloride concentration, passive film becomes more unstable, which brings about more serious crevice corrosion [4]. In contrast to chloride ions, nitrate ions act as a passivating agent to inhibit pitting corrosion [5].

By trapping chloride deposits on the surface of canisters, crevice corrosion becomes one of the promoting factors of the so-called atmospherically-induced SCC (AISCC). The initiation of crevice corrosion is particularly dependent on aggressive Cl^- ions, crevice geometry, and temperature. In canister systems, crevice conditions occur where the canister contacts the support structure of a storage module [6].

In terms of austenitic stainless steel, it is subject to SCC, especially in an aggressive environment, such as stress corrosion cracking due to Cl^- ions [7–9]. An increased volume of corrosion products possibly creates localized stress at the local sites of crevices to induce the occurrence of the stress corrosion cracking. In addition, Tani et al. reported that crevice corrosion causes the initiation of SCC on the surfaces beneath sea salt particles [10].

The following mechanisms are usually used to explain the behavior of crevice corrosion: The mechanism of critical crevice solution potential drop [3], in which the oxygen depleted in the crevice leads to acidification of the crevice solution, which generates breakdown of the passive film and initiation of the crevice corrosion. This mechanism places emphasis on aggressive ions, such as chloride ions accumulated in the crevice, and, following depassivation, results from active dissolution of the base metal [11]. As for the mechanism of IR drop, when the value of the IR exceeds a critical threshold, crevice corrosion occurs by means of transiting the potential from the passive state to the active state. By increasing the corrosion current, aggressive ions (e.g., Cl^-) facilitate crevice corrosion, which causes an increase in the IR value [12].

The areal density of the salt deposit and local environmental conditions determine the chloride concentration present on the canister surface. The Southwest Research Institute (SwRI) [13] conducted experiments at a chloride concentration of 0.1–10 g/m^2 of synthetic sea salts deposited on 304, 304L, and 316L stainless steel U-bend specimens. Dong et al. investigated the SCC susceptibility of deposited $MgCl_2$ on 316L stainless steel at 75 °C with a relative humidity of 70% for 480 h. They suggested that fine transgranular SCC (TGSCC) appeared from the side of the pits of the weld region [14]. Corrosion pits can also be connected to tensile stress. Higher tensile stress is prone to increasing the pitting density in the heat-affected zone (HAZ) [15,16]. Furthermore, the morphology of a crevice-like attack or pit could impact the susceptibility of the pit developing into a crack due to the morphology being able to affect the distribution of stress/strain [17]. Guo and coworkers reported atmospheric corrosion of 304 stainless steel under mixed salts of $MgCl_2$ and NaCl at 21 °C and 45% RH. They suggested that dish-shaped pits can be observed in mixed salts solutions. Moreover, crevice corrosion occurs under NaCl crystals [18]. Shoji et al. proposed systematic research of the AISCC phenomenon in terms of chloride ions, relative humidity, and temperature, and illustrated its happening in 316L and 304L austenitic stainless steel. It was reported that $MgCl_2$ is the constituent of sea salt responsible for enhancing low-temperature AISCC in AISI 316L and AISI 304L austenitic stainless steel [19] because of its low deliquescence point [20]. Scatigno et al. investigated the impact of salt on chloride-induced SCC in AISI 304L stainless steel at a 70% relative humidity and 90 °C. It was reported that AISCC was found in the environment with a chloride concentration of 1.7 g/m^2 of $MgCl_2$ [21]. Engelberg and coworkers provided an order of susceptibility for the occurrence of atmospherically induced SCC of 316L and 304L austenitic stainless steel. Moreover, the RH ranges over which AISCC took place at a chloride concentration of 25 g/m^2 of $MgCl_2$ and sea water were summarized, based on the work of Shoji and coworkers [22]. Prosek et al. suggested that the corrosivity of deposited chloride under specific exposure situations was in the descending order of: $CaCl_2 > MgCl_2 > NaCl$. Due to interaction of the given salt with the water vapor in the air, it was controlled by the chloride concentration on the surface [23]. Furthermore, the threshold of the chloride concentration for the SCC initiation of 304L steel after 10,000 h of testing at 45 °C was between 0.1 and 1 g/m^2 [24]. Moreover, Ornek et al. reported that based on salt loading tests for the 304L stainless steel, the chloride concentration at the crack tips was saturated and, hence, the bulk concentration of chloride at the surface did not affect the crack propagation rates [25].

Masuda et al. presented the SCC phenomenon of AISI 304 steel with regard to the growth of a pit, surface potential distribution, and slip deformation. They conducted SCC tests at 343 K and a relative humidity of 28% with $MgCl_2$ droplets deposited on specimens. Discrete cracks were usually found near the region of the crack tip [26]. Qiao et al. investigated the SCC phenomenon of 321 stainless steel in an $MgCl_2$ solution. They reported that the interaction between the main crack and discontinuous microcracks may promote the effective stress intensity factor and enhance the coalescence of the crack, resulting in mechanical fracture of the ligaments between the cracks in stainless steel [27].

Dust accumulation is prone to being affected by the roughness of the material surface. Higher surface roughness insinuates the presence of deeper grooves on the material surface, where a higher rate of dust accumulation and a higher concentration of ions occur, causing early initiation of corrosion [28].

The aim of this research was to survey the behavior of the crevice corrosion of AISI 304L steel under the combined effect of accumulated dust, different chloride concentrations, and relative humidity, and the experimental results of this research are expected to contribute to evaluating the durability of 304L canisters located in coastal areas. 304L stainless steel specimens with as-machined surfaces were adopted in this work to accelerate the effect of crevice corrosion for dust accumulated on the specimens' surfaces. Furthermore, this research offers greater knowledge of the initiation of stress corrosion cracking at sites of crevice corrosion.

2. Experimental Procedure

The shape and dimensions of the test specimens are presented in Figure 1. White emery was adopted in this work to simulate the dust accumulated on the as-machined specimen surface. The as-machined surface roughness (Ra) of the specimen was 4.9 μm. The chemical compositions of the AISI 304L steel used in the work are listed in Table 1. The test samples were manufactured from AISI 304L austenitic stainless steel, and white emery was made from alumina. Duplicate tests were conducted for each experimental condition.

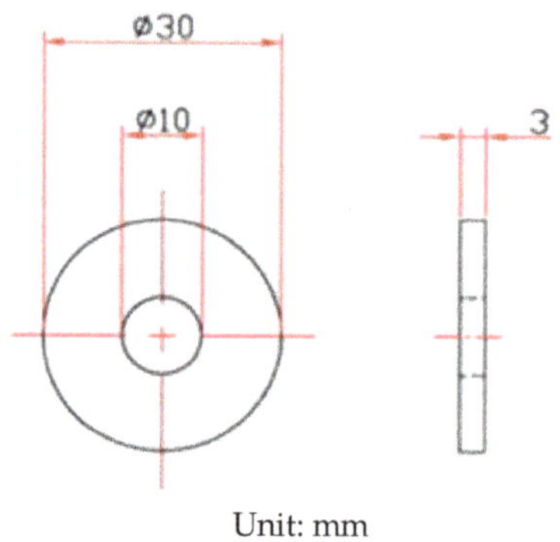

Figure 1. Dimensions of the specimen used for crevice corrosion test.

Table 1. Chemical composition of the 304L stainless steel used in this study.

Element	C	S	Si	Ni	Cr	Mn	Fe
wt.%	0.017	0.0290	0.450	9.000	18.000	1.540	Bal.

The chemical composition of sea salt used in the present study, based on ASTM D 1141-98 (13) Formula a, are listed in Table 2. A schematic diagram of the test specimen sprayed with synthetic sea water and white emery is presented in Figure 2. First of all, the samples were sprayed with synthetic sea water of 3.5 wt.% and subsequently dried on a hot plate at 65 °C for 20 min. After the specimens were dried, the mass change of specimens was measured to calculate the salt loadings in g/m^2. The concentration of chloride for the samples was 0.1 g/m^2 and 1 g/m^2. Finally, 0.5 g of white emery was sprayed uniformly in a semicircle of the sample. The constant humidity and temperature chambers were maintained at 45 °C with 45%, 55%, and 70% RH for 7000 h. Microstructural examination was acquired through a proper metallographic preparation process of the samples. The samples were washed ultrasonically in deionized water and dried carefully, then mounted in resin. The mounted specimens were polished using a powder of aluminum oxide. The microstructures and the surface features of the specimens were characterized with an optical microscope (OM) and the scanning electron microscope (SEM). The compositions of the corrosion products were determined by the technique of energy-dispersive spectrometer

(EDS). Furthermore, a scanning electron microscope equipped with an electron backscatter diffraction (EBSD) detector was used to inspect the stress corrosion cracking features of the specimens. Moreover, a kernel average misorientation (KAM) map and strain countering map for samples were acquired by inputting the electron back scatter diffraction map into the HKL-Channel 5 software for conducting data processing. The crack morphologies of the samples were examined by SEM.

Table 2. Chemical composition of the sea salt used in this study.

Composition	$NaCl$	Na_2SO_4	$MgCl_2$	KCl	$CaCl_2$	$NaHCO_3$	KCl	KBr	$SrCl_2$	H_3BO_3
wt.%	58.490	9.750	26.460	1.645	2.765	0.477	1.645	0.238	0.095	0.071

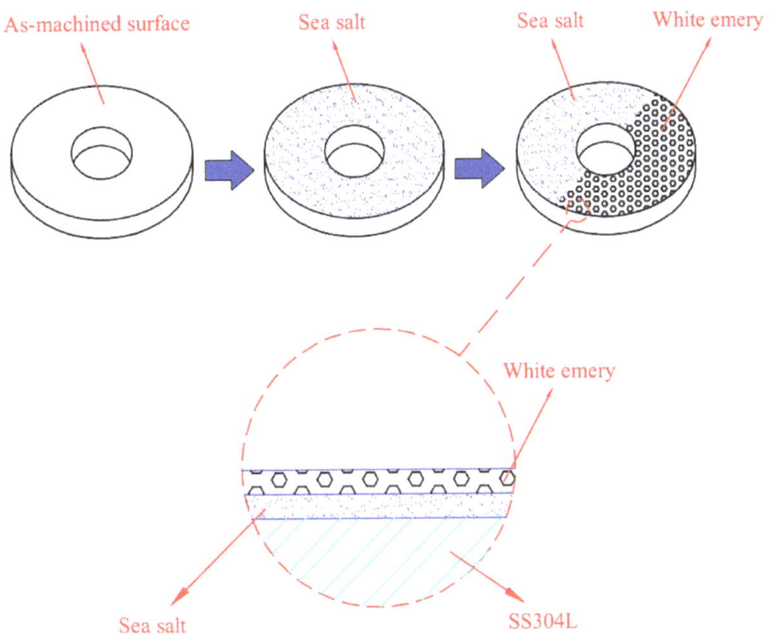

Figure 2. Schematic diagram of the specimen sprayed with synthetic sea water and white emery.

3. Results and Discussion

3.1. Surface Morphology Analysis

Figures 3 and 4 illustrate macrographs of the AISI 304L steel samples with a 1 g/m^2 concentration of chloride after testing for 7000 h at a relative humidity of 70%. Figure 3 reveals that significant crevice corrosion occurred in corroded areas. Figure 4a illustrates that obvious crevice corrosion was only observed in the regions deposited with white emery, while Figure 4b demonstrates a clear image of the crevice corrosion after the specimen was cleaned. Figure 5 reveals macrographs of the AISI 304L stainless steel samples with a 0.1 g/m^2 chloride concentration at three different relative humidity levels. A trace of crevice corrosion was discerned with the specimens. The rusted areas of the specimens at a relative humidity of 70% seem to be larger than the specimens at relative humidities of 45% and 55%. Figure 6 shows SEM micrographs of the samples deposited with a chloride concentration of 0.1 g/m^2 at various levels of relative humidity for 7000 h. The rust observed on the specimens presents proof of crevice corrosion, which was induced by chloride. The morphologic features of the crevice corrosion changed with RH. Figure 6a reveals some small corrosion pits on the specimens. Furthermore, there are some rust spots on the specimens shown in Figure 6b,c, with shallow corrosion existing beneath the rust spots. Figure 6c shows that the specimens subjected to a relative humidity of 70% present slightly more corrosion than those exposed to relative humidities of 45% and 55%, as shown in Figure 6a,b. Figure 7 demonstrates macrographs of the AISI 304L steel samples deposited with a chloride concentration of 1 g/m^2 at three different relative humidity levels. Figure 7a–c demonstrates that there is no obvious dissimilarity in the corroded regions on the samples at relative humidities of 45% and 55%. However, the corroded areas increased significantly at a relative humidity of 70%. In addition, the samples with a chloride concentration of 1 g/m^2, shown in Figure 7, reveal more severe corrosion at all three RH levels than that with a chloride concentration of 0.1 g/m^2, as illustrated in Figure 5. Figure 8 reveals SEM micrographs of the samples deposited with a chloride concentration of 1 g/m^2 at various levels of RH for 7000 h. As for those samples deposited with a chloride concentration of 0.1 g/m^2, the rust on the samples with a chloride concentration of 1 g/m^2 is proof of crevice corrosion, which was induced by chloride. Figure 8a–c illustrate several rust spots that occurred on the specimen and shallow corrosion beneath the rust spots. The samples subjected to a relative humidity of 70% (Figure 8c) reveal more serious corrosion than those exposed to relative humidities of 45% and 55% (Figure 8a). Moreover, those samples with a chloride concentration of 1 g/m^2, as illustrated in Figure 8, reveal more serious corrosion at all three RH levels than those with a chloride concentration of 0.1 g/m^2, as shown in Figure 6.

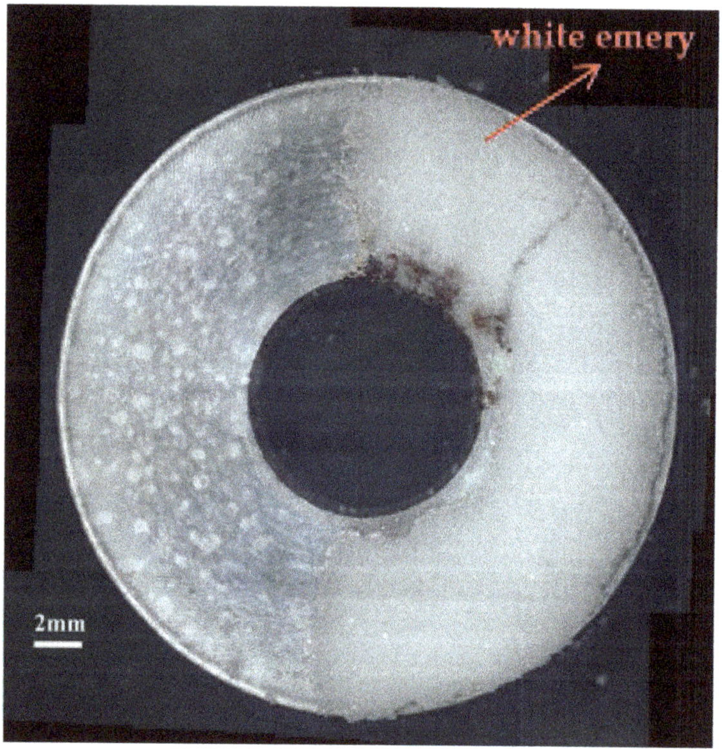

Figure 3. Macrographs of the specimens with a 1 g/m^2 chloride concentration after 7000 h of testing a relative humidity (RH) of 70%.

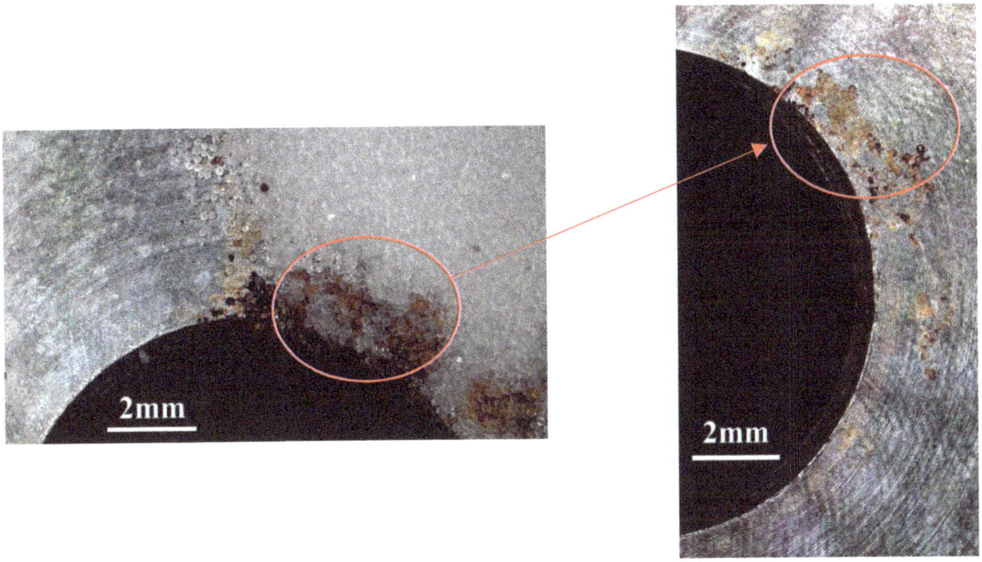

(**a**) (**b**)

Figure 4. Macrographs of the corrosion regions of those samples with a chloride concentration of 1 g/m^2 at RH = 70%: (**a**) Specimen before cleaning and (**b**) specimen after cleaning.

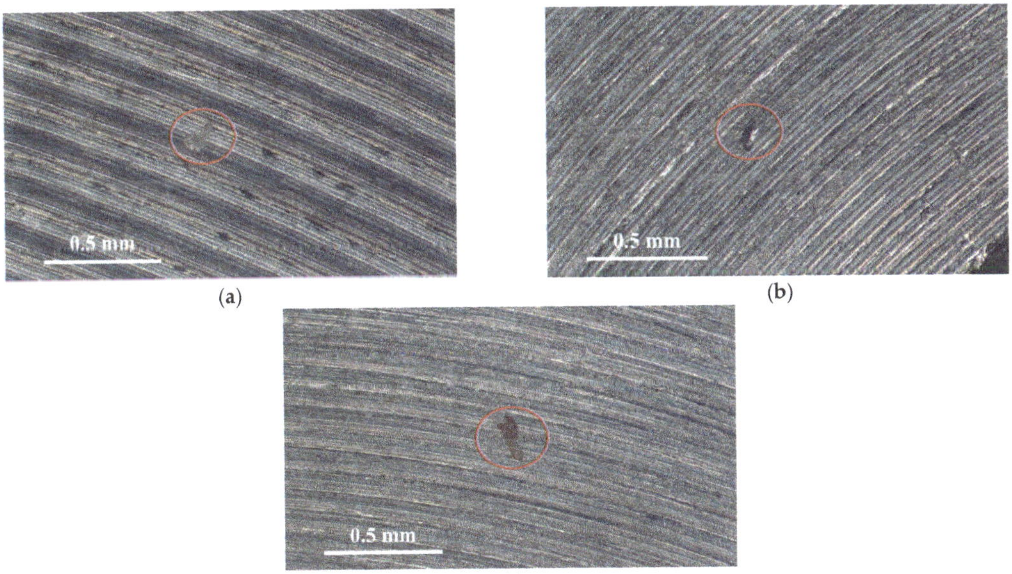

(**a**) (**b**) (**c**)

Figure 5. Macrographs of the corrosion regions of those samples with a 0.1 g/m^2 chloride concentration at: (**a**) Relative humidity = 45%, (**b**) relative humidity = 55%, and (**c**) relative humidity = 70%.

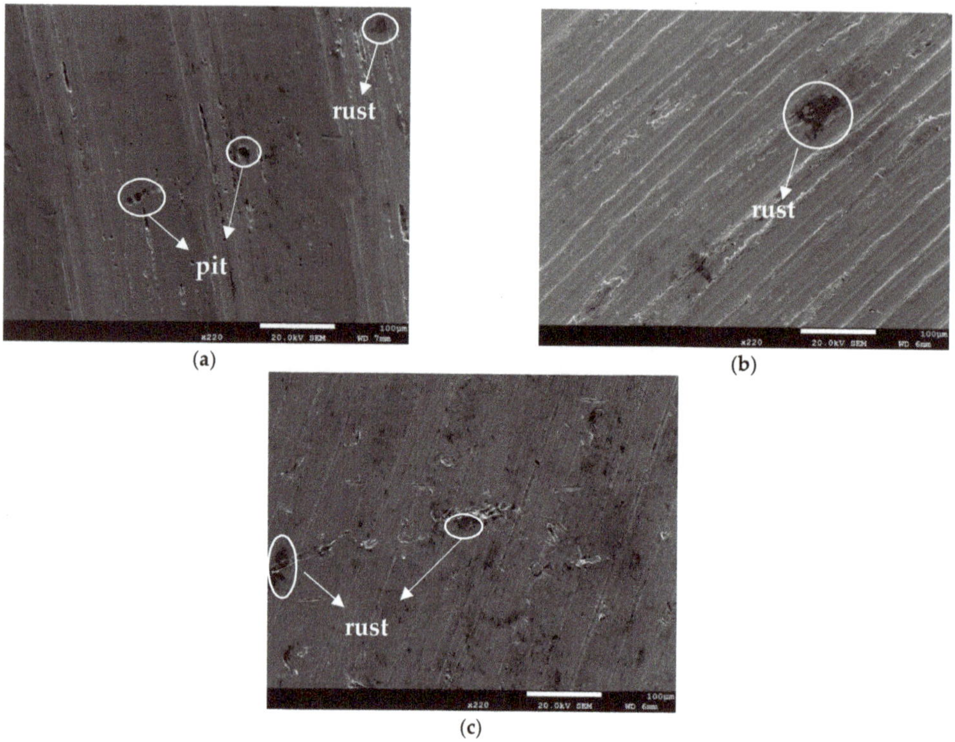

Figure 6. SEM micrographs of those specimens with a 0.1 g/m^2 chloride concentration at: (**a**) Relative humidity = 45%, (**b**) relative humidity = 55%, and (**c**) relative humidity = 70%.

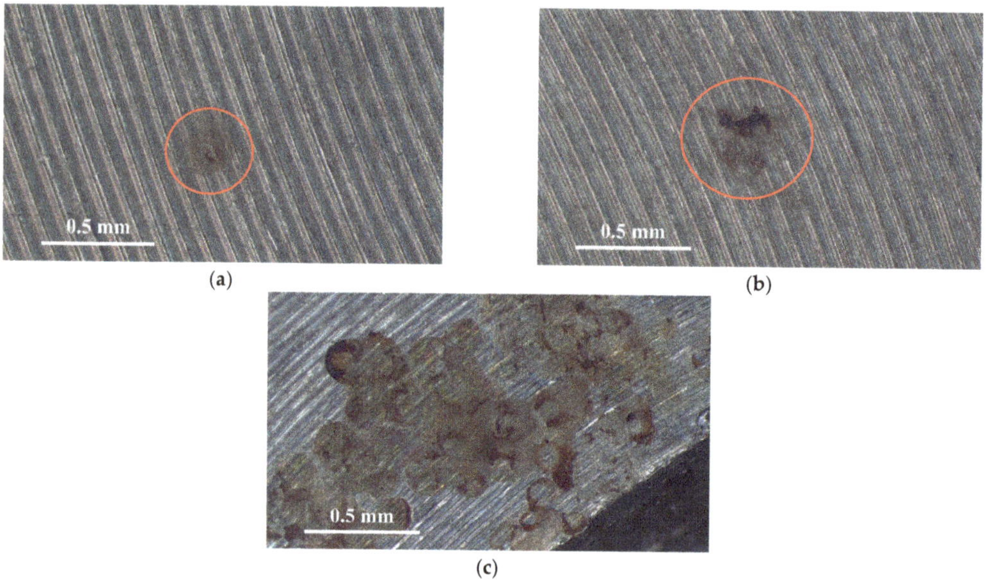

Figure 7. Macrographs of the corroded regions of those samples with a 1 g/m^2 chloride concentration at: (**a**) Relative humidity = 45%, (**b**) relative humidity = 55%, and (**c**) relative humidity = 70%.

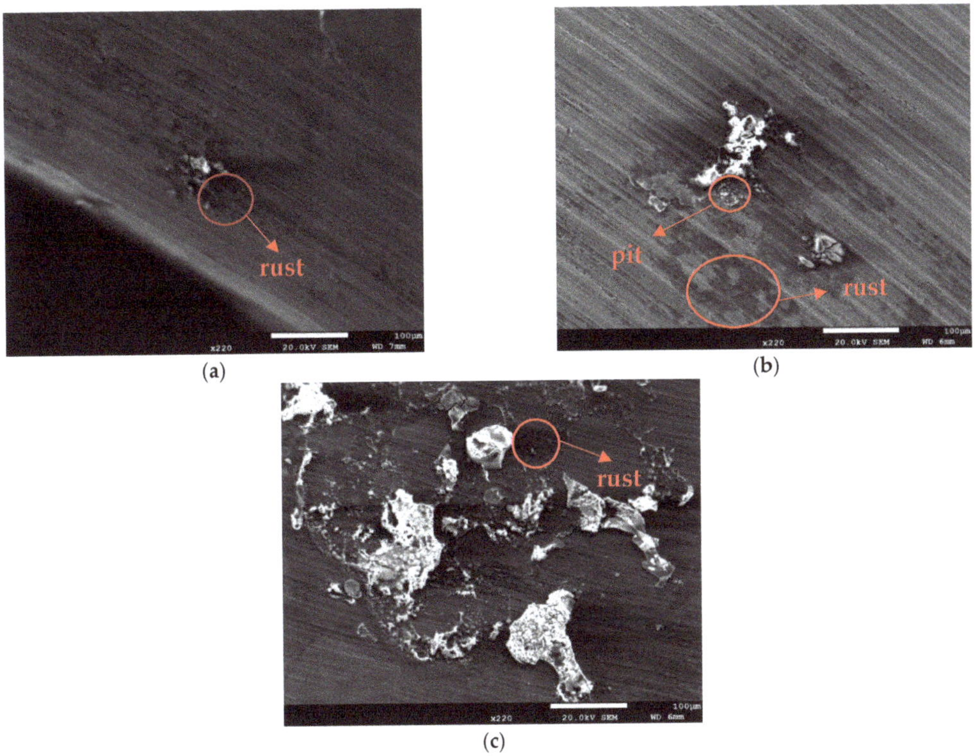

Figure 8. SEM micrographs of those samples with a chloride concentration of 1 g/m^2 after 7000 h of testing at: (**a**) RH = 45%, (**b**) RH = 55%, and (**c**) RH = 70%.

3.2. EDS and EBSD Analyses

Figure 9 depicts EDS examinations of the corroded regions on test specimens deposited with a chloride concentration of 1 g/m^2 at 55% RH. The results of the EDS analysis of the corroded regions are shown in Table 3, revealing that the sulfur and chlorine contents of Point A, which is the matrix of stainless steel, were so low that they could not be detected. Points B, C, and D, located in the corrosion products, were inspected to have higher contents of sulfur and chlorine, and the rust at Point E contained trace amounts of sulfur and chlorine. Furthermore, Points B, C, D, and E had higher contents of sulfur in comparison to Point A, which is probably explained by the sulfates contained in the synthetized sea water.

Table 3. Results of the EDS analysis of the corroded regions (wt.%).

Location	O	S	Cl	Mn	Cr	Ni	Fe
A	0.70	0.00	0.00	1.70	18.6	8.50	70.4
B	17.5	0.80	1.20	1.50	23.5	7.20	48.3
C	42.6	0.90	4.10	1.30	10.6	6.10	34.4
D	30.0	1.80	3.60	2.20	39.1	1.90	21.3
E	15.9	0.30	0.20	1.90	16.2	7.30	58.3

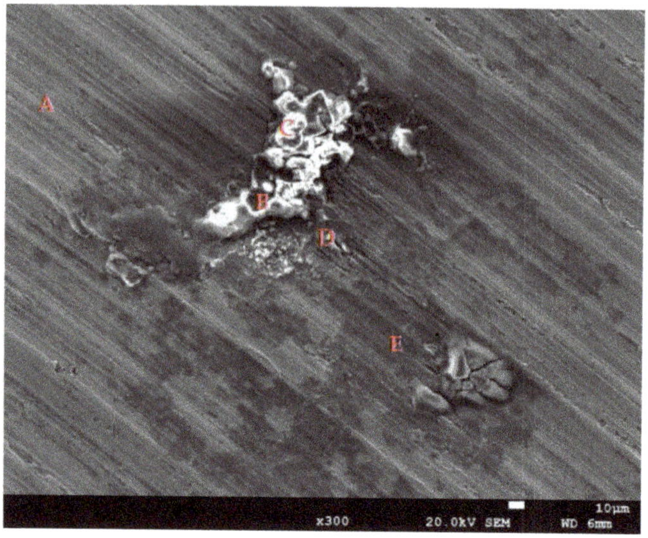

Figure 9. Energy dispersive X-ray spectrometry (EDS) analysis of the corroded regions on those specimens with a 1 g/m^2 chloride deposit after testing at 55% relative humidity for 7000 h.

Cl^- ions are indispensable for the initiation of crevice corrosion. The CCS mechanism, owes the initiation of crevice corrosion to the accumulation of the aggressive ions, particularly Cl^- in the crevice, brings about the appearance of highly aggressive localized corrosion that destroys the passive film of stainless steel [29]. Point B has a higher content of iron but a lower content of oxygen compared to Points C and D, which indicates that the oxide product was peeled off to expose the metal surface. Moreover, Point E, located in the rust region, has a higher Fe content but lower Cl and O contents compared to Points B, C, and D.

Figures 10 and 11 are SEM micrographs demonstrating the morphology of the SCC of the samples deposited with chloride concentrations of 0.1 g/m^2 and 1 g/m^2 at various RH levels, respectively. For samples deposited with a 0.1 g/m^2 chloride concentration, no SCC cracks were observed on those specimens exposed to the 45% and 55% relative humidities, as shown in Figure 6a,b, whereas Figure 10 demonstrates discontinuous SCC cracks on those specimens tested at a 70% relative humidity. For samples deposited with a 1 g/m^2 chloride concentration, no SCC cracks were observed on those specimens exposed to a 45% relative humidity, as exemplified in Figure 8a. Meanwhile, Figure 11a shows discrete SCC cracks on those samples exposed to a 55% RH, whereas Figure 11b reveals continuous SCC cracks on the specimens tested at a 70% relative humidity. Shoji et al. demonstrated that $MgCl_2$ is the sea salt constituent responsible for promoting low-temperature atmospherically induced SCC in AISI 316L and 304L steel. They inspected the phenomenon of atmospherically induced SCC of samples deposited with a 25 g/m^2 chloride concentration using sea water [19]. Prosek and coworkers investigated the phenomenon of atmospherically induced SCC of AISI 316L and 304 stainless steel with a 260 g/m^2 chloride concentration acquired by $MgCl_2$ droplets [23]. The maximum 1 g/m^2 chloride concentration used in this research is much lower compared to that reported in the literature [19,22,23]. Figures 10 and 11a demonstrate short and shallow cracks on those samples deposited with 0.1 g/m^2 of chloride at a 70% RH and 1 g/m^2 of chloride at a 55% RH, respectively, as opposed to Figure 11b, in which long and deep cracks appear on those specimens deposited with 1 g/m^2 of chloride at a relative humidity of 70%. It is conjectured that microcracks nucleate discontinuously at favorable sites in the first place and, under some favorable situations, microcracks grow and then coalesce by means of breaking ligaments between them to connect into one main crack [26–28,30]. On the basis of this hypothesis, some of the shallow and short cracks present on those samples with a 0.1 g/m^2 chloride concentration at a 70% RH and a 1 g/m^2 chloride concentration at a 55% RH could be attributed to finite amounts of Cl^- ions transmitted to the crevice sites of the samples. As a result of the low concentration of chloride of the samples tested, it is essential to have higher RH environments to promote the transmission of adequate amounts of Cl to crevice sites for promoting crack nucleation. The Cl maps obtained by EDS mapping are proof of the aforementioned argument. There was no SCC at RH = 45% and 55% whereas specimens exposed at RH = 70%, the highest RH of the three, cracked with those specimens deposited with a chloride concentration of 0.1 g/m^2, as shown in Figure 6a,b and Figure 10. This is probably because the chloride concentration of 0.1 g/m^2 was so low that it could not initiate stress corrosion cracking in the specimens tested at lower levels of relative humidity (RH = 45% and 55%). Moreover, when the chloride concentration for the specimens was increased to 1 g/m^2, there were still no SCC cracks with specimens exposed to RH = 45% whereas those exposed to RH = 55% and 70% cracked, as illustrated in Figures 8a and 11a,b. This can be possibly interpreted as finite chloride transmitted into crevice sites at the lowest relative humidity (relative humidity = 45%) in the specimens deposited with 1 g/m^2 of chloride. Therefore, the synergistic effect of the chloride concentration and relative humidity could account for the SCC initiation of the 304L stainless steel. From the experimental results of the work, it can be concluded that the relative humidity threshold for the SCC initiation of AISI 304L stainless steel with an as-machined surface deposited with a chloride concentration of 0.1 g/m^2 at 45 °C is between 55% and 70% RH, whereas the relative humidity threshold for those specimens deposited with a chloride concentration of 1 g/m^2 is between 45% and 55% RH. The increased volume of corrosion products possibly creates localized stress beneath white emery, which acts like a crevice former, leading to the occurrence of SCC. Furthermore, with the increase in relative humidity, the length of the cracks was observed to increase when comparing Figure 11a to Figure 11b.

Figure 10. SEM morphology of the SCC in those samples with 0.1 g/m^2 of chloride deposited after 7000 h of testing at RH = 70%.

(a) (b)

Figure 11. SEM morphology of the SCC in those samples with a chloride concentration of 1 g/m^2 after 7000 h of testing at: (**a**) RH = 55% and (**b**) RH = 70%.

Figure 12 demonstrates EDS mapping for the crack regions of those samples deposited with a 1 g/m^2 chloride concentration tested at a 70% RH. The crack region is obviously enriched with chlorine and oxygen, as respectively demonstrated by Figure 12e,f, but depleted with iron, as manifested in Figure 12b. Figure 12c,d reveal enriched manganese and sulfur, respectively, on the center bottom and left top of the figure.

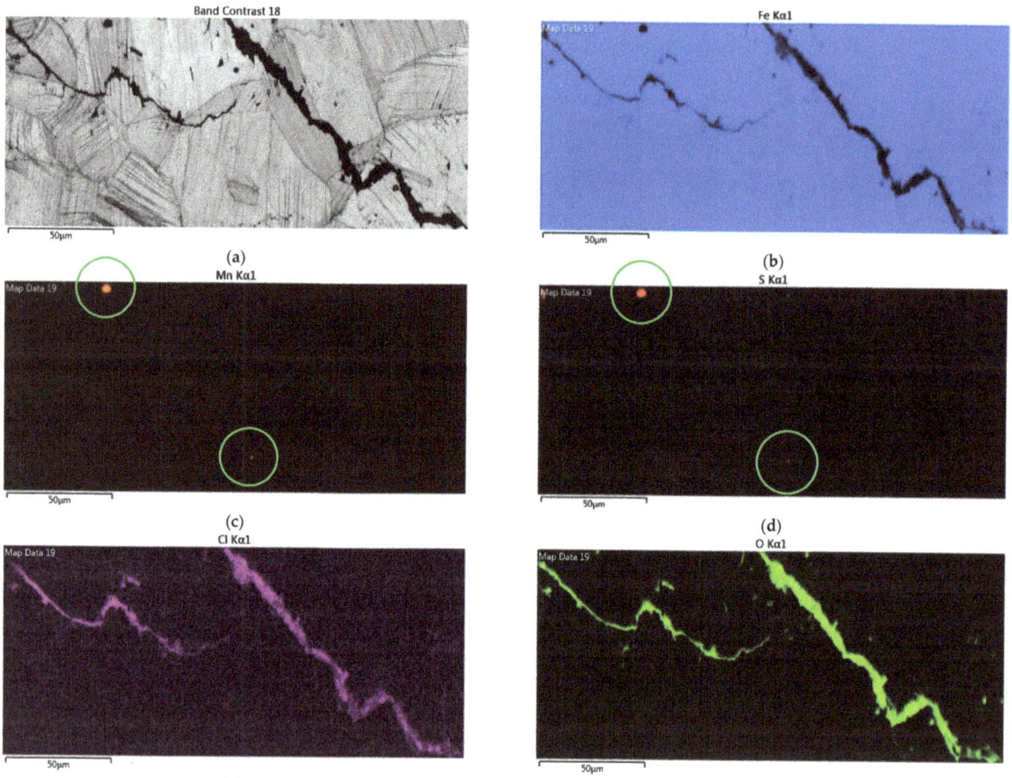

(a) (b) (c) (d) (e) (f)

Figure 12. EDS mapping for the crack regions of those samples with a chloride concentration of 1 g/m^2 tested for 7000 h at a relative humidity of 70%: (**a**) Band contrast image, (**b**) Fe mapping, (**c**) Mn mapping, (**d**) S mapping, (**e**) Cl mapping, and (**f**) O mapping.

Figure 13 demonstrates EBSD maps for the crack regions of those specimens deposited with 1 g/m^2 of chloride at a 70% RH. Figure 13a,d present samples cracked by a transgranular stress corrosion cracking (TGSCC) mode, which is in good agreement with those findings on transgranular stress corrosion cracking of stainless steel induced by chlorides when the temperature was above 50 °C [10]. The strain countering map for the specimens tested (Figure 13b) illustrates that the stress is more concentrated around the crack regions. Moreover, the kernel average misorientation (KAM) map for the specimens tested (Figure 13c) demonstrates that the high plastic strain, located around the crack regions, is probably linked to the plastic strain at the crack tip.

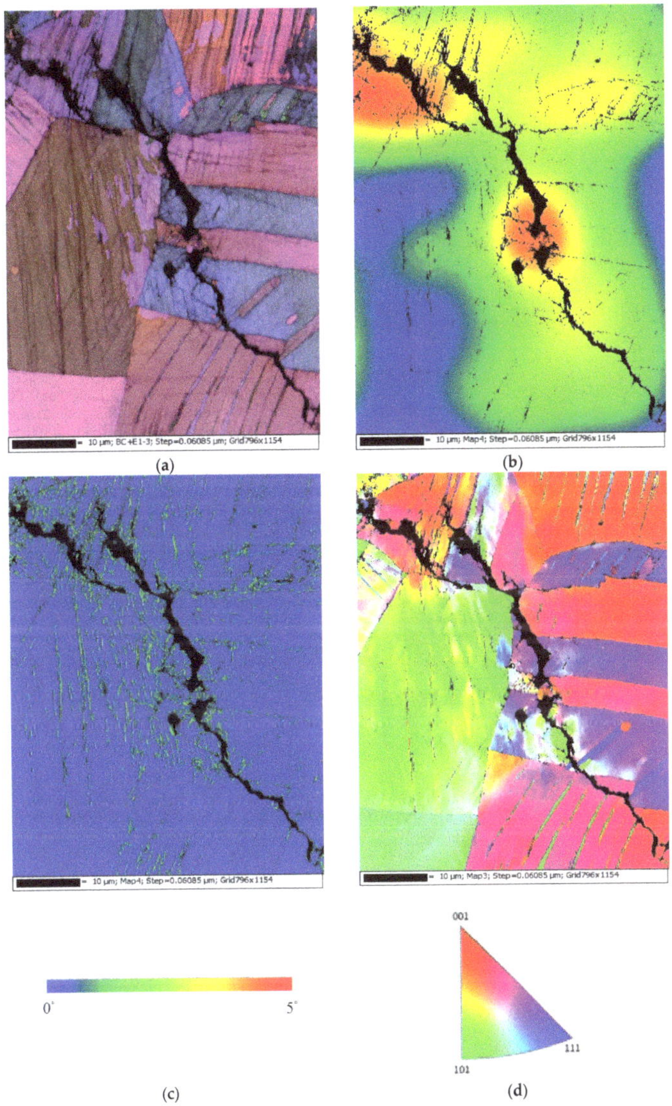

Figure 13. Electron back scatter diffraction (EBSD) maps for the crack region of the specimens deposited with a 1 g/m^2 chloride concentration tested for 7000 h at 70% RH: (**a**) Euler map, (**b**) strain countering map, (**c**) kernel average misorientation (KAM) map, and (**d**) inverse pole figure (IPF) map.

4. Conclusions

In this work, the crevice corrosion behavior of 304L stainless steel with white emery deposited on the as-machined surface was inspected by testing the samples at 45 °C under a combination condition of 0.1 g/m^2 or 1 g/m^2 chloride concentration and a RH of 45%, 55%, or 70% RH. The conclusions of this work are given as below:

(1) No cracks were observed on the samples deposited with a 0.1 g/m^2 chloride concentration at a 45% and 55% RH. However, discontinuous cracks were found on those samples exposed to a 70% relative humidity, which is the highest RH of the three. This could be accounted for by the fact that the chloride concentration of 0.1 g/m^2 was so low that it could not initiate stress corrosion cracking in the specimens tested at lower relative humidity levels (45% and 55% RH).

(2) When the chloride concentration for the specimens was increased to 1 g/m^2, there were still no SCC cracks on those specimens exposed to RH = 45% whereas those exposed to 55% and 70% RHs cracked. Furthermore, discrete SCC cracks were found on those specimens exposed to a 55% RH, whereas continuous SCC cracks were observed on those exposed to a relative humidity of 70%. This could be interpreted by finite chloride being transported into the crevice sites at a 45% RH for the specimens deposited with 1 g/m^2. The observation that the crack region had obviously been enriched with chloride provides evidence to verify the aforementioned argument.

(3) The results suggest that the relative humidity threshold was between 55% and 70% RH for the SCC initiation of 304L stainless steel with an as-machined surface and a 0.1 g/m^2 chloride concentration at 45 °C, whereas the relative humidity threshold decreased to between 45% and 55% RH when the chloride concentration for the specimens increased to 1 g/m^2.

(4) The specimens with a chloride concentration of 1 g/m^2 had obvious crevice corrosion only at those regions deposited with white emery.

(5) The 304L stainless steel specimens with an as-machined surface tested at 45 °C were cracked with the TGSCC mode, which was substantiated by the results of EBSD.

Author Contributions: Conceptualization, C.-P.Y. and K.-C.T.; formal analysis, C.-P.Y.; investigation, C.-P.Y. and K.-C.T.; project administration, J.-Y.H.; writing—original draft, C.-P.Y.; writing—review and editing, C.-P.Y. and J.-Y.H. All authors have read and agreed to the published version of the manuscript.

Funding: This research received no external funding.

Institutional Review Board Statement: Not applicable.

Informed Consent Statement: Not applicable.

Data Availability Statement: Data sharing is not applicable to this article.

Conflicts of Interest: The authors declare no conflict of interest.

References

1. Saegusa, T.; Yagawa, G.; Aritomi, M. Topics of research and development on concrete cask storage of spent nuclear fuel. *Nucl. Eng. Des.* **2008**, *238*, 1168–1174. [CrossRef]
2. Aoyama, T.; Sugawara, Y.; Muto, I.; Hara, N. In situ monitoring of crevice corrosion morphology of type 316 L stainless steel and repassivation behavior induced by sulfate ions. *Corros. Sci.* **2017**, *127*, 131–140. [CrossRef]
3. Li, Y.Z.; Wang, X.; Zhang, G.A. Corrosion behavior of 13 Cr stainless steel under stress and crevice in 3.5 wt.% NaCl solution. *Corros. Sci.* **2020**, *163*, 108290. [CrossRef]
4. Han, D.; Jiang, Y.; Shi, C.; Deng, B.; Li, J. Effect of temperature, chloride ion and pH on the crevice corrosion behavior of SAF 2205 duplex stainless steel in chloride solutions. *J. Mater. Sci.* **2012**, *47*, 1018–1025. [CrossRef]
5. Cook, C.; Padovani, C.; Davenport, A.J. Effect of nitrate and sulfate on atmospheric corrosion of 304 L and 316 L stainless steels. *J. Electron. Soc.* **2017**, *164*, C146–C163. [CrossRef]
6. Meyer, R.M.; Pardini, A.; Cuta, J.; Adkins, H.; Casella, A.; Qiao, A.; Larche, M.R.; Diaz, A.; Doctor, S.R. *NDE to Manage Atmospheric SCC in Canister for Dry Storage of Spent Fuel: An Assessment; PNNL-22495 401001060*; Pacific Northwest National Laboratory: Richland, WA, USA, 2013.

7. Schoell, R.; Xi, L.; Zhao, Y.; Wu, X.; Yu, Z.; Kenesei, P.; Almer, J.; Shayer, Z.; Kaoumi, D. In situ synchrotron X-ray tomography of 304 stainless steels undergoing chloride-induced stress corrosion cracking. *Corros. Sci.* **2020**, *170*, 108687. [CrossRef]
8. Wu, X. On residual stress analysis and microstructural evolution for stainless steel type 304 spent nuclear fuel canisters weld joint: Numerical and experimental studies. *J. Nucl. Mater.* **2020**, *534*, 152131. [CrossRef]
9. Yeom, H.; Dabney, T.; Pocquette, N.; Ross, K.; Pfefferkorn, F.E.; Sridharan, K. Cold spray deposition of 304 L stainless steel to mitigate chloride-induced stress corrosion cracking in canisters for used nuclear fuel storage. *J. Nucl. Mater.* **2020**, *538*, 152254. [CrossRef]
10. Tani, J.I.; Mayuzumi, M.; Hara, N. Stress corrosion cracking of stainless-steel canister for concrete cask storage of spent fuel. *J. Nucl. Mater.* **2008**, *379*, 42–47. [CrossRef]
11. Machuca, L.L.; Bailey, S.I.; Gubner, R.; Watkin, E.L.; Ginige, M.P.; Kaksonen, A.H.; Heidersbach, K. Effect of oxygen and biofilms on crevice corrosion of UNS S31803and UNS N08825 in natural seawater. *Corros. Sci.* **2013**, *67*, 242–255. [CrossRef]
12. Shojaei, E.; Mirjalili, M.; Moayed, M.H. The influence of the crevice induced IR drop on polarization measurement of localized corrosion behavior of 316L stainless steel. *Corros. Sci.* **2019**, *156*, 96–105. [CrossRef]
13. Ahn, T.; Oberson, G.; Depaula, S. Chloride-induced stress corrosion cracking of austenitic stainless steel used for dry storage of spent nuclear fuel. *ECS Trans.* **2013**, *50*, 211–226. [CrossRef]
14. Dong, P.; Scatigno, G.G.; Wenman, M.R. Effect of salt composition and microstructure on stress corrosion cracking of 316 L austenitic stainless steel for dry storage canisters. *J. Nucl. Mater.* **2021**, *545*, 152572. [CrossRef]
15. Wu, G.; Singh, P.M. Effect of elastic stresses on pitting behavior of stainless steel 304. *J. Electron. Soc.* **2019**, *166*, C209–C216. [CrossRef]
16. Islami, N.; Rashid, S.; Ariffin, A.K.; Nuawi, M.Z. Stress corrosion damage on austenitic stainless steel in sodium chloride. *Int. J. Automot. Mech. Eng. A* **2017**, *14*, 3824–3836. [CrossRef]
17. Guo, L.; Zhou, S.; Crocker, L.; Turnbull, A. Initiation sites for cracks developed from pits in a shot-peened 12 Cr martensitic stainless steel. *Int. J. Fatigue* **2017**, *98*, 195–202. [CrossRef]
18. Guo, L.; Mi, N.; Mohammed-Ali, H.; Ghahari, M.; Plessis, A.D.; Cook, A.; Street, S.; Reinhard, C.; Atwood, R.C.; Rayment, T.; et al. Effect of mixed salts on atmospheric corrosion of 304 stainless steel. *J. Electrochem. Soc.* **2019**, *166*, C3010–C3014. [CrossRef]
19. Shoji, S.; Ohnaka, N.; Furutani, Y.; Saitoh, T. Effects of relative humidity on atmospheric stress corrosion cracking of stainless steels. *Corros. Eng.* **1986**, *35*, 559–565. [CrossRef]
20. Scatigno, G.G.; Ryan, M.P.; Giuliani, F.; Wenman, M.R. The effects of prior cold work on the chloride stress corrosion cracking of 304 L austenitic stainless steel under atmospheric condition. *Mater. Sci. Eng. A* **2016**, *668*, 20–29. [CrossRef]
21. Scatigno, G.G.; Dong, P.; Ryan, M.P.; Wenman, M.R. The effects of salt loading on chloride-induced stress corrosion cracking of 304 L austenitic stainless steel under atmospheric conditions. *Materialia* **2019**, *8*, 100509. [CrossRef]
22. Cook, A.B.; Lyon, S.B.; Stevens, N.P.C.; Gunther, M.; McFiggans, G.; Newman, R.C.; Engelberg, D.L. Assessing the risk of under-deposit chloride-induced stress corrosion cracking in austenitic stainless steel nuclear waste containers. *Corros. Eng. Sci. Technol.* **2014**, *49*, 529–534. [CrossRef]
23. Prosek, T.; Iverson, A.; Taxsen, C.; Thierry, D. Low temperature on stress corrosion cracking of stainless steels in the atmosphere in the presence of chloride deposits. *Corrosion* **2009**, *65*, 105–117. [CrossRef]
24. Yeh, C.P.; Tsai, K.C.; Huang, J.Y. Influence of chloride concentration on stress corrosion cracking and crevice corrosion of austenitic stainless steel in saline environments. *Materials* **2020**, *13*, 5640. [CrossRef] [PubMed]
25. Ornek, C.; Engelberg, D.L. Toward understanding the effects of strain and chloride deposition density on atmospheric chloride-induced stress corrosion cracking of type 304 austenitic stainless steel under $MgCl_2$ and $FeCl_3$: $MgCl_2$ droplets. *Corrosion* **2019**, *75*, 167–182. [CrossRef]
26. Masuda, H. SKFM observation of SCC on SUS304 stainless steel. *Corros. Sci.* **2007**, *49*, 120–129. [CrossRef]
27. Qiao, L.J.; Gao, K.W.; Volinsky, A.A.; Li, X.Y. Discontinuous surface cracks during stress corrosion cracking of stainless steel single crystal. *Corros. Sci.* **2011**, *53*, 3509–3514. [CrossRef]
28. Acharyya, S.G.; Khandelwal, A.; Kain, V.; Kumar, A.; Samajdar, I. Surface working of 304 L stainless steel: Impact on microstructure, electrochemical behavior and SCC resistance. *Mater. Charact.* **2012**, *72*, 68–76. [CrossRef]
29. Li, J.X.; Chu, W.Y.; Wang, Y.B.; Qiao, L.J. In situ TEM study of stress corrosion cracking of austenitic stainless steel. *Corros. Sci.* **2003**, *45*, 1355–1365. [CrossRef]
30. Zhu, L.K.; Yu, Y.; Li, J.X.; Qiao, L.J.; Volinsky, A.A. Stress corrosion cracking under low stress: Continuous or discontinuous cracks? *Corros. Sci.* **2014**, *80*, 350–358. [CrossRef]

materials

Article

Ex Situ Examination of Matrix and Inclusions of API-X100 before and after Exposure to Bitumen at Elevated Temperature

Hongxing Liang * and Edouard Asselin

Department of Materials Engineering, The University of British Columbia, Vancouver, BC V6T 1Z4, Canada; edouard.asselin@ubc.ca
* Correspondence: hongxingliang314@gmail.com

Abstract: The corrosivity of bitumen at 60 and 120 °C was examined by exposing American Petroleum Institute (API) X100 (yield strength 100 ksi, 690 MPa) pipeline steel to bitumen in an autoclave for 30 days. Prior to the autoclave measurements, the inclusion types in the API-X100 pipeline steel were characterized by scanning electron microscopy (SEM), and four types of inclusions were identified, according to their elemental compositions. The four types of inclusions and the surrounding matrix were characterized by ex situ SEM before and after exposure to bitumen. The results show that no obvious corrosion occured at the inclusions or the matrix after exposure at 60 and 120 °C.

Keywords: API-X100 pipeline; bitumen; corrosion; inclusion

check for updates

Citation: Liang, H.; Asselin, E. Ex Situ Examination of Matrix and Inclusions of API-X100 before and after Exposure to Bitumen at Elevated Temperature. *Materials* **2021**, *14*, 5007. https://doi.org/10.3390/ma14175007

Academic Editors: Claudio Gennari, Luca Pezzato and Amir Mosta-faei

Received: 25 July 2021
Accepted: 29 August 2021
Published: 2 September 2021

1. Introduction

According to the National Energy Board of Canada, by 2040, oil sands bitumen production will reach 4.5 million barrels per day (MMb/d), which is 2.3 times the 2012 production level [1]. By 2040, an increase in bitumen production is also predicted to occur in Venezula (from 0.6 to 2.1 MMb/d), Madagascar (from 0.03 to 0.38 MMb/d), the United States (from 0.005 to 0.035 MMb/d), and the Republic of Congo [2]. Bitumen is typically diluted with a diluent (a natural-gas condensate, such as naphtha) to reduce its viscosity: in this form, it is commonly referred to as "dilbit". The reduced viscosity enables transportation in pipelines, which is cost effective at large scales [3]. The capacity of existing pipelines is limited, and the construction of new pipelines has been proposed to transport the increasing bitumen production to domestic markets in Canada or new international markets, such as Asia [4]. However, new pipeline proposals result in wide-ranging policy disputes involving both key stakeholders and the general population [5]. For example, the Northern Gateway Pipeline, which planned to ship dilbit from Alberta to British Columbia, was cancelled in 2016 [6]. Similarly, after years of changing status and dispute, the Keystone XL project was cancelled in 2021 [7]. Dusyk et al. [8] found that the dispute was primarily framed as an issue of economic benefit versus environmental risk through the analysis of 2097 articles about the Northern Gateway Pipeline between 2008 and 2014. Among the different environmental risks, terrestrial and aquatic spills of dilbit were common concerns. Dilbit spills resulting from pipeline ruptures or leaks can be related to pipeline corrosion [9]. An insufficient understanding of corrosion of dilbit pipelines, especially their internal corrosion, exacerbates the debate around new pipeline construction.

Though the roles of chloride droplets and deposits in the internal corrosion of dilbit pipelines were reported in [10] and in our previous studies [10–14], the corrosion effect of bitumen itself on the pipeline has been in dispute. Corrosion has been identified in dilbit pipelines near over-bends where deposits coupled with chloride droplets accumulate on the pipeline surface [10]. In separate studies, we used paraffin-covered or dilbit-covered droplets to simulate the corrosive environment in dilbit pipelines. We identified that the distribution of corrosion penetration under the oil-covered droplet is uneven, with the anode regions showing deeper penetration, due to either lower pH [15], enhanced chloride concentration [13], the possible direct reduction of corrosion products [13] or the presence of silica deposition [11]. Furthermore, as summarized in the reports [9,16] or literature [17], and in contrast to other heavy oils, bitumen may lead to faster pipeline corrosion because of the higher concentration of naphthenic acids, sulfur compounds, and asphaltenes that it contains. However, in [18] it was found that only under high temperatures (230–460 °C) could the naphthenic acid compounds in bitumen be corrosive. Thus, one may not expect these compounds to be corrosive at the normal operating temperatures (45–70 °C) of dilbit pipelines [19]. McIntyre, et al., [20] also concluded that bitumen was not corrosive. Via autoclave corrosion measurements of API-X65 (yield strength 65 ksi, 450 MPa) pipeline steel exposed to dilbit and synbit (a mixture of bitumen and synthetic crude oil) at 65 °C for 14 days, they showed that no apparent pitting corrosion occurred on the specimens. Thus, there are conflicting reports on the corrosive nature of bitumen, and this conflict engenders further conflict at the regulatory and project approval stages. In addition, since corrosion around inclusions can be a concern for many materials, it is of added interest to study the potential for corrosion at these microscopic features. To the best of our knowledge, the stability vis-à-vis corrosion of steel inclusions immersed in bitumen has not yet been studied.

Therefore, in order to evaluate the corrosive nature of bitumen, ex situ SEM characterization was used in this work to evaluate the same area on API-X100 pipeline steel before and after exposure to bitumen at 60 and 120 °C for 30 days. These two temperatures were selected because some dilbit pipelines are operated at 60 °C (the now abandoned Keystone XL pipeline) [19], while some pipelines (for example, a test line at MacKay River, AB, Canada) are designed to transport bitumen without dilution at 120 °C [21]. The objective of this work was to see whether bitumen at 60 or 120 °C would cause the corrosion of either the matrix or the inclusions of API-X100 steel. On the one hand, correctly evaluating the corrosion rate of pipeline steel in bitumen provides information for policy actors, key stakeholders and general citizens, who would determine whether to build new bitumen or dilbit pipelines. On the other hand, such work ensures that operators of existing dilbit pipelines use appropriate information for corrosion mitigation, which can provide environmental and economic benefits.

2. Materials and Methods

2.1. Materials

API-X100 (Evraz Inc., Chicago, IL, USA) samples (10 mm × 10 mm × 6 mm in dimension) were cut from a 6 mm thick sheet with a composition provided in Table 1. The specimens were then ground successively, using 120, 320, 600, and 1200 grit silicon carbide papers. After grinding, 6 and 1 μm diamond suspensions were used to polish the steel surfaces. The microstructure of the steel consists of acicular (needle-like) ferrite, polygonal ferrite, and bainite, as reported in our previous study [11]. Grid marks were made on the specimen surface, using a sharp tweezer and ruler without etching in order to locate the same areas by SEM prior to and after autoclave immersion in bitumen, as shown in Figure 1.

Table 1. Chemical composition (assay from the steel manufacturer) of the API-X100 pipeline steel samples (unit: wt.%).

Element	C	Cu	Mn	V	Cr	Nb	Ni	Al	Mo	Ti
API-X100 (wt.%)	0.1	0.25	1.66	0.003	0.016	0.043	0.13	0.02	0.19	0.02

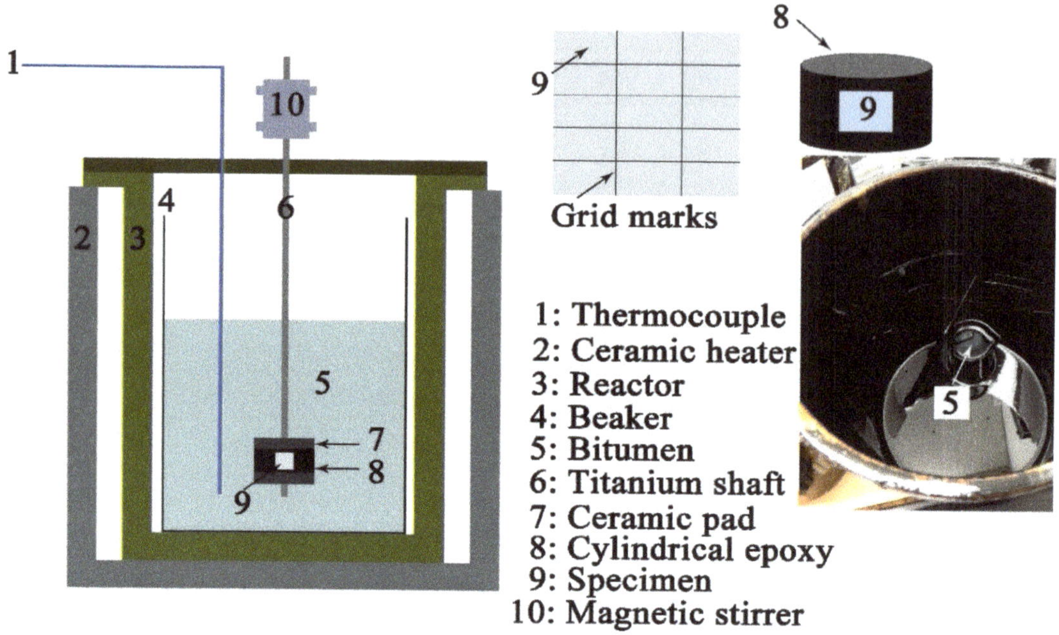

Figure 1. Schematic of the autoclave setup used to examine the corrosion of API-X100 pipeline steel in bitumen at 60 and 120 °C. Bitumen is shown at right in an unrelated separate bucket.

2.2. Bitumen and Organic Solvent

Bitumen was supplied by a large Canadian producer. The properties and elemental constituents of the bitumen were given in our previous study [13]. Bitumen is a heavy oil with an atomic C/H ratio of 0.72. Importantly, the sulfur concentration in bitumen (4.09 wt.%) is higher than that in crude oil (~3 wt.%) [19]. After the autoclave immersions, the specimens were cleaned using toluene ($C_6H_5CH_3$, VWR International, Mississauga, ON, Canada).

2.3. Autoclave Exposure Experiments

Figure 1 presents the autoclave setup used to assess the potential corrosivity of bitumen. The temperatures were set to either 60 or 120 °C. The rotational speed was controlled at 720 r/min, and the achieved surface velocity (144 cm/s) across the specimen was equal to the flow rate in a real dilbit pipeline [20]. The steel specimen was mounted in the cylindrical epoxy resin (M.G. Chemicals Ltd., Surrey, BC, Canada) with a temperature tolerance of 250 °C. The diameter of the cylindrical epoxy resin was 3.80 cm. After 30 days of exposure, the steel specimen was cleaned as noted above. At least 3 different API-X100 pipeline steel specimens were exposed in the autoclave at each temperature in order to obtain reliable results.

2.4. Surface Morphology

The morphologies of the matrix and inclusions in the steel prior to and after exposure to bitumen for 30 days at 60 and 120 °C were imaged by a field emission SEM (Zeiss Germany, Oberkochen, Germany). SEM equipped with an EDAX Genesis (Zeiss Germany, Oberkochen, Germany) integrated energy dispersive X-ray spectroscopy (EDS) was used to study the elemental distribution of the inclusions. An accelerating voltage of 20 kV was selected to obtain the images of EDS and secondary electrons. At least 3 different positions in the matrix and 50 different inclusions (the identification of inclusions were conducted using SEM EDS area mapping) were examined to ensure reliable results.

3. Results

3.1. Inclusions in API-X100 Pipeline Steel

At least 50 inclusions randomly selected on the steel surface were observed by microscopy in order to obtain statistically relevant results. Based on the composition measured by EDS, the inclusions were classified into four different types, namely, Mg−Al−Si−O−Ca−S (Figure 2a), Mg−Al−O−Ca−S (Figure 2b), Mg−Al−O−Ca−Mn−S (Figure 2c), and Al−O−Ca−S (Figure 2d). Of the 50 inclusions, these 4 types of inclusions represented 58%, 30%, 10%, and 2% of the total number, respectively.

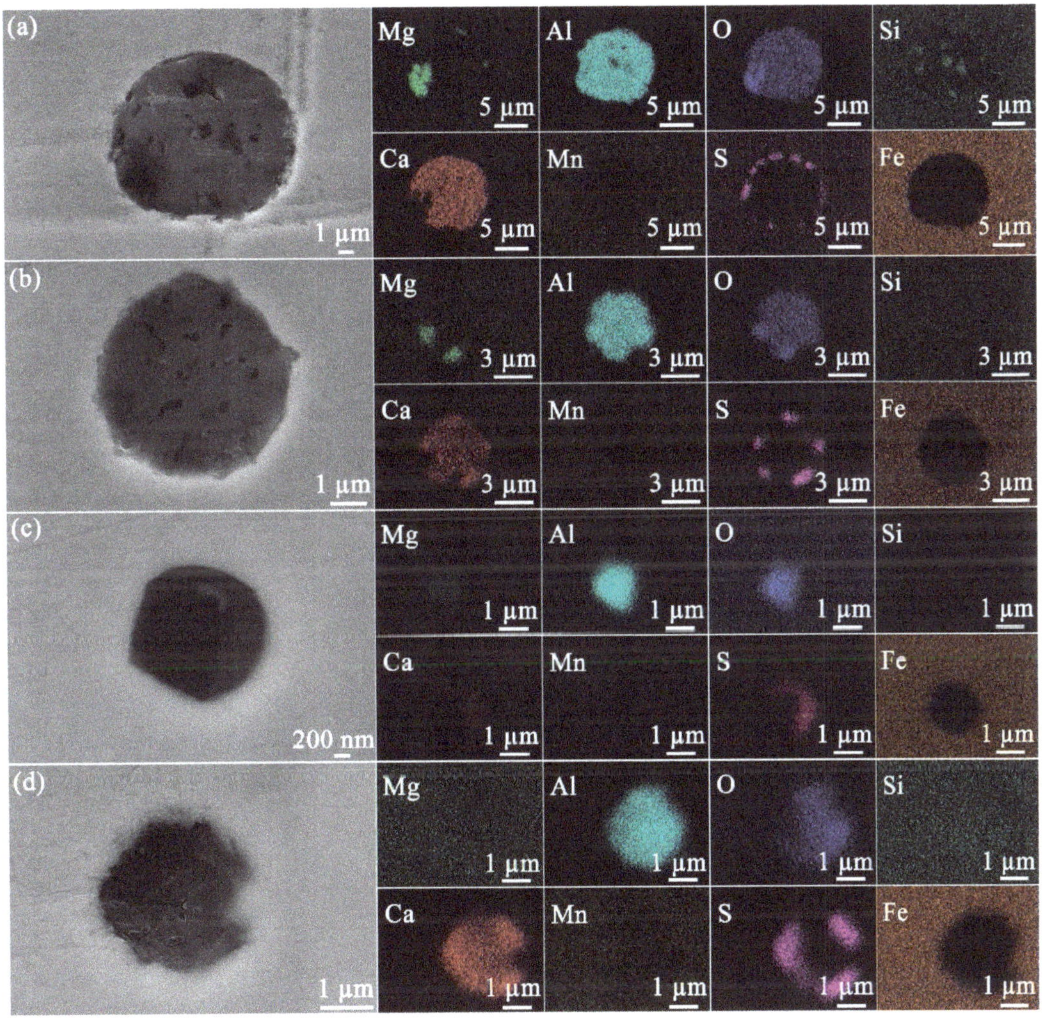

Figure 2. SEM/SE (secondary electron) images of inclusions containing (**a**) Mg−Al−Si−O−Ca−S, (**b**) Mg−Al−O−Ca−S, (**c**) Mg−Al−O−Ca−Mn−S, and (**d**) Al−O−Ca−S on unexposed steel. The elemental distributions of the inclusions are given at right.

3.2. Evaluation of the Inclusions after Exposure to Bitumen at 60 and 120 °C

Ex situ SEM characterization of identical areas and inclusions before and after exposure to bitumen avoids providing misleading results. The morphologies of the four types of inclusions prior to and after exposure to bitumen at 60 °C are given in Figure 3. After an exposure of 30 days, no obvious corrosion could be found to occur at the inclusions. Furthermore, the API-X100 pipeline steel matrix (Figure 4) also did not show any signs of corrosion: no trace of uniform or pitting corrosion was detected.

At 120 °C, the four types of inclusions also remained inert after exposure for 30 days (Figure 5). In addition, neither uniform corrosion nor pitting corrosion was observed in the steel matrix (Figure 6). The current results indicate that the bitumen tested here, without diluent, and even at elevated temperature, was not corrosive to API-X100 pipeline steel.

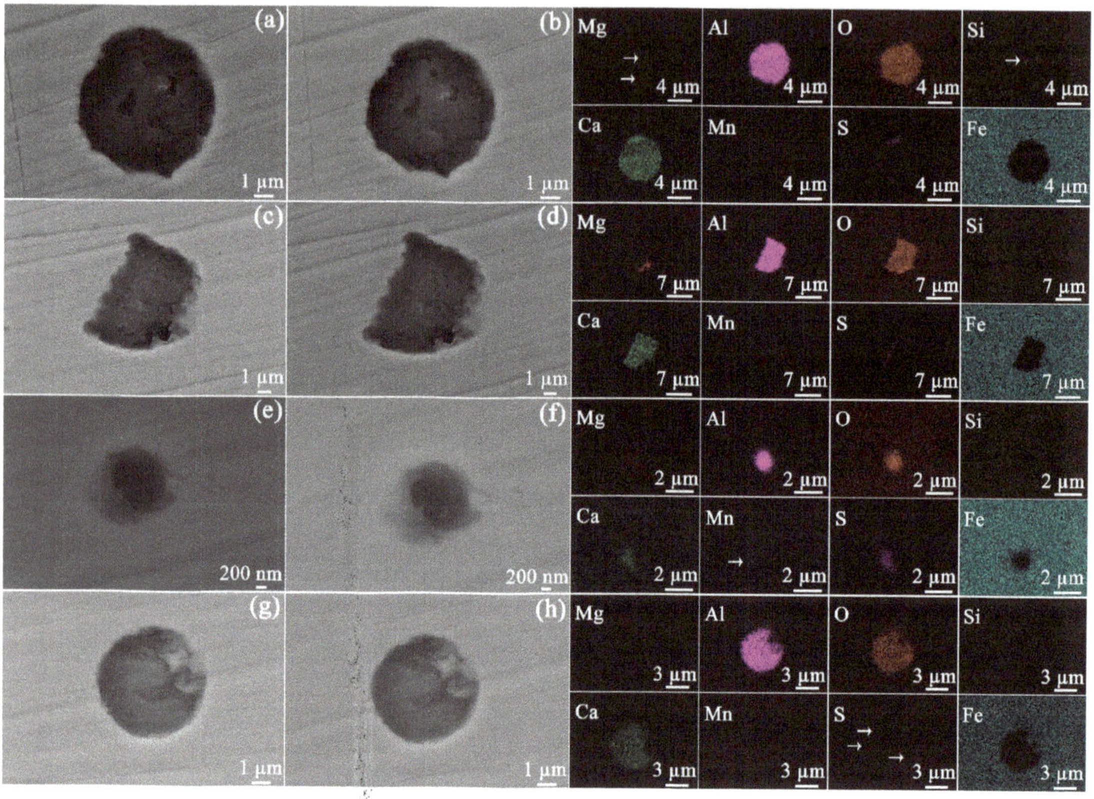

Figure 3. SEM SE images of the inclusion containing Mg−Al−Si−O−Ca−S before (**a**) and after (**b**) exposure to bitumen at 60 °C, of the inclusion containing Mg−Al−O−Ca−S before (**c**) and after (**d**) exposure to bitumen at 60 °C, of the inclusion containing Mg−Al−O−Ca−Mn−S before (**e**) and after (**f**) exposure to bitumen at 60 °C, and of the inclusion containing Al−O−Ca−S before (**g**) and after (**h**) exposure to bitumen at 60 °C. The elemental distributions of the inclusions are given at right. In the EDS maps given at right, the areas with low signal intensity are marked with arrows.

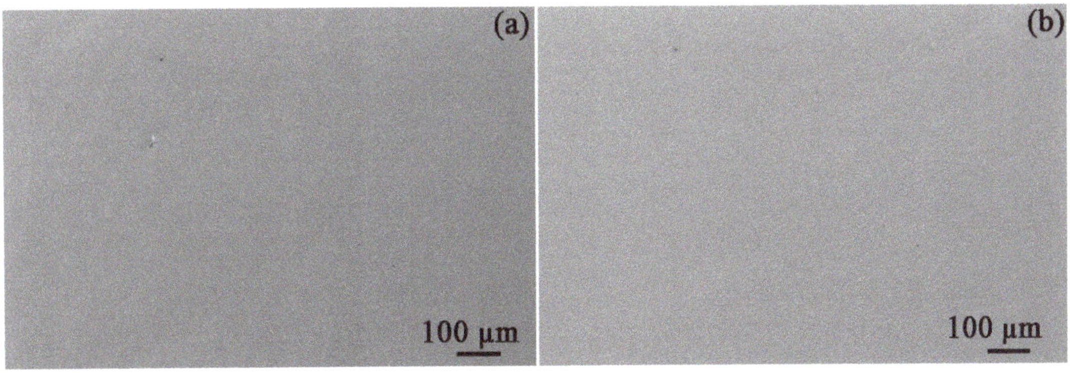

Figure 4. SEM SE images of the API-X100 pipeline steel matrix before (**a**) and after (**b**) exposed to bitumen at 60 °C.

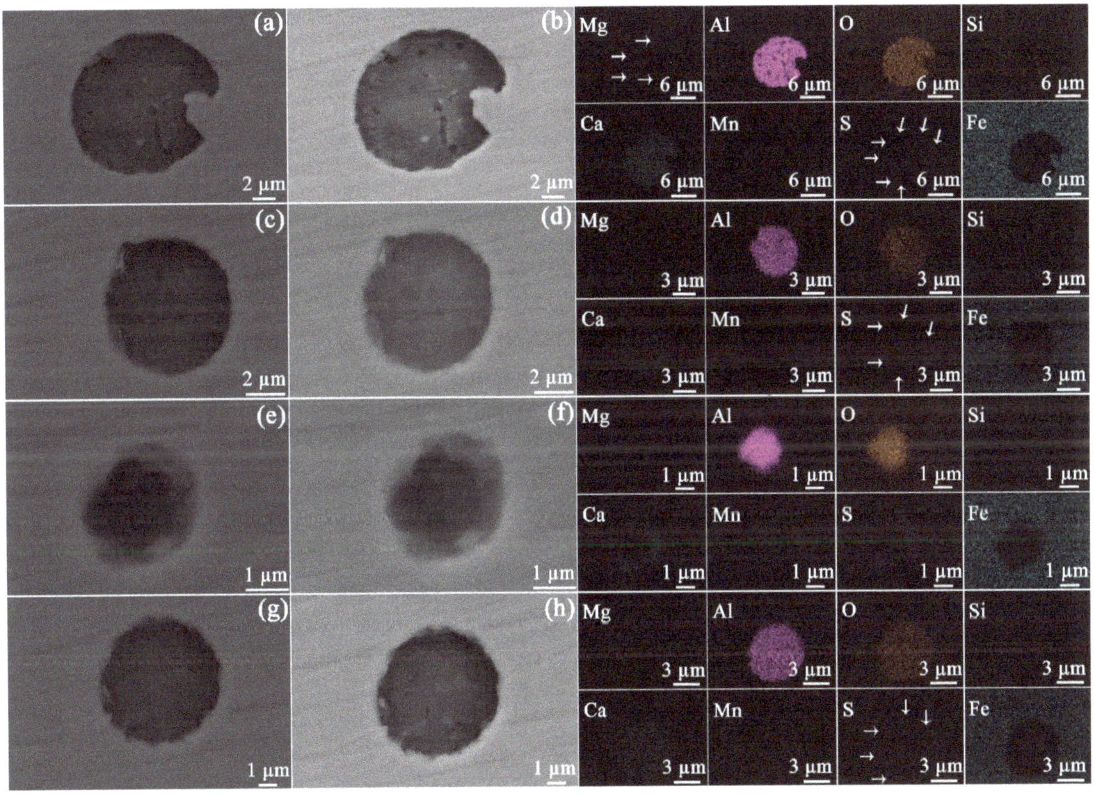

Figure 5. SEM SE images of the inclusion containing Mg−Al−Si−O−Ca−S before (**a**) and after (**b**) exposed to bitumen at 120 °C, of the inclusion containing Mg−Al−O−Ca−S before (**c**) and after (**d**) exposed to bitumen at 120 °C, of the inclusion containing Mg−Al−O−Ca−Mn−S before (**e**) and after (**f**) exposed to bitumen at 120 °C, and of the inclusion containing Al−O−Ca−S before (**g**) and after (**h**) exposed to bitumen at 120 °C. The elemental distributions of the inclusions are given at right. In the EDS maps given at right, the areas with low signal intensity are marked with arrows.

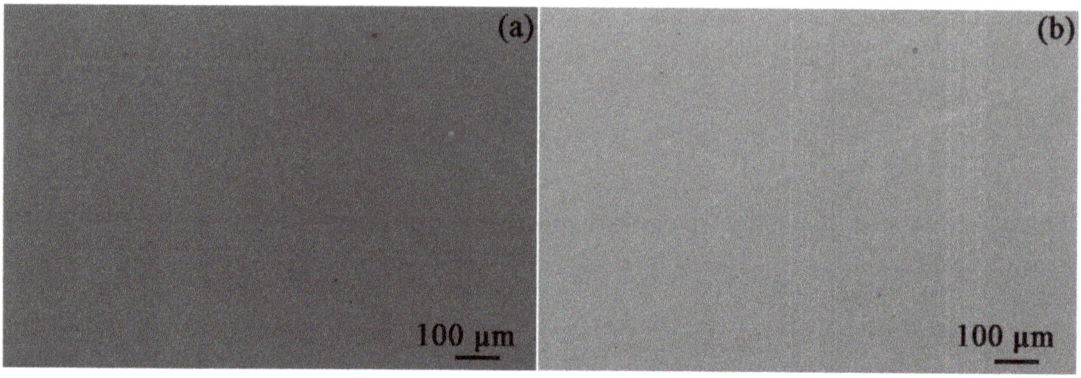

Figure 6. SEM SE images of the API-X100 pipeline steel matrix before (**a**) and after (**b**) exposed to bitumen at 120 °C.

4. Discussion

4.1. Inclusions in the Steel

In Figure 2, four types of inclusions were identified. The inclusions that are frequently detected in carbon steel include magnesium oxide (MgO), calcium oxide (CaO), silicon dioxide (SiO_2), aluminum oxide (Al_2O_3), calcium sulfide (CaS), and manganese sulfide (MnS) [22]. Further, it is reported that the S in API-X100 is bound in CaS [23]. It is inferred that the inclusion in Figure 2a consists of MgO, Al_2O_3, SiO_2, CaO, and CaS. Compared to the inclusion in Figure 2a, the inclusion of Figure 2b does not show SiO_2. Further, neither MgO nor SiO_2 were observed in Figure 2d. As shown in Figure 2c, the Al and Mg are bound with O, whereas the Ca and Mn are bound with S, indicating a separation within the same particle, i.e., into (Ca, Mn)S and (Mg, Al)O rich portions. During refining, the improvement of steel cleanliness and modification of Al_2O_3 and MnS are achieved by the addition of Ca [24]. The sulfur in inclusions exists as CaS with the CaS formed by the direct reaction of S and Ca as well as the reaction of CaO with dissolved Al and S in liquid steel [24].

API-X100 pipeline steels from different manufacturers may show different inclusions, which are dependent on the manufacturing, deoxidation, and desulfurization processes [22]. Dong et al. [25] concluded that the primary inclusions in this steel are Al_2O_3, titanium dioxide (TiO_2) and ferric carbide (Fe_3C). Jin and Cheng reported that four types of inclusions (Si-enriched, Al-enriched, microvoid, and carbide inclusions) were identified in API-X100, and the electrochemical activity of Si-enriched inclusions is higher than that of the matrix in a near-neutral pH bicarbonate solution purged with 5% carbon dioxide (CO_2) [26]. In another independent study from the same research group, five types of inclusions, including MnS, MnS/Al_2O_3, Ca/Al/Mg/O, Al_2O_3, and Si-enriched particles were identified in API-X100 [27]. However, Li et al. [28] concluded that the Si-enriched inclusions are inert during the exposure to near neutral pH bicarbonate solution (purged with 5% CO_2) and that SiO_2, Al–Si–S–Ca–O, Al–Si–Ca–O, and Al–Ca–O inclusions were observed in the steel. Moreover, Li et al. did not observe the MnS inclusion in API-X100 by SEM. Arafin and Szpunar [29] disclosed that the inclusions in API-X100 pipeline steel can be classified into three types (Si-based, Al–Mg–S–Ca–O, and Al–Mg–O). The findings in the current work are different from those studies. Al_2O_3, CaO, and CaS were identified in all four types of inclusions (Figure 2) and (Mg, Al)O/(Ca, Mn)S (Figure 2c) was detected instead of MnS/Al_2O_3. Neither SiO_2 nor MgO were detected in the inclusions (Figure 2d). Because the SEM can only observe a limited area, other types of inclusions may exist in the steel samples used here. However, these 4 types of inclusions represent the entire spectrum of the 50 inclusions imaged herein.

4.2. After Exposure to Bitumen at 60 and 120 °C

The different inclusions were inert in bitumen for 30 days at 60 and 120 °C. It was reported that the corrosion initiation of API-X100 in a near-neutral pH bicarbonate solution is associated with localized dissolution at or near inclusions [27]. When X70 steel is exposed to near-neutral pH bicarbonate solution, (Ca, Mn)S particles, which have a lower resistance to corrosion in comparison with the matrix and (Mg, Al)O particles, can dissolve preferentially [30]. However, in bitumen, the CaMnS particles within the API-X100 remained inactive at both temperatures (Figure 3e,f and Figure 5e,f) even after 30 days.

The API-X100 steel matrix also remained inactive (Figures 4 and 6) in the bitumen. Derungs [31] indicated that the corrosion caused by naphthenic acids cannot occur below 220 °C. The sulfur-containing compounds can be separated into two categories, namely, polar (mercaptans) and non-polar (sulfide, thiophenes) forms [32]. Ayello et al. [32] confirmed that 1-tetradecanethiol (a mercaptan) can inhibit the corrosion of carbon steel in chloride solution purged by CO_2 at room temperature, whereas dibenzothiophene (a sulfide) and dioctyl-sulfide (a thiophene) have no significant effect on carbon steel corrosion. Morales et al. [33] reported that asphaltene can deposit on a pipeline surface and behave as a physical barrier that prohibits corrosive species from contacting the pipeline surface. If a paraffinic solvent diluent is used with bitumen, asphaltene deposition may occur coupled with the accumulation of solid deposits, forming a sludge, which can result in the corrosion of pipelines. However, the main objective of this work was to verify the corrosivity of bitumen without consideration of the diluents.

Although bitumen itself does not trigger obvious corrosion of pipelines, that does not indicate that the transportation of dilbit by pipelines is absolutely safe. As mentioned in the introduction, corrosion of pipelines can occur under deposits that may contain aqueous chloride droplets. It is still necessary to improve the understanding of dilbit pipeline corrosion, especially its internal corrosion, and to consider other factors, such as microbially induced corrosion.

5. Conclusions

In this study, four types of inclusions (Mg−Al−Si−O−Ca−S, Mg−Al−O−Ca−S, Mg−Al−O−Ca−Mn−S, and Al−O−Ca−S) were detected in the API-X100 pipeline steel based on the elemental composition of the inclusions. The Mg−Al−Si−O−Ca−S inclusions consist of MgO, Al_2O_3, SiO_2, CaO, and CaS, whereas SiO_2 is not observed in Mg−Al−O−Ca−S. The Ca and Mn in Mg−Al−O−Ca−Mn−S are bound with S, while the Al and Mg are bound with O, demonstrating a separation within the same particle, i.e., into (Ca, Mn)S and (Mg, Al)O. Further, Al−O−Ca−S does not show either MgO or SiO_2. An ex situ SEM/EDS examination of an identical area on API-X100 pipeline steel before and after exposure to bitumen indicates that neither the matrix nor the inclusions corrode at 60 or 120 °C. Insufficient understanding of the corrosivity of bitumen exacerbates the debate around new pipeline construction. The finding herein of the non-corrosive nature of bitumen in dilbit pipeline conditions provides information to policy makers, key stakeholders and citizens, who determine whether to build new dilbit pipelines.

Author Contributions: Investigation, H.L.; Methodology, H.L.; Project administration, E.A.; Supervision, E.A.; Writing—original draft, H.L.; Writing—review & editing, E.A. All authors have read and agreed to the published version of the manuscript.

Funding: UBC Pipeline Integrity Institute.

Institutional Review Board Statement: Not applicable.

Informed Consent Statement: Not applicable.

Data Availability Statement: All the data is available within the manuscript.

Acknowledgments: The authors acknowledge the financial support from the UBC Pipeline Integrity Institute. The corresponding author wishes to thank the China Scholarship Council for the financial support.

Conflicts of Interest: The authors declare no conflict of interest.

References

1. National Energy Board. Market Snapshot: Oil Sands Bitumen Production Will Continue to Grow to 2040. 2018. Available online: https://www.nebone.gc.ca/nrg/ntgrtd/mrkt/snpsht/2018/01-05lsndsbtmn-eng.html (accessed on 30 September 2019).
2. Rosa, L.; Davis, K.F.; Rulli, M.C.; D'Odorico, P. Environmental consequences of oil production from oil sands. *Earth's Future* **2017**, *5*, 158–170. [CrossRef]
3. Verma, A.; Nimana, B.; Olateju, B.; Rahman, M.; Radpour, S.; Canter, C.; Subramanyam, V.; Paramashivan, D.; Vaezi, M.; Kumar, A. A techno-economic assessment of bitumen and synthetic crude oil transport (SCO) in the Canadian oil sands industry: Oil via rail or pipeline? *Energy* **2017**, *124*, 665–683. [CrossRef]
4. Vaillancourt, K.; Alcocer, Y.; Bahn, O. An analysis of the impacts of new oil pipeline projects on the Canadian energy sector with a TIMES model for Canada. In *Informing Energy and Climate Policies Using Energy Systems Models*; Springer: Cham, Swizerland, 2015; pp. 247–260.
5. Bowles, P.; MacPhail, F. The town that said "No" to the Enbridge Northern Gateway pipeline: The Kitimat plebiscite of 2014. *Extr. Ind. Soc.* **2017**, *4*, 15–23. [CrossRef]
6. Hoberg, G. The Political Economy of Pipelines: The Battle over Oil Sands Access to Tidewater. *Can. Public Policy* **2013**, *39*, 371–391. [CrossRef]
7. Corkal, V. *Pipelines or Progress: Government Support for Oil and Gas Pipelines in Canada*; International Institute for Sustainable Development: Geneva, Switzerland, 2021.
8. Dusyk, N.; Axsen, J.; Dullemond, K. Who cares about climate change? The mass media and socio-political acceptance of Canada's oil sands and Northern Gateway Pipeline. *Energy Res. Soc. Sci.* **2018**, *37*, 12–21. [CrossRef]
9. Bakker, G. *The Corrosive Nature of Diluted Bitumen and Crude Oil Literature Review*; MR Gordon and Associates, Ltd.: North Vancouver, BC, Canada, 2011.
10. Place, T.; Holm, M.R.; Cathrea, C.; Ignacz, T. Understanding and Mitigating Under-Deposit Corrosion in Large Diameter Crude Oil Pipelines: A Progress Report. In Proceedings of the 2008 7th International Pipeline Conference, Calgary, AB, Canada, 29 September–3 October; American Society of Mechanical Engineers: New York, NY, USA, 2018; pp. 809–821.
11. Liang, H.; Schaller, R.F.; Asselin, E. Three phase corrosion of pipeline steel: Size effects of deposited solids under water droplets and an oil diffusion barrier. *J. Pipeline Sci. Eng.* **2021**, *1*, 137–147. [CrossRef]
12. Liang, H.; Schaller, R.F.; Asselin, E. The Effects of Chloride Droplet Properties on the Underoil Corrosion of API X100 Pipeline Steel. *Corrosion* **2019**, *75*, 1051–1064. [CrossRef]
13. Liang, H.; Liu, J.; Alfantazi, A.; Asselin, E. Corrosion behaviour of X100 pipeline steel under a salty droplet covered by simulated diluted bitumen. *Mater. Lett.* **2018**, *222*, 196–199. [CrossRef]
14. Liang, H.; Schaller, R.F.; Asselin, E. Aqueous Corrosion of Deformed Steel Under Simulated Diluted Bitumen. *Corrosion* **2019**, *75*, 1194–1206. [CrossRef]
15. Liang, H.; Liu, J.; Schaller, R.F.; Asselin, E. A New Corrosion Mechanism for X100 Pipeline Steel Under Oil-Covered Chloride Droplets. *Corrosion* **2018**, *74*, 947–957. [CrossRef]
16. Swift, A.; Shope, E.; Casey-Lefkowitz, S.; Club, S. *Tar sands Pipelines Safety Risks*; Natural Resources Defense Council: New York, NY, USA, 2011.
17. Warmore, S. Tar Sends Oil and Pipeline Safety: Examining Regulatory Shortcomings. *Wayne L. Rev.* **2013**, *59*, 175.
18. Slavcheva, E.; Shone, B.; Turnbull, A. Review of naphthenic acid corrosion in oilrefining. *Br. Corros. J.* **1999**, *34*, 125–131. [CrossRef]
19. Dilbit Corrosivity. Canadian Energy Pipelines Association. *State of the Art Report*; Document Number: 12671-RPT-001 REV 1; Penspen Integrity: Newcastle, UK, 2013; pp. 1–39.
20. McIntyre, D.R.; Achour, M.; Scribner, M.E.; Zimmerman, P.K. Laboratory tests comparing the corrosivity of dilbit and synbit with conventional crudes under pipeline conditions. In Proceedings of the Corrosion 2014, San Antonio, TX, USA, 9–13 March 2014. Paper no. 3824, NACE.
21. Winter, C.H.; Sillers, R.; Glowach, A.M. High-temp., insulated coating aids construction of Alberta bitumen pipeline. *Oil Gas J.* **2003**, *101*, 56.
22. Atkinson, H.; Shi, G. Characterization of inclusions in clean steels: A review including the statistics of extremes methods. *Prog. Mater. Sci.* **2003**, *48*, 457–520. [CrossRef]
23. Al-Mansour, M.; Alfantazi, A.; El-Boujdaini, M. Sulfide stress cracking resistance of API-X100 high strength low alloy steel. *Mater. Des.* **2009**, *30*, 4088–4094. [CrossRef]
24. Zhu, H.; Zhao, J.; Li, J.; Hu, Q.; Peng, C. Evolution of nonmetallic inclusions in pipeline steel during LF and VD refining process. *High Temp. Mater. Process.* **2020**, *39*, 424–432. [CrossRef]
25. Dong, C.; Liu, Z.; Li, X.; Cheng, Y. Effects of hydrogen-charging on the susceptibility of X100 pipeline steel to hydrogen-induced cracking. *Int. J. Hydrog. Energy* **2009**, *34*, 9879–9884. [CrossRef]
26. Jin, T.; Cheng, Y. In situ characterization by localized electrochemical impedance spectroscopy of the electrochemical activity of microscopic inclusions in an X100 steel. *Corros. Sci.* **2011**, *53*, 850–853. [CrossRef]
27. Peng, X.Y.; Liang, G.C.; Jin, T.Y.; Cheng, Y.F. Correlation of initiation of corrosion pits and metallurgical features of X100 pipeline steel. *Can. Met. Q.* **2013**, *52*, 484–487. [CrossRef]

28. Li, Y.; Liu, J.; Deng, Y.; Han, X.; Hu, W.; Zhong, C. Ex situ characterization of metallurgical inclusions in X100 pipeline steel before and after immersion in a neutral pH bicarbonate solution. *J. Alloys Compd.* **2016**, *673*, 28–37. [CrossRef]
29. Arafin, M.; Szpunar, J. Effect of bainitic microstructure on the susceptibility of pipeline steels to hydrogen induced cracking. *Mater. Sci. Eng. A* **2011**, *528*, 4927–4940. [CrossRef]
30. Yang, Z.; Kan, B.; Li, J.; Su, Y.; Qiao, L.; Volinsky, A.A. Pitting Initiation and Propagation of X70 Pipeline Steel Exposed to Chloride-Containing Environments. *Materials* **2017**, *10*, 1076. [CrossRef] [PubMed]
31. Derungs, W.A. Naphthenic Acid Corrosion—An Old Enemy of the Petroleum Industry. *Corrosion* **1956**, *12*, 41–46. [CrossRef]
32. Ayello, F.; Robbins, W.; Richter, S.; Nešić, S. Crude oil chemistry effects on inhibition of corrosion and phase wetting. In Proceedings of the CORROSION 2011, Houston, TX, USA, 13–17 March 2011. Paper no. 11060, NACE.
33. Ramirez, M.; Perdomo, J.; Morales, J.L.; Viloria, A. Effect of crude oil contaminants on the internal corrosion in gas pipelines. In Proceedings of the CORROSION 2000, Orlando, FL, USA, 26–31 March 2000. Paper no. 40, NACE.

materials

MDPI

Article

Microstructure, Mechanical Properties, and Corrosion Behavior of Ultra-Low Carbon Bainite Steel with Different Niobium Content

Yun Zong [1,2,3,*] and Chun-Ming Liu [1,3,*]

1 Key Laboratory for Anisotropy and Texture of Materials, Ministry of Education, Northeastern University, Shenyang 110819, China
2 School of Mechanical and Automobile Engineering, Qilu University of Technology (Shandong Academy of Sciences), Jinan 250353, China
3 School of Materials Science and Engineering, Northeastern University, Shenyang 110819, China
* Correspondence: yun_zong@qlu.edu.cn (Y.Z.); cmliu@mail.neu.edu.cn (C.-M.L.)

Abstract: Four types of ultra-low carbon bainite (ULCB) steels were obtained using unified production methods to investigate solely the effect of niobium content on the performance of ULCB steels. Tensile testing, low-temperature impact toughness testing, corrosion weight-loss method, polarization curves, electrochemical impedance spectroscopy (EIS), and the corresponding organizational observations were realized. The results indicate that the microstructure of the four steels comprise granular bainite and quite a few martensite/austenite (M/A) elements. The niobium content affects bainite morphology and the size, quantity, and distribution of M/A elements. The elongation, yield strength, and tensile strength of the four types of ULCB steels are above 20%, 500 MPa, and 650 MPa, respectively. The impact toughness of the four types of ULCB steels at −40 °C is lower than 10 J. Steel with Nb content of 0.0692% has better comprehensive property, and maximum charge transfer resistance in 3.5 wt.% NaCl solution at the initial corrosion stage. The corrosion products on the surface of steel with higher niobium content are much smoother and denser than those steel with lower niobium content after 240 h of corrosion. The degree of corrosion decreases gradually with the increase of niobium content at the later stage of corrosion.

Keywords: ultra-low carbon bainite steel; niobium microalloying; mechanical property; corrosion resistance; corrosion behavior

Citation: Zong, Y.; Liu, C.-M. Microstructure, Mechanical Properties, and Corrosion Behavior of Ultra-Low Carbon Bainite Steel with Different Niobium Content. *Materials* **2021**, *14*, 311. https://doi.org/10.3390/ma14020311

Received: 19 November 2020
Accepted: 4 January 2021
Published: 9 January 2021

1. Introduction

Ultra-low carbon bainite (ULCB) steel is a kind of multipurpose steel that has attracted wide attention owing to its outstanding comprehensive properties such as high strength and toughness, better weldability, and other characteristics [1–8]. The performance of the steel is directly relevant to the composition, production process, and corresponding microstructure [9]. The composition of ULCB steel does not require more carbon and alloying elements to maintain strength. Its superior performance can be obtained by a low-carbon microalloying composition and advanced production methods such as controlled rolling [10,11]. The strength of steel mainly depends on dislocation strengthening in bainite structure, fine crystal strengthening, precipitation strengthening and micro-alloy strengthening. The design of lower carbon content and the lower total amount of alloys results in an excellent comprehensive performance and overcomes the contradiction between material performance, cost, profit, and energy consumption. Currently, low-carbon bainitic steel is the choice of steel for offshore platforms [12].

When ULCB steel is used in the construction of offshore platforms under marine environment or at low temperatures, its superior mechanical and welding properties as well as certain seawater corrosion resistance are necessary. Some research has indicated that the strength, hardness, and plasticity of steel can be significantly improved, and the corrosion rate can be remarkably reduced by the addition of micro-alloyed elements [13,14]. Generally, appropriate amounts of copper, nickel, and chromium are added to ULCB steel for effectively improving the corrosion resistance [15–19]]. Niobium, an important micro-alloyed element, is also added for strength and improvement of the low-temperature toughness in steel for both marine engineering structures and liquefied petroleum gas (LPG) transport ship hulls [20]. Niobium is considered particularly critical because of its multiple effects on ULCB steel. Trace niobium can increase the strength, toughness, and ductility of steel by multiple mechanisms [21–24]. However, it has been reported that the effect of niobium content on steel weldability is on both sides [20,21]. The addition of niobium must be moderate in range. Moreover, the change in niobium content causes a change in the chemical composition, which significantly affects the phase transition behavior and the corresponding microstructure. The composition and microstructure play an important role for determining the corrosion susceptibility of steel [25–27]. However, the effect of niobium content on the seawater corrosion resistance of ULCB steel is still unclear. Because there are complex interactions between chemical composition, deformation, the phases formed, and microstructural unit sizes and homogeneity, the properties of steel are affected by several factors. Thus far, there have been relatively few studies on the influence of Nb content on the microstructure and properties of ULCB steel used in offshore platforms, excluding other factors. In addition, the correlation between niobium content and the extent to which it affects performance, particularly corrosion resistance, is ambiguous; furthermore, the niobium content appropriate for a favorable comprehensive performance in offshore platform steel remains unclear. The cost-effective design and production of value-added niobium-bearing offshore platform steel with an excellent comprehensive performance is in progress.

To investigate the microstructure, mechanical properties, seawater corrosion resistance, and corrosion behavior of ULCB steel with various niobium contents and to reveal the correlation between them, the design of low-carbon micro-alloyed composition with different Nb contents is adopted in this study. All steel samples were obtained by unified production and the preparation methods were intended to investigate the effect of niobium content only on the properties, excluding other factors. Then, the strength, plasticity, and impact toughness at −40 °C, as well as corrosion resistance in three and a half percent mass fraction (3.5 wt.%) NaCl solution were examined. In addition, the corresponding microstructure and surface morphology were observed.

This study provides a beneficial reference for the design and manufacture of low-carbon bainite steel and its application in full immersion marine environments.

2. Materials and Methods

2.1. Materials

Based on the low carbon content, trace niobium, titanium, copper, nickel, chromium, and molybdenum were added to ensure the acquisition of bainite structure and excellent comprehensive performance. To investigate the effect of Nb content on the microstructure and properties of low-carbon bainitic steel, the content of Nb in four types of steels was 0.035%, 0.062%, 0.084%, and 0.107%, respectively. Other elements were present in same quantities in all four samples. The chemical compositions of the four types of experimental steels with different niobium contents are listed in Table 1.

Table 1. The chemical compositions of the four types of experimental steels with different niobium contents (wt.%).

Steel	C	Si	Mn	P	S	Nb	Ni	Cr	Mo	Ti	Cu
No.8	0.047	0.20	1.59	0.009	0.006	0.035	0.615	0.413	0.249	0.015	0.29
No.9	0.05	0.21	1.55	0.008	0.006	0.069	0.596	0.410	0.252	0.014	0.30
No.10	0.05	0.19	1.54	0.008	0.007	0.084	0.600	0.430	0.240	0.011	0.343
No.11	0.05	0.197	1.56	0.009	0.007	0.107	0.582	0.450	0.241	0.013	0.336

The experimental steel was smelted in a 50 kg vacuum furnace (Made in Liaoning Jinzhou North Electric Furnace Equipment Co. Ltd, Jinzhou, China) and rolled on a 300 mill. The size of ingots was 80 mm × 80 mm ×200 mm, and the thickness of rolled plates was 14 mm. Multiple passes of high temperature cumulative deformation at 1050–950 °C were adopted, and the total reduction ratio was 82.5%. After rolling, the plates were cooled at 850–550 °C with a fast-cooling process, and then cooled to room temperature (25 ± 1 °C) in air.

2.2. Methods

Samples were taken from the four rolled steel plates. Tension samples and impact samples were machined into the size, as shown in Figure 1.

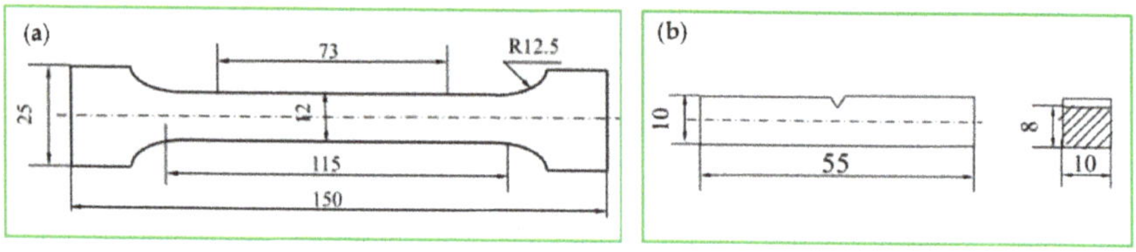

Figure 1. Size of samples for (**a**) tensile test and (**b**) impact test (mm).

Uniaxial tension tests were performed with a CMT5105 SANS microcomputer control electronic universal testing machine (New Sans (Shanghai) Enterprise Development Co., Ltd., Shanghai, China). The impact toughness at −40 °C was tested based on the standard of the American Society for Testing Material (ASTM) E2298 [28]. The Vickers hardness (HV) was measured using HVS-50 (Laizhou Huayin Experimental Instrument Co., Ltd., Laizhou, China) under 98 N load with 15 s dwell time according to ASTM E384 [29]. Three parallel samples were utilized to test the same observation point.

The sample size for the polarization curve and impedance testing was machined into 10 mm × 10 mm × 2 mm specimen. The sample size for the corrosion weight-loss experiment was machined into 25 mm × 10 mm × 2 mm. All samples were ground up to 1500 grit, washed with deionized water and alcohol by ultrasonic sound, degreased with acetone, and dried in a desiccator.

The electrochemical tests were conducted using a three electrode system. The experimental steel sample, platinum-coated niobium mesh electrode, and saturated calomel electrode (SCE) were the working, auxiliary, and reference electrodes, respectively. A copper wire was welded on the back of each polarization curve and impedance test sample as an electrical connection, exposing a square (10 mm × 10 mm) surface on the front. The rest of the sample surface was sealed with paraffin. Samples were completely immersed and suspended in 3.5 wt.% NaCl solution at room temperature (25 ± 1 °C). The electrochemical tests were conducted using an electrochemical workstation (Interface 1000, Gamry Instruments, Inc., Warminster, PA, USA). The electrochemical impedance spectroscopy (EIS) results were obtained by employing an AC voltage with an amplitude of 10 mV in the frequency range 0.01–100,000 Hz. The test parameters of the polarization curve ranged from −1 mV to +1 mV, and the scanning rate was 0.5 mV·s^{-1}.

The corrosion weight-loss method was used to determine the corrosion rates and study the kinetics of corrosion. Corrosion samples of all four types of steels underwent 24 h, 72 h, 168 h, 240 h, and 360 h of corrosion, respectively. The corrosion samples were grouped according to the experimental scheme. The corrosion samples were grouped according to the experimental scheme. Four samples were taken from each observation point, three of which were used for the weight loss test, and the other one was used for scanning electron microscopy to observe the surface morphology of rust layer. The samples were weighed before soaking. Then, the sample was suspended vertically in a solution with an insulating plastic cord. The samples were separated from each other at a certain distance, each sample was kept on almost the same level and immersion time was recorded. During corrosion, the macroscopic morphological changes of the sample surface were observed. After the samples were taken out, the loose corrosion products on the sample surface were brushed with a bristle brush. Then, three parallel samples were selected from each group and polished with 800# sandpaper. The remaining corrosion products on the surface were removed without damaging the sample matrix. After the corrosion products were removed, ultrasonic cleaning with deionized water was performed for five minutes to remove surface impurities. They were then cleaned and dehydrated with anhydrous ethanol. Finally, the samples were blown dry and weighed.

The corrosion rate can be calculated as follows [30,31]:

$$v^{-} = (m_0 - m_1)/S \times t \quad (1)$$

where v^{-} is the corrosion rate of the experimental steel ($g \cdot m^{-2} \cdot h^{-1}$); m_0 denotes the weight of the sample before corrosion (g); m_1 refers to the weight of the sample after rust removal (g); t is the corrosion time (h); and S is the area of sample surface in corrosion test (m^2).

The metallographic structures and surface morphology of the corrosion products on the experimental steel were made by optical microscopy LEICA DM2700M (Leica Microsystems Inc., Wetzlar, GER) and SEM QUANTA 200 (FEI Company of the Netherlands, Eindhoven, The Netherlands).

3. Results

3.1. Microstructure of Four Types of Experimental Steels

Metallographic structures of the four types of niobium bearing steels are shown in Figure 2. The microstructures of the four kinds of ULCB steel mainly consist of granular bainite, and a certain number of martensite/austenite (M/A) components are distributed on the matrix. With increase in niobium content, the grain shape of bainite changes from a multilateral shape to an angular shape, and the size reduces. The content, size, and distribution of M/A elements are different for the four types of steels. In steel with lower niobium content, the M/A components are less in number and larger in size. However, the M/A components in steel with higher Nb content, are more abundant and smaller in size.

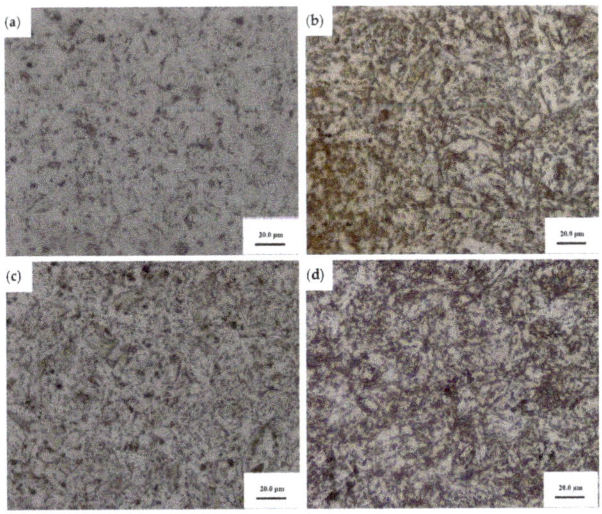

Figure 2. Metallographic structures of the four types of niobium bearing steels (**a**) No. 8 steel (**b**) No. 9 steel (**c**) No. 10 steel (**d**) No. 11 steel.

The content, size, and distribution of the M/A components depend on the composition of steel, cooling temperature, cooling strength, and other deformation processes. A large amount of M/A structures in low-alloy steel destroys the integrity of steel, affects the properties of steel, and is detrimental to the toughness and plasticity of steel. In addition, the M/A components also affect the corrosion resistance of bare steel.

3.2. Mechanical Properties of Four Types of Experimental Steels

The mechanical properties of the four types of ULCB steels are shown in Figure 3. The four types of ULCB steels in as-rolled state have higher strength and better plasticity but lower impact toughness at −40 °C. The elongation (δ), yield strength (Rel), and tensile strength (R_m) of the three performance indexes are above 20%, 500 MPa, and 650 MPa, respectively. However, the Charpy V-Notch (CVN) impact toughness at −40 °C (A_{KV} (−40 °C)) of the four types of experimental steels in the as-rolled state did not vary significantly with the change in niobium content and is lower than 10 J. The yield strength of No. 8 steel with an average of 507 MPa is the lowest. The yield strength of No. 11 steel is the highest, reaching 555 MPa. The yield strength of No. 10 steel is similar to that of No. 9, which is approximately 535–537 MPa. This suggests that the Rel of all experimental steels increases with niobium content. No. 10 steel has the highest tensile strength, exceeding 700 MPa. The tensile strength of steel No. 8, 9, and 10 increases in turns. However, the tensile strength of steel No. 11 with a niobium content of 0.107% is lower than that of No. 10 with a niobium content of 0.084%.

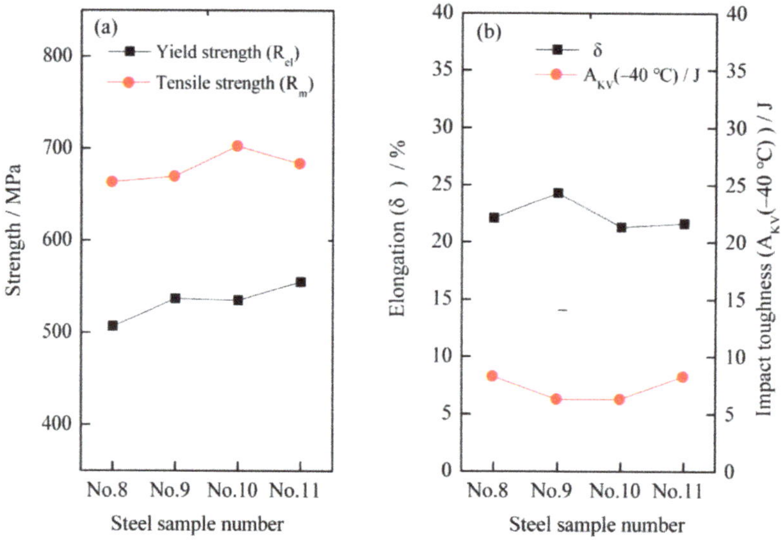

Figure 3. Mechanical properties of four types of ULCB steels (**a**) Yield strength (R_{el}) and Tensile Scheme (R_m) (**b**) Elongation (δ) and −40 °C impact toughness (A_{KV} (−40 °C)).

The elongation of steel No. 8 and 9 with lower niobium content is higher than that of No. 10 and 11 with higher niobium content. The elongation of steel No. 9 exceeds 24%, which is the largest among the four types of experimental steels.

Evidently, the niobium content has a vital influence on the strength and plasticity of steel; however, the effect of niobium content on the strength and plasticity is not monotonic, but complex. Therefore, an optimal value of niobium content exists for both strength and plasticity. Among the four experimental steels, No. 9 with 0.0692% niobium content has a better comprehensive property of strength and elongation.

The low-temperature toughness of the low-carbon bainite steel in the as-rolled condition is not ideal in a low-temperature resistance environment. To obtain a perfect comprehensive performance, appropriate heat treatment must be adopted, particularly for steel plates with large thicknesses. The reason is that if only thermal mechanical rolling is used to produce this type of steel, it is liable to be limited by the capacity of the rolling mill equipment and the cooling capacity of the cooling equipment. Under these circumstances, it is difficult to ensure sufficient pass pressure and cooling rate. Consequently, it can easily cause uneven microstructure in the thickness and damage the strength, lamellar tearing resistance, and low temperature toughness. Therefore, the application of reasonable heat treatment is indispensable to produce large-thickness ULCB steel plates.

3.3. The Initial Corrosion Behavior and Corrosion Resistance of Four Experimental Steels in 3.5 wt.% NaCl Aqueous Solution

The polarization curves of the four types of experimental steels exposed to 3.5 wt.% NaCl aqueous solution in the initial soaking stage are shown in Figure 4. The polarization curves of the four types of steels have similar characteristics. With the increase in current density, the anode precipitation potential increases and the cathode precipitation potential decreases.

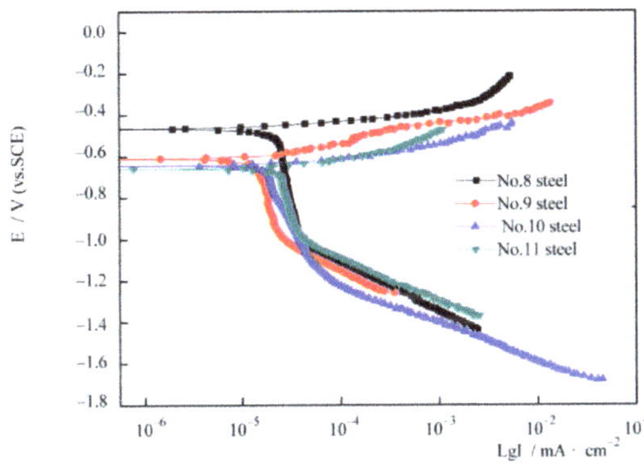

Figure 4. Polarization curves of four types of experimental steels exposed to 3.5 wt.% NaCl aqueous solution.

The polarization curve was fitted using Tafel's equation [31]. The electrochemical parameters obtained by fitting are listed in Table 2. It is revealed that the corrosion potential (E_0) becomes more negative with the increase in Nb content. The values of corrosion current density (I_0) of the four types of steels are of the same magnitude order, and the difference is not evident. The I_0 values of steel No. 8 and 11 are slightly smaller than those of steel No. 9 and 10. Steel No. 9 with 0.0692% Nb has the smallest corrosion current density compared to other experimental steels in the initial corrosion stage.

Table 2. Fitting results of polarization curves.

Steel No.	No.8	No.9	No.10	No.11
I_0 (Amp/cm^2)	2.32×10^{-5}	1.57×10^{-5}	1.71×10^{-5}	2.28×10^{-5}
E_0 (V)	−0.4578	−0.6084	−0.645	−0.665

The cathodic part of the polarization curve is due to the limiting current density determined by the oxygen reduction reaction. Oxygen plays an important role in cathodic corrosion [30].

There are two evident inflection points on the cathode section of the polarization curve. The cathode oxygen reduction is divided into three stages. In the first stage, the potential of the metal is positive, and the cathode reaction is controlled by the activation polarization of oxygen. In the second stage, the metal potential is negative, and the cathode reaction changes from the oxygen activation polarization control to oxygen diffusion control. The rate of corrosion is limited by the rate of oxygen diffusion. In the third stage, the potential of the metal is more negative, and the cathode reaction of the metal material changes from the oxygen concentration diffusion control to hydrogen evolution activation control.

In the initial stage, the corrosion rate of bare steel is higher. This is because there is abundant dissolved oxygen in the corrosion solution, and the bare steel surface is in direct contact with the corrosion solution. At this stage, the electrochemical reaction is active, so the corrosion rate is higher. Then, the electrochemical reaction proceeds continuously. The dissolved oxygen is constantly consumed and reduces gradually. Thus, the cathode reaction is inhibited, and the corrosion rate falls rapidly. After that, more corrosion products are produced on the bare steel surface that prevent the sample surface from contacting the corrosion solution, which further inhibits the corrosion [19]. Therefore, the corrosion rate decreases gradually.

3.4. Electrochemical Impedance Spectroscopy of Four Types of ULCB Steels in 3.5 wt.% NaCl Solution

EIS was applied to explore the corrosion kinetics of the four types of ULCB steels. The Nyquist and Bode plots of the four types of ULCB steels are presented in Figure 5. As shown in the Nyquist plots, the impedance spectra of the four types of ULCB steels have similar shapes, demonstrating that they have the same corrosion mechanisms and processes in the initial stage. All the Nyquist plots consist of a clear capacitive reactance arc. This indicates that activation controls the corrosion reaction. The high-frequency part in the capacitive reactance arc reflects the impedance information between the corrosion solution and corrosion products generated on the steel surface, which corresponds to the double layer behavior and the change in solution resistance (R_s) at the original reaction interface of the four types of steel substrates. The low-frequency part in the capacitive reactance arc shows the impedance information of the steel electrode surface; hence, the low-frequency part reflects the corrosion behavior of the electrode surface.

Among the four types of ULCB steels, the steel with the smallest capacitive arc reactance radius is the steel No. 8, which has the lowest niobium content, and the steel with the largest capacitive arc reactance radius is the steel No. 9. Generally, a large capacitive arc diameter represents a large charge transfer resistance (R_t). R_t describes the interfacial resistance between the steel and the corrosion product layer, which is the final barrier that prevents chlorine ions from reaching the metal surface [32]. A large capacitive arc diameter means that corrosion products have a strong protective ability, and the degree of metal dissolution reaction is weak [19,33–35]. Steel No. 9 with 0.0692% niobium has the largest arc reactance radius compared to other experimental steels in the initial corrosion stage. Consequently, the corrosion resistance of steel No. 9 is the best in the initial corrosion stage.

The Bode plots reflect the relationship between the impedance modulus, phase angle, and frequency. Figure 5b shows the characteristics of the relationship between the impedance modulus and frequency. The impedance modulus measured in the highest frequency region in the Lg Z—Lg ω Bode plots represents R_s. The impedance modulus measured in the lowest frequency region represents the sum of R_s and R_t. The information shown in Figure 5b is consistent with that shown in Figure 5a. The solution resistance R_s changes slightly; however, the sum of R_s and R_t gradually increases with increasing niobium content, except for steel No. 9. The (R_s + R_t) of steel No. 9 is the largest. R_t is inversely proportional to the degree of electrochemical reaction [30,36]. Therefore, Figure 5b also shows that, in steel No. 8, 10, and 11, the charge transfer resistance of the corrosion reaction increases, and the degree of corrosion reaction decreases with increasing niobium content.

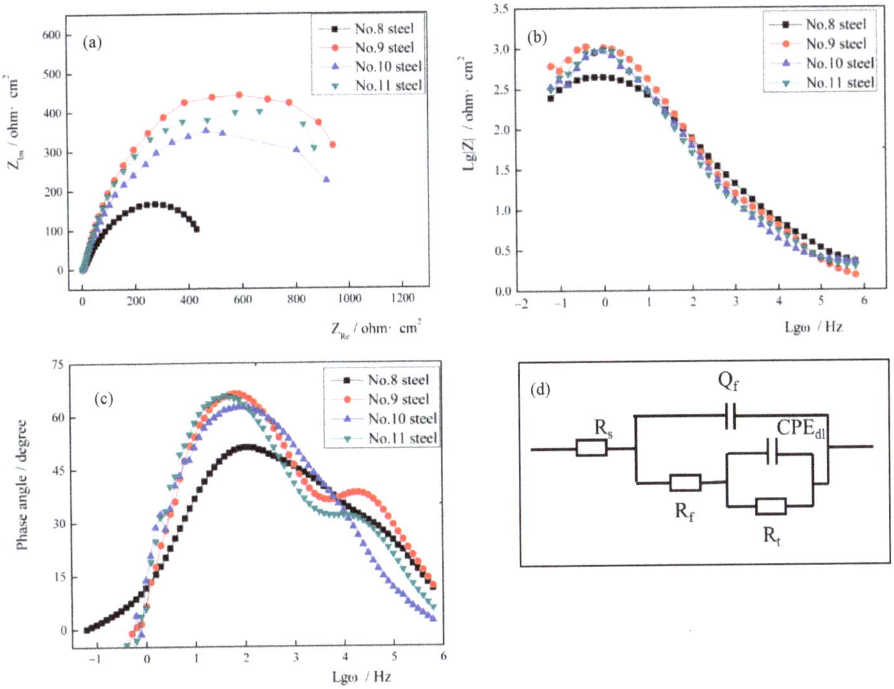

Figure 5. (**a**) Nyquist plots, (**b**) Lg Z—Lg ω Bode plots and (**c**) phase angle—Lg ω Bode plots of four types ULCB steels samples with different content of niobium in 3.5 wt.% NaCl solution and (**d**) equivalent circuit to fitting EIS data (Here, Rs represents solution resistance; Cc and Rc refer to the capacitance of corrosion products film and its resistance, respectively; Cd is the capacitance of double electrode layer, and Rt stands for charge transfer resistance).

The phase angle—Lg ω curves of the four ULCB steels still have similar characteristics. A distinct peak exists in the low-frequency part and a small hump in the middle and high frequency parts. The small hump of steel No. 9 and 11 in the middle and high frequency areas are more evident than those of steel No. 8 and steel No. 10. The corrosion process appears as two-time constants. This implies that the electrochemical reaction of the four types of steels was affected by two state variables, the double electric layer and corrosion products. The equivalent circuit for the test steel is illustrated in Figure 5d. Here, R_s represents the solution resistance; C_c and R_c refer to the capacitance of the corrosion product film and its resistance, respectively; C_d is the capacitance of the double electrode layer; and R_t stands for the charge transfer resistance.

The charge transfer resistances of steel No. 8, 9, 10, and 11 calculated analytically after fitting are 523.3, 1148.0, 1090.4, and 1116 ohm·cm^2, respectively. The fitting results are consistent with the curve characteristics presented in the Bode plots. R_t values increase with increasing niobium content. It is further proved that, except for steel No. 9, the charge transfer resistance of other steels increases, and the degree of corrosion tends to decrease with increasing niobium content. Steel No. 9 with 0.0692% Nb has the largest charge transfer resistance compared to other experimental steels in the initial corrosion stage. This conclusion is consistent with the conclusion drawn from the polarization curve.

3.5. Corrosion Rate Curves and Surface Corrosion Morphology of Four Experimental Steels

The weight-loss method is the most basic and effective quantitative evaluation method in the study of corrosion resistance of materials. The corrosion rate curves of the four types of bainite steels in 3.5 wt.% NaCl solution based on gravimetric experiment results are demonstrated in Figure 6a. The corrosion phenomenon observed in steel No.10 and 11 was soaked in 3.5 wt.% NaCl solution for 168 h is shown in Figure 6b,c. As shown in Figure 6a, the corrosion rates of the four types of ULCB steels are approximately the same in general trend. Within 0–360 h of corrosion time, the corrosion rate of the four types of steels is relatively high during the initial immersion stage. With the increase in immersion time, the corrosion rate gradually decreases and becomes stable; then, the corrosion rate increases slightly. Although the overall corrosion rate trend is similar, the time nodes of the corrosion rate valley values of the four types of steels are different.

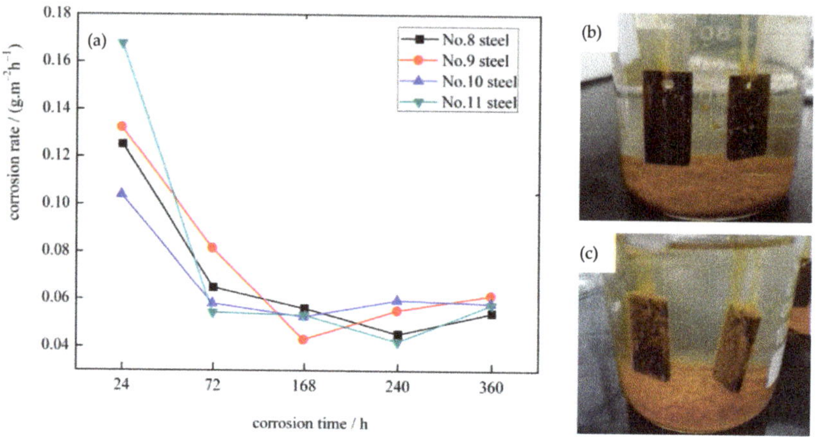

Figure 6. (**a**) The corrosion rates curves of four types bainite steels in 3.5 wt.% NaCl aqueous solution and corrosion phenomenon observed of (**b**) No. 10 steel and (**c**) No. 11 steel was soaked in 3.5 wt.% NaCl solution with 168 h.

Within the immersion time range of this experiment, the corrosion rates of the four types of steels reached the maximum after soaking for 24 h. At this point, the corrosion rate of steel No.11 was the highest, and that of No.10 was the lowest. After soaking for 72 h, the corrosion rates of the four types of steels decreased significantly, among which the lowest one was steel No.11. The corrosion rates of steel No.9 and 10 reached the minimum after 168 h of immersion. After soaking for 240 h, the corrosion rates of steel No.8 and 11 reached the minimum. Meanwhile, the corrosion rates of steel No.9 and 10 increased slightly. After soaking for 360 h, the corrosion rates of steel No.8 and 11 increased slightly, and those of steel No.9 and 10 tended to be stable. The corrosion rates of the four types of steels showed slight difference when immersed for 360 h.

During the experiment, it was observed that the solution in the beaker turned slightly yellow after the sample was soaked for approximately 1 h. After 24 h of corrosion, corrosion precipitation clearly contained in the beaker. The solution had completely turned yellow. Distinct corrosion occurred on the surface of the sample. After 168 h of corrosion, the surface of the sample was full of corrosion products, and the bottom of the beaker was full of corrosion precipitate, as shown in Figure 6b,c. If the beaker shakes slightly, some corrosion products on the sample surface fall off and leak out the corrosion morphology of the sample surface.

The surface corrosion morphologies of the four types of steel samples after 240 h of corrosion are shown in Figure 7. It is observed that the surface corrosion products of steel No.10 and 11 with higher niobium content are much smoother and denser than those of steel No.8 and 9 with lower niobium content when the experimental steels undergo corrosion for 240 h, on the whole. In particular, the white corrosion products on the surface of steel No.8 are uniformly distributed as islands and are denser. There is more granular material, and the protrusion is more evident, as shown in Figure 7a. After further magnification, the white corrosion products of steel No.8 are in the form of debris, and there are evident cracks in the middle of the corrosion products with a large number of cracks and local corrosion pits, as shown in Figure 7e. In steel No.9, there are significantly fewer white corrosion products and lower convex height of granular material compared to steel No.8. A small number of short shallow cracks and local corrosion pits are present in the corrosion products, as shown in Figure 7b,f. As shown in Figure 7c,g, the surface of the corrosion products of steel No.10 is relatively smooth and dense, and cracks are almost invisible; the number and size of white corrosion products are small, and the protrusions are not evident. The corrosion product of steel No.11 is slightly loose than that of steel No.10. As shown in Figure 7d,h, there are no large white corrosion products on the surface of steel No.11; only a few small white spots exist, and no cracks can be seen in the corrosion products.

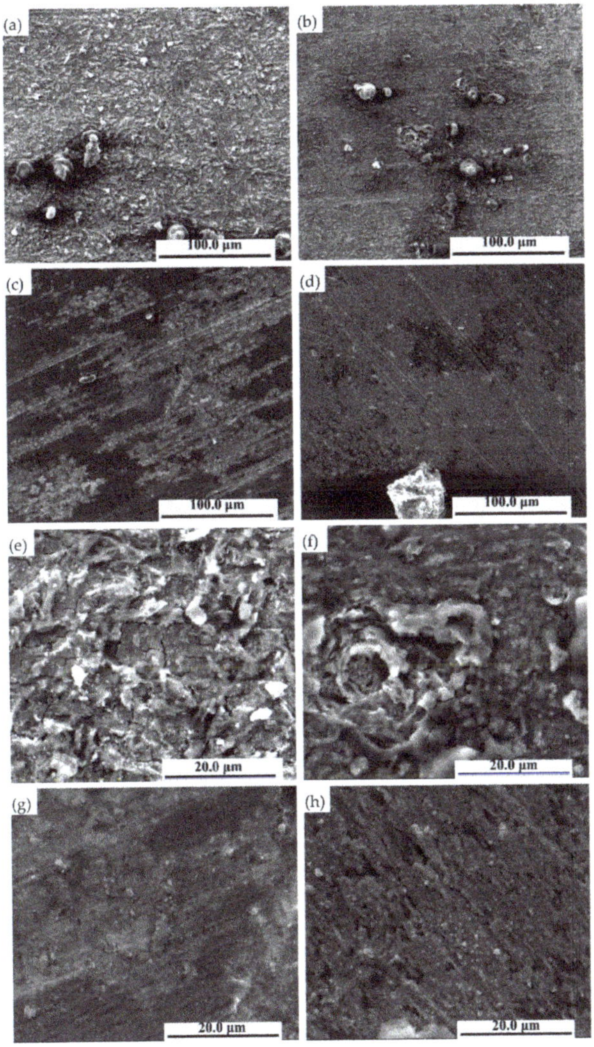

Figure 7. Surface corrosion morphology of No. 8 steel (**a**,**e**), No. 9 steel (**b**,**f**), No. 10 steel (**c**,**g**) and No. 11 steel (**d**,**h**) after 240 h corrosion.

4. Discussion

4.1. Correlation between Microstructure, Mechanical Properties, Corrosion Behavior of ULCB Steel, and Nb Content

From the experimental results, it can be concluded that niobium content plays a vital role in the microstructure, mechanical properties, and corrosion properties of ULCB steel.

Nb microalloying can be beneficial for achieving finer grain structure. The equilibrium dissolution temperature of the carbonitride of niobium precipitation can be calculated by the following equation [37]:

$$\log [\mathrm{Nb}][\mathrm{C} + 12\mathrm{N}/14] = 2.26 - 6\,770/\mathrm{T} \tag{2}$$

where [Nb], [C], and [N] are the elemental contents of niobium, carbon, and nitrogen (mass %), respectively, and T refers to the temperature (K).

The calculated equilibrium dissolution temperature of the carbonitride of niobium precipitation of the four types of steels is approximately 1076.3 °C, 1160.5 °C, 1186.6 °C, and 1220.8 °C, respectively. The calculated results exceed the initial rolling temperature of 1050 °C. Niobium is precipitated in steel in the form of carbonitrides. Carbonitride of niobium precipitation can control the austenite grain size during reheating, retard recrystallization during deformation, and provide precipitation strengthening during cooling. In a certain range of niobium content, with the increase in niobium content, the precipitates increase and the effect of strengthening and toughening is excellent. However, too much niobium content leads to the coarsening of precipitates, which is not conducive to the improvement of strengthening and toughening.

ULCB steel with higher Nb content has evident advantages in inhibiting the coarsening of the original austenite grains and maintaining the grain uniformity, which is beneficial to the subsequent phase transition. Moreover, Nb addition is also more conducive to the generation of bainite microstructure. As the niobium content increases, the microstructure of ULCB steel undergoes transition from granular bainite to Lath bainite. The content and size of the M/A component in the microstructure also changes with niobium content.

These multiplex effects decrease the grain size of the microstructure after transformation and improve the strength and toughness of the steel [25]. It also compensates for the loss of strength arising from the lower carbon content of the ULCB steel.

Moreover, by adding trace Nb to improve the strength and toughness of materials, the use of other elements can be reduced. The carbon equivalent decreases accordingly. A low carbon equivalent ensures better toughness and excellent weldability. The influence of Nb content on the steel weldability is on both sides [21,22], and the appropriate content of niobium is critical. The amount of niobium added needs to be moderate in range. It may refer to the American Petroleum Institute (API) Specification (Spec) 5 L standard [38] for the amount of niobium, vanadium, and titanium in micro-alloyed pipeline steel. The maximum Nb content in micro-alloyed pipeline steel can reach 0.15% in the absence of vanadium and titanium. However, it has also been suggested that the niobium content limit in offshore steel is extremely strict compared to pipeline steel [20].

Furthermore, Nb addition can reduce the carbon content of ULCB steel, change the chemical composition, and affect the microstructure, thus directly or indirectly affecting the corrosion performance.

When the steel is immersed in seawater, galvanic corrosion is formed between the ferrite (anode) and cementite (cathode) phases in the steel matrix. The reduction of carbon content can hinder the formation of cementite (Fe_3C). Correspondingly, the associated galvanic corrosion associated with Fe_3C reduces, resulting in a lower corrosion rate. In particular, for offshore platform steels, the addition of alloying elements such as Cu, Ni, and Cr can effectively increase the corrosion resistance [16,17,19]. Changes in chemical composition can change the transformation behavior of steel and the corresponding microstructure. Composition and microstructure play an extremely important role in determining the corrosion susceptibility of steel [11,26,39]. Metal composition can influence the first kinetically controlled corrosion phase [40].

Trace Nb affects the corrosion process of ULCB steel at different stages. At the initial stage of corrosion, a certain amount of Nb can improve the corrosion resistance of low-carbon bainite steel. This is because Nb can improve the uniformity of steel and increase the grain boundary with a small angle. However, the corrosion process of steel is essentially an electrochemical process; therefore, the reaction rate depends largely on the potential difference between the anode and cathode. Improving the uniformity of the steel structure is equivalent to reducing the potential difference between the anode and cathode of the corroded battery, thus reducing the corrosion rate. In contrast, low-carbon bainite steel has low interfacial energy and relatively low impurity concentration; therefore, there is no evident interfacial corrosion between the laths.

At the beginning of bare steel corrosion, except for steel No.9, the steel No.8, 10, and 11, exhibited the trend of increase in arc radius of the Nyquist plots increased with increasing niobium content. This indicates that the charge transfer resistance in the corrosion reaction increases and the current generated in the corrosion reaction decreases with increasing niobium content.

With the increase of corrosion time, ULCB steel with higher niobium content has a more denser rust layer, which can hinder the migration of ions and slow down the corrosion of the internal matrix, thus improving the corrosion resistance.

The reason for steel No.9 to have the largest capacitive reactance radius may be that the corrosion reaction is related not only to the chemical composition of the steel itself, but also to the structure, surface, and environmental media of the steel. When the surface state and environmental medium are similar, the main influencing factors are the composition and microstructure. Steel in NaCl solution has both macro- and micro-corrosion activities. The change in microstructure has a great influence on the micro-corrosion. In the initial corrosion stage without effective rust layer protection, micro-corrosion accounts for a large proportion. Theoretically, in ULCB steel, ultra-low carbon content can decrease corrosion rates, as mentioned above. However, specific to different micro-alloyed steel, the change of trace niobium content in their composition is not only the change of composition itself, but also the corresponding change of steel microstructure. The most evident changes are the microstructure type, grain size and shape, precipitation, size and distribution of the carbo-nitrate, as well as the shape, size, and distribution of M/A components. These changes undoubtedly affect the micro-corrosion. Therefore, the effect of the niobium content change is not linear, but complex.

In summary, there is no doubt that the change in niobium content can significantly affect the mechanical properties and corrosion resistance at the same time when other technological factors remain unchanged during the steel production.

4.2. Relationship between Corrosion Behavior of the Four Types ULCB Steels with Different Nb Content and the Evolution of the Corrosion Product Layer

The corrosion behavior of the four types of steels is closely related to the evolution of the rust layer. As mentioned previously, the corrosion rate of the four types of steels is relatively high in the initial immersion stage and then gradually decreases and becomes stable. The rust layer generated on the metal surface and the thickness, density of the rust layer, and crack in layer can be responsible for interpreting this trend. The effect of corrosion products on the macroscopic corrosion process can be divided into two stages: at the early corrosion stage, the thin and loose corrosion products on the steel surface cannot protect the matrix effectively, corrosion rate increases with corrosion time. At the later corrosion stage, the formation of a compact, continuous, and uniform corrosion product layer can protect the steel substrate from direct contact with the corrosion solution.

When the experimental steel was immersed in the NaCl solution, a galvanic corrosion cell was formed. The metal dissolution reaction occurred in the anode region, and a low potential region was formed on the metal surface; oxygen reduction and reduction reaction of corrosion products occurred in the cathode region, and a high potential region was formed. At the early stage of corrosion, there was no protective layer on the surface, and the exposed steel plate contacted the corrosion liquid rich in oxygen and Cl^- directly.

Consequently, the dissolution of iron is fast, and the corrosion rate is exceedingly high. Then, a layer of corrosion products starts to form on the surface of the steel as the corrosion reaction progresses. At an early stage, the corrosion product layer is quite thin and loose. Dissolved oxygen in seawater passes through the corrosion product layer to reach the corrosive interface readily. At this point, the corrosion process is predominantly determined by the concentration of dissolved oxygen transferred from the solution medium to the metal surface. As more dissolved oxygen enters the metal interface to participate in the cathode reduction, the corrosion rate is elevated. The corrosion driving force is the potential difference between the anode and cathode on the surface at this stage. Therefore, the corrosion rate of the experimental steel in the initial immersion stage is relatively high.

When steel No.9 and 10 were soaked for 168 h, steel No.8 and 11 soaked for 240 h, the corrosion rate increased slightly. This phenomenon can be attributed to two reasons. One reason is that the rust layer formed at this time is too thin to prevent the entry of dissolved oxygen; inversely, its porous and loose structure provides a channel for dissolved oxygen to contact the metal matrix, so the corrosion rate increases. The other reason is the cracks generated in the corrosion products layer. At the early corrosion stage, the generated iron oxide generally maintains a certain coherent relationship with the steel matrix, reducing the surface energy and increasing the elastic strain energy of the interface. However, the corrosion product layer itself is brittle and difficult to deform. Therefore, cracks easily occur in the corrosion-producing layer, which provides a convenient channel for dissolved oxygen and Cl^- to further erode the matrix. Consequently, the thin, porous, and cracked rust layer has no protective effect, which is the main reason for an increase in the corrosion rate at this stage.

Combined with the surface corrosion morphology analysis, the sample steel No.8 showed a rougher surface and more white corrosion products than other steels after 240 h of corrosion in the simulated seawater, indicating that steel No.8 underwent serious corrosion in the corrosive liquid. From the corrosion morphology, the degree of corrosion of steel with higher niobium content is lower after 240 h of corrosion.

From the corrosion rate curve, steel No.11 with the highest niobium content has a faster corrosion rate at the initial corrosion stage. The main reason may be that steel No.11 has a small grain size and higher interfacial energy; thus, it shows a higher corrosion rate in the early stage. However, it can form a denser film that can inhibit further increases in the corrosion rate in a short time. The existing data indicate that the initial corrosion characteristics of steel are correlated with long-term corrosion behavior. If the bare steel corrodes uniformly and slowly, it may be beneficial to the tightness and stabilization of the corrosion product layer; that is, the corrosion behavior of bare steel is still transmissible after the formation of the rust layer. This may explain the reason for the highest initial corrosion rate of steel No.11 that reduces later.

For steel No.8 and 9, the corrosion rates of steel No.8 and 9 decreased rapidly after the formation of the corrosion product protective film at the early corrosion stage. The corrosion rates of steel No.8 and 9 minimized after 240 h and after 168 h, respectively, and there was a slight increase in the later stage. This can be explained by the sparse and scattered distribution of corrosion products and the existence of pores and cracks between the corrosion products, as shown in Figure 7.

The corrosion rates of the four types of bainitic steels gradually become stable after 360 h immersion, which is due to the formation of a protective corrosion product layer. With the increase in soaking time, the rust layer gradually covered the entire sample surface and the thickness gradually increased. The rust layer prevented Cl^- from contacting the metal matrix by setting obstacles in the transmission of dissolved oxygen to the matrix. Oxygen depolarization corrosion has been slightly inhibited. At the same time, the existence of a rust layer also prevents the iron ion from diffusing outward from the corrosion boundary. The increasing thickness of the surface rust slows down the diffusion process, eventually leading to a gradual decrease in the corrosion rate and a gradual reduction in the potential difference between the anode and cathode. The corrosion reaction rate of the corrosive liquid and bainite structure reached a dynamic balance with the inhibition rate of the corrosion layer, which inhibited further deepening of the corrosion phenomenon. At this time, a tight structure was beneficial for isolating the steel matrix from the corrosion solution.

The novelty of this work is the effect of niobium content on the microstructure, mechanical properties, and corrosion behavior of ULCB steel. It is well understood that both the composition and microstructure have a strong influence on the mechanical properties and corrosion behavior of steel. The present work studied the effect of niobium content on the microstructure, mechanical properties, and corrosion behavior of steel separately without the influence of the chemical composition of the steel and processing.

5. Conclusions

Four types of ULCB steels with different Nb contents were designed and produced. The correlation between the microstructure, mechanical properties, corrosion behavior of ULCB steel, and Nb content were investigated. The main conclusions drawn are as follows:

The microstructure of the four types of ULCB steels is mainly granular bainite with some martensite/austenite (M/A) components. The niobium content affects bainite morphology and the size, quantity, and distribution of M/A group elements.

All experimental steels in the as-rolled state have better plasticity and higher strength but lower impact toughness at −40 °C. The yield strength (R_{el}), tensile strength (R_m), and elongation (δ) of the four types of steels are all above 500 MPa, 650 MPa, and 20%, respectively. However, the impact toughness at −40 °C (A_{KV}) of the four types of experimental steels in the as-rolled state is lower than 10 J. Steel No.9 with 0.0692% niobium has a better comprehensive property of elongation and strength.

Based on the weight-loss experiment results, the corrosion rate of the four types of steels is relatively high during the early immersion stage. With the increase in corrosion time, the corrosion rate gradually decreases and becomes stable; then, the corrosion rate increases slightly.

Steel No.9 with 0.0692% niobium has the largest charge transfer resistance compared to other experimental steels at the initial corrosion stage according to the polarization curve and impedance spectrum of bare steel. The surface of steel with higher niobium content has smoother and denser corrosion products compared to the steel with lower niobium content after 240 h of corrosion. The degree of corrosion decreases gradually with the increase in niobium content at the later stage of corrosion can be drawn by combining the corrosion rate curve and the morphology of the corrosion products.

Author Contributions: Conceptualization: Y.Z.; methodology: Y.Z.; validation: Y.Z. and C.-M.L.; formal analysis: Y.Z.; writing–original draft preparation: Y.Z.; writing–review and editing: Y.Z. and C.-M.L.; supervision: C.-M.L.; project administration: Y.Z.; funding acquisition: Y.Z. All authors have read and agreed to the published version of the manuscript.

Funding: This work was funded by the National Natural Science Foundation of China (51401112).

Institutional Review Board Statement: Not applicable.

Informed Consent Statement: Not applicable.

Data Availability Statement: The data presented in this study are available on request from the corresponding author.

Acknowledgments: We gratefully acknowledge the support and help of Senior engineer Zhao Peilin, who is working at the Technical center of Shandong Laiwu Steel Group. We also acknowledge all those who have helped us during the writing of this thesis.

Conflicts of Interest: The authors declare no conflict of interest.

References

1. Wang, S.; Yang, S.; Gao, K.; Shen, X.; He, X. Corrosion Behavior and Variation of Apparent Mechanical Property of One Newly-Developed Low Carbon Bainitic Steel in Environment Containing Chloride Ion. *Acta Metall. Sin.* **2008**, *44*, 1116–1124.

2. Wang, S.C.; Kao, P.W. The Effect of Alloying Elements On the Structure and Mechanical Properties of Ultra Low Carbon Bainitic Steels. *J. Mater. Sci.* **1993**, *28*, 5169–5175. [CrossRef]
3. He, X.L.; Shang, C.J.; Yang, S.W.; Wang, X.M.; Guo, H. (Eds.) *High Performance Low-Carbon Bainite Steel: Composition, Process, Organization, Performance and Application*; Metallurgical Industry Press: Beijing, China, 2008; pp. 8–18.
4. Lan, L.; Qiu, C.; Zhao, D.; Gao, X.; Du, L. Microstructural Characteristics and Toughness of the Simulated Coarse Grained Heat Affected Zone of High Strength Low Carbon Bainitic Steel. *Mater. Sci. Eng. A* **2011**, *529*, 192–200. [CrossRef]
5. Song, F.Y.; Zhang, P.Y.; Wang, P.; Zhu, F.X. Effects of the Small Amount of Ferrite in Bainite Laths on the Strength and Toughness of High Strength Steels. *J. Northeast. Univ. Nat. Sci.* **2014**, *35*, 686–689. [CrossRef]
6. Li, D.Z.; Cui, T.X.; Wang, Y.T.; Fu, P.M.; Meng, C.F. Status and Development of Low Carbon Microalloying Bainitic High Strength Steel. *J. Iron Steel Res.* **2013**, *25*, 1–5.
7. Citti, P.; Giorgetti, A.; Millefanti, U. Mechanical Characterization of a New Low Carbon Bainitic Steel for High Performance Crankshaft. *Procedia Struct. Integr.* **2018**, *12*, 438–447. [CrossRef]
8. He, X.L.; Shang, C.J. Microstructure Fining Theory of Low-carbon Bainitic Steel. In *Ultra-Fine Grained Steels*; Weng, Y.Q., Ed.; Metallurgical Industry Press: Beijing, China, 2009; pp. 235–299. [CrossRef]
9. Apostolopoulos, C.A.; Demis, S.; Papadakis, V.G. Chloride-Induced Corrosion of Steel Reinforcement–Mechanical Performance and Pit Depth Analysis. *Constr. Build. Mater.* **2013**, *38*, 139–146. [CrossRef]
10. Cui, Z.; Patel, J.; Palmiere, E.J. Thermomechanical processing of structural steels with dilute niobium additions. In *HSLA Steels 2015, Microalloying 2015 & Offshore Engineering Steels 2015*; Springer: Cham, Switzerland, 2016; pp. 281–287. [CrossRef]
11. Cizek, P.; Wynne, B.P.; Davies, C.H.J.; Hodgson, P.D. The effect of simulated thermomechanical processing on the transformation behavior and microstructure of a low-carbon Mo-Nb linepipe steel. *Metall. Mate. Trans. A* **2015**, *46*, 407–425. [CrossRef]
12. Yan-Lei, Z.; Jun, C.; Zhen-Yu, L. Corrosion Behavior of Rusted 550 MPa Grade Offshore Platform Steel. *J. Iron. Steel. Res. Int. (Engl. Lett.)* **2013**, *020*, 66–73. [CrossRef]
13. Chen, Y.; Chen, L.; Zhou, X.; Zhao, Y.; Zha, X.; Zhu, F. Effect of Continuous Cooling Rate on Transformation Characteristic in Microalloyed Low Carbon Bainite Cryogenic Pressure Vessel Steel. *Trans. Indian Inst. Met.* **2016**, *69*, 817–821. [CrossRef]
14. Balart, M.J.; Davis, C.L.; Strangwood, M. Cleavage Initiation in Ti–V–N and V–N Microalloyed Ferritic–Pearlitic Forging Steels. *Mater. Sci. Eng. A* **2000**, *284*, 1–13. [CrossRef]
15. Katiyar, P.K.; Misra, S.; Mondal, K. Corrosion Behavior of Annealed Steels with Different Carbon Contents (0.002, 0.17, 0.43 and 0.7% C) in Freely Aerated 3.5% NaCl Solution. *J. Mater. Eng. Perform.* **2019**, *28*, 4041–4052. [CrossRef]
16. Sun, F.; Li, X.; Cheng, X. Effects of Carbon Content and Microstructure on Corrosion Property of New Developed Steels in Acidic Salt Solutions. *Acta Metall. Sin. (Engl. Lett.)* **2014**, *27*, 115–123. [CrossRef]
17. Cleary, H.J.; Greene, N.D. Corrosion Properties of Iron and Steel. *Corros. Sci.* **1967**, *7*, 821–831. [CrossRef]
18. Zhang, D.; Gao, X.; Su, G.; Du, L.; Liu, Z.; Hu, J. Corrosion Behavior of Low-C Medium-Mn Steel in Simulated Marine Immersion and Splash Zone Environment. *J. Mater. Eng. Perform.* **2017**, *26*, 2599–2607. [CrossRef]
19. Chen, X.; Dong, J.; Han, E.; Ke, W. Effect of Ni On the Ion-Selectivity of Rust Layer On Low Alloy Steel. *Mater. Lett.* **2007**, *61*, 4050–4053. [CrossRef]
20. Suzuki, S.; Wang, H.X. Niobium in structural steel. *World Iron Steel (Chin. Lett.)* **2012**, *12*, 27–41. [CrossRef]
21. Zhang, Y.Q.; Zhang, H.Q.; Jin-Fu, L.I.; Liu, W.M. Effect of Heat Input on Microstructure and Toughness of Coarse Grain Heat Affected Zone in Nb Microalloyed HSLA Steels. *J. Iron Steel Res. Int.* **2009**, *16*, 73–80. [CrossRef]
22. Hausmann, K.; Krizan, D.; Spiradek-Hahn, K.; Pichler, A.; Werner, E. The Influence of Nb On Transformation Behavior and Mechanical Properties of TRIP-assisted Bainitic–Ferritic Sheet Steels. *Mater. Sci. Eng. A* **2013**, *588*, 142–150. [CrossRef]
23. Hannula, J.; Kömi, J.; Porter, D.A.; Somani, M.C.; Kaijalainen, A.; Suikkanen, P.; Yang, J.; Tsai, S. Effect of Boron on the Strength and Toughness of Direct-Quenched Low-Carbon Niobium Bearing Ultra-High-Strength Martensitic Steel. *Metall. Mate. Trans. A* **2017**, *48*, 5344–5356. [CrossRef]
24. Misra, D.; Jansto, S.G. Niobium-based alloy design for structural applications: Processing-structure-property paradigm. In *HSLA Steels 2015, Microalloying 2015 & Offshore Engineering Steels 2015*; Springer: Cham, Switzerland, 2016; pp. 261–266. [CrossRef]
25. Katiyar, P.K.; Misra, S.; Mondal, K. Comparative Corrosion Behavior of Five Microstructures (Pearlite, Bainite, Spheroidized, Martensite, and Tempered Martensite) Made from a High Carbon Steel. *Metall. Mate. Trans. A* **2019**, *50*, 1489–1501. [CrossRef]
26. Souza, E.C.D.; Rossitti, S.M.; Fortulan, C.A.; Rollo, J.M.D.D. Influence of Ferrite Phase Content on the Electrochemical Properties of Duplex Stainless Steels. *Mat. Res.* **2017**, *20*, 21–29. [CrossRef]
27. Zhou, Y.; Chen, J.; Xu, Y.; Liu, Z. Effects of Cr, Ni and Cu on the Corrosion Behavior of Low Carbon Microalloying Steel in a CF-Containing Environment. *J. Mater. Sci. Technol. (Engl. Lett.)* **2013**, *29*, 168–174. [CrossRef]
28. ASTM E2298-18. *Standard Test Method for Instrumented Impact Testing of Metallic Materials*; ASTM International, West Conshohocken: Montgomery County, PA, USA, 2018.
29. ASTM E384-17. *Standard Test Method for Microindentation Hardness of Materials*; ASTM International, West Conshohocken: Montgomery County, PA, USA, 2017.
30. Saleem Khan, M.; Xu, D.; Liu, D.; Lekbach, Y.; Yang, K.; Yang, C. Corrosion Inhibition of X80 Steel in Simulated Marine Environment with Marinobacter aquaeolei. *Acta Metall. Sin. (Engl. Lett.)* **2019**, *32*, 1373–1384. [CrossRef]
31. Wang, F.P.; Jing, H.M.; Xin, C.M. *Corrosive Electrochemistry*, 2nd ed.; Chemical Industry Press: Beijing, China, 2017; p. 133.

32. Yao, S.; Liu, S.; Zeng, G.; Li, X.; Lei, T.; Li, Y.; Du, Y. Effect of Manganese on Microstructure and Corrosion Behavior of the Mg-3Al Alloys. *Metals* **2019**, *9*, 460. [CrossRef]
33. Nady, H.; El-Rabiei, M.M.; Samy, M. Corrosion Behavior and Electrochemical Properties of Carbon Steel, Commercial Pure Titanium, Copper and Copper–Aluminum–Nickel Alloy in 3.5% Sodium Chloride Containing Sulfide Ions. *Egypt J. Petrol.* **2016**, *26*, 79–94. [CrossRef]
34. Cheng, Z.J.; Song, D.; Jiang, J.Y.; Jiang, J.H.; Xiao-Long, M.A.; You, K.; Ai-Bin, M.A. Microstructure Characteristic and Electrochemical Corrosion Behavior of Surface Nano-Crystallization Modified Carbon Steel. *J. Iron Steel Res. Int.* **2016**, *23*, 1281–1289. [CrossRef]
35. Seikh, A.H.; Halfa, H.; Baig, M. Microstructure Characterization and Corrosion Resistance Behavior of New Cobalt-Free Maraging Steel Produced through ESR Techniques. *J. Mater. Eng. Perform.* **2017**, *26*, 1589–1597. [CrossRef]
36. Dong, Z.H.; Liu, T.; Liu, H.F. Influence of EPS Isolated From Thermophilic Sulphate-Reducing Bacteria On Carbon Steel Corrosion. *Biofouling* **2011**, *27*, 487–495. [CrossRef]
37. Palmiere, E.J.; Garcia, C.I.; Ardo, A.J.D. Compositional and Microstructural Changes Which Attend Reheating and Grain Coarsening in Steels Containing Niobium. *Metall. Mate. Trans. A* **1994**, *25*, 277–286. [CrossRef]
38. API SPEC 5L-2012. *Specification for line pipe FORTY-FIFTH EDITION*; American Petroleum Institute: Washington, DC, USA, 2012.
39. Melchers, R.E. Effect On Marine Immersion Corrosion of Carbon Content of Low Alloy Steels. *Corros. Sci.* **2003**, *45*, 2609–2625. [CrossRef]
40. Melchers, R.E. Effect of Small Compositional Changes on Marine Immersion Corrosion of Low Alloy Steels. *Corros. Sci.* **2004**, *46*, 1669–1691. [CrossRef]

materials

MDPI

Article

Effect of Different Austempering Heat Treatments on Corrosion Properties of High Silicon Steel

Mattia Franceschi [1], Luca Pezzato [1,*], Alessio Giorgio Settimi [1], Claudio Gennari [1], Mirko Pigato [1], Marina Polyakova [2], Dmitry Konstantinov [2], Katya Brunelli [1] and Manuele Dabalà [1]

1 Department of Industrial Engineering, University of Padua, Via Marzolo 9, 35131 Padova, Italy; mattia.franceschi@phd.unipd.it (M.F.); alessiogiorgio.settimi@unipd.it (A.G.S.); claudio.gennari@unipd.it (C.G.); mirko.pigato@unipd.it (M.P.); katya.brunelli@unipd.it (K.B.); manuele.dabala@unipd.it (M.D.)

2 Department of Mechanical Engineering and Metallurgical Technologies, Nosov Magnitogorsk State Technical University, pr. Lenina, 38, 455000 Magnitogorsk, Russia; m.polyakova-64@mail.ru (M.P.); const_dimon@mail.ru (D.K.)

* Correspondence: luca.pezzato@unipd.it; Tel.: +39-049-827-5498

Abstract: A novel high silicon austempered (AHS) steel has been studied in this work. The effect of different austenitizing temperatures, in full austenitic and biphasic regime, on the final microstructure was investigated. Specimens were austenitized at 780 °C, 830 °C, 850 °C and 900 °C for 30 min and held isothermally at 350 °C for 30 min. A second heat treatment route was performed which consisted of austenitizing at 900 °C for 30 min and austempering at 300 °C, 350 °C and 400 °C for 30 min. Scanning electron microscopy (SEM) and X-ray diffraction (XRD) have been used to evaluate the microstructural evolution. These techniques revealed that the microstructures were composed of carbide-free bainite, ferrite, martensite and retained austenite (RA) in different volume fractions ($V\gamma$). An aqueous borate buffer solution with 0.3 M H_3BO_3 and 0.075 M $Na_2B_4O_7\partial 10H_2O$ (pH = 8.4) was used for corrosion tests in order to evaluate the influence of the different volume fractions of retained austenite on the corrosion properties of the specimens. The results showed that when increasing the austenitization temperatures, the volume fractions of retained austenite reached a maximum value at 850 °C, and decrease at higher temperatures. The corrosion properties were investigated after 30 min and 24 h immersion by means of potentiodynamic polarization (after 30 min) and electrochemical impedance spectroscopy (after both 30 min and 24 h) tests. The corrosion resistance of the samples increased with increases in the volume fraction of retained austenite due to lower amounts of residual stresses.

Keywords: high-silicon steel; austempering; retained austenite; corrosion resistance

Citation: Franceschi, M.; Pezzato, L.; Settimi, A.G.; Gennari, C.; Pigato, M.; Polyakova, M.; Konstantinov, D.; Brunelli, K.; Dabalà, M. Effect of Different Austempering Heat Treatments on Corrosion Properties of High Silicon Steel. *Materials* **2021**, *14*, 288. https://doi.org/10.3390/ma14020288

Received: 14 December 2020
Accepted: 5 January 2021
Published: 8 January 2021

1. Introduction

In the latest years, significant interest has been dedicated to austempered high-silicon steels (AHS) [1,2]. High strength, toughness and ductility characterize these steel grades, making them very attractive materials. Their technological application can be in several industrial fields, such as automotive, mining, machinery, etc. [3].

Silicon is the most important element for these alloys, and it determines the final microstructural constituents and their amount [4]. It inhibits carbides precipitation, in particular cementite [5]. The suppression of cementite precipitation allows the partitioning of carbon into austenite during bainitic transformation, increasing its carbon content. This increases the stability of austenite, favoring its retention at room temperature.

Zhu et al. [6] observed the strong silicon contribution to the recrystallization processes; silicon delayed dynamic recrystallization and FCC iron grain growth, further favoring austenite retention.

In order to obtain the best microstructure, these steels have to be subjected to a particular heat treatment (HT) consisting of several steps. A heating phase in the dual phase, austenitic–ferritic, or in the austenitic region, represents the first step of the treatment. Austenitizing in a dual phase regime should be preferable because it contributes to carbon partitioning and retained austenite (RA) stabilization. In fact, newly formed ferrite rejects carbon that diffuses into austenite, which becomes more stable. After the austenitizing step, the material is cooled to the austempering temperature (AT) that is between bainite start (Bs) and martensite start (Ms). Once the selected AT is reached, the material is isothermally held and then cooled to room temperature [4,7].

The main purpose of this HT is to obtain a microstructure consisting of RA and free carbide bainite, which exhibits a high strength and is tougher than martensite [8]. In several cases, it is preferable to avoid the presence of RA, because it leads to a strong reduction in the mechanical properties, as reported by Pezzato et al. [9]. However, in modern AHSS-retained austenite has great importance due to its capability to undergo martensitic transformation when subjected to mechanical stresses. This phenomenon is called the transformation-induced plasticity (TRIP) effect, and enhances mechanical properties (i.e., strength, fracture strain, uniform strain, work hardening rate, etc.) [3,4,10,11].

The second aim of the treatment is precisely the stabilization of austenite at room temperature. Austenite's stability is influenced by several factors, such as carbon content, grain size, orientation and residual stresses [12–14]. Austenite's stability increases with its carbon content, reducing the grains size; moreover, non-preferred orientation and the absence of textures favor its retention.

RA is characterized by two morphologies: (a) film-like austenite, and (b) blocky austenite [15]. Filmy austenite is located between bainitic ferrite laths, and its presence is attributed to silicon, which prevents the precipitation of cementite. Supersaturated ferrite rejects the carbon that enriches austenite, which becomes stable at room temperature. Films of austenite have enhanced strength, toughness and wear resistance due to their low tendency to exhibit the TRIP effect, and their being an obstacle for cracks propagation [4,15,16].

The carbon content in blocky austenite is lower than in film-like RA and is inhomogeneous. These characteristics make RA more susceptible to martensitic transformation, and it transforms at lower deformation levels in comparison with filmy austenite. The carbon inhomogeneity inside RA blocks provokes a gradual transformation: the poorest zones, within the grain, transform before the richer ones. Austenite blocks also have lower mechanical stability because they contain a large number of nucleation sites for martensite formation and less constraints for the transformation [15,16].

The microstructure and the final mechanical properties of steel depend on the heat treatment parameters [17]. Concerning the HT parameters effects on microstructure, in the literature there are numerous works and results focused on the soaking step at the AT temperature. Palaksha et al. [18] observed that when increasing the austempering temperature from 275 to 350 °C, the final volume fraction of retained austenite increases, then a further temperature increase leads to a decrease in RA stabilization. Such an effect has been observed also by Putatunda [2], who demonstrated that the same trend can be observed for carbon content in RA and material fracture toughness. Son et al. [19] observed on a 0.9C-2.3Si steel an increase in both ferritic and austenitic grain size, and RA final amount. Kumar et al. [20] demonstrated that the bainite volume fraction increases with a decreasing austempering temperature, and that RA decreases with holding time. The reduction in the final volume fraction of the retained austenite's increasing isothermal bainitic transformation (IBT) time was also observed by Acharya et al. [21]. Concerning the effect of austenitizing temperature, Zhao et al. observed that increasing austenitizing temperatures led to coarser prior austenite grain size (PAGB) and bainitic packets, coarser RA blocks, and a higher final volume fraction of the retained austenite [22].

The authors in a previous work [17] reported results for the effect of different austenitizing conditions; as the austenitization temperature increased, a grain coarsening was observed. Moreover, this led to the presence of bainitic ferrite (BF), promoted by element diffusion [23].

Microstructural modification induced by heat treatments could influence corrosion resistance, as observed by several authors both in low-alloyed carbon steels [24] and in high-alloyed steel, such as stainless steel [25] and duplex stainless steel (DSS) [26].

Concerning the effect of retained austenite on corrosion resistance, in the literature there are several studies. Bignozzi et al. [27], studying a high-chromium martensitic steel, observed that when increasing the volume fraction of RA, the corrosion rate decreased due to the low amount of internal stress in the retained austenite [28,29]. Additionally, Han et al. [30] observed that when increasing the volume fraction of austenite and the nodularity of graphite in cast iron, the corrosion rate decreased.

In this work, the effects of different austenitizing temperatures at fixed AT and the effects of different ATs at the fixed austenitizing temperature were investigated. Microstructural evolution was evaluated with electron microscopy and X-ray diffraction.

In addition, given the lack of data in the literature regarding the corrosion resistance of these steel grades, the effect of the microstructural constituents and RA content on the corrosion resistance of the material was investigated in borate buffer solution [31].

2. Materials and Methods

High silicon steel was supplied by Nosov Magnitogorsk State Technical University (NMSTU) in the form of 8 mm diameter wires. Chemical composition, given in weight percentage (wt.%), is summarized in Table 1.

Table 1. High silicon steel chemical composition (wt.%).

Fe	C	Si	Mn	P	S	Cr	Ni	Cu	Mo	Ti	V	Al
Balance	0.430	3.260	2.720	0.010	0.008	0.043	0.074	0.060	0.022	0.001	0.005	0.105

Carbolite tubular and Nabetherm 3000 electric furnaces were used for the realization of austempering heat treatments. The first was used for the normalizing treatment (900 °C and calm air-cooling) and for the austenitizing step. The second was used for the austempering step, i.e., soaking at AT for isothermal bainitic transformation (IBT). An air-cooling system was used for the cooling step from the austenitization temperature to the AT temperature, and a water-cooling system for the final cooling to room temperature.

The subsequent treatments were performed (summarized in Table 2):

- *Normalizing:* heating at 900 °C at 1 °C/s, 30 min holding time and air-cooling (10 °C/s);
- *Austempering (Route 1)*: Pre-normalization treatment from 900 °C (30 min) and water-cooling. Heating at 780, 830, 850 °C for 30 min at 1 °C/s, air cooling at 10 °C/s to 350 °C and holding for 30 min followed by water-cooling to room temperature at 40 °C/s (Figure 1a);
- *Austempering (Route 2)*: pre-normalization treatment from 900 °C (30 min) and water-cooling. Heating at 900 °C for 30 min at 1 °C/s, air cooling at 10 °C/s to 300, 350 and 400 °C and holding for 30 min followed by water-cooling to room temperature at 40 °C/s (Figure 1b).

Table 2. Scheduled heat treatment.

Heat Treatment Route	Treatment	Austenitizing (°C) (30 min)	Austempering (°C) (30 min)
-	Normalizing	900	-
1	780 + 350	780	350
	830 + 350	830	350
	850 + 350	850	350
2	900 + 300	900	300
	900 + 350	900	350
	900 + 400	900	400

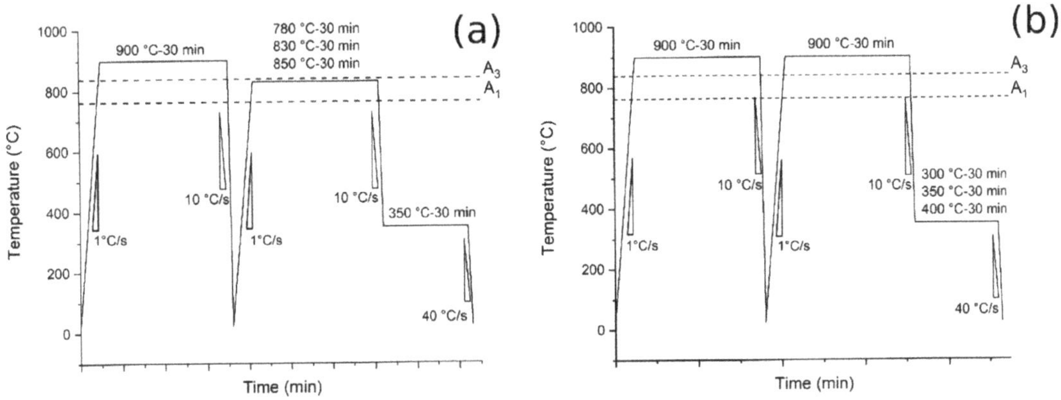

Figure 1. Heat treatments scheme: (**a**) the tests performed at different austenitizing and fixed austempering temperatures (route 1); (**b**) the tests performed at fixed austenitizing and different austempering temperatures (route 2).

Microstructure evolution was investigated with a scanning electron microscope (SEM, Leica Cambridge Leo Stereoscan 440, Leica Microsystems S.r.l., Milan, Italy), operating with a secondary electron (SE) at 15 kV and in backscattered electron mode (BSE) at 25 kV. Heat-treated specimens were cut along the cross section with lubricated SiC disks, mounted using phenolic resin, ground with abrasive SiC papers (from 320 to 1200 grit) and polished using clothes and polycrystalline diamond suspensions of 6, 3 and 1 μm. In order to reveal microstructures, Nital 2 (2 mL of nitric acid in 98 mL of ethanol) etchant was used.

Phase identification and quantification were performed with X-ray diffraction, using a Siemens D500 X-ray diffractometer (Siemens, Munich, Germany), equipped with a monochromator on the detector side and a Cu radiation tube, working at 40 kV and 30 mA. The $2\theta = 40–105°$ angular range was investigated with a step scan of 0.025° and a counting time of 6 s per step. The evaluation of the volume fraction of the phases was performed by means of Rietveld analysis with Maud software (Lutterotti, University of Trento, Department of Material Engineering, Trento, Italy).

The corrosion performances were investigated by potentiodynamic polarization tests (PDP) at room temperature. Polarization tests were performed in a borate buffer solution containing 0.3 M H_3BO_3 and 0.075 M $Na_2B_4O_7 \bullet 10H_2O$ (pH = 8.4) [31], using an AMEL 2549 Potentiostat (Amel Electrochemistry S.r.l., Milan, Italy). The reference electrode was a saturated calomel electrode (SCE), the counter electrode was a platinum electrode, and the scan rate was 0.5 mVs^{-1} in a potential range between OCP−0.2V (0.2 V below the measured OCP) and +2.5 V. The electrolyte was chosen in order to simulate a non-strongly aggressive environment to underline the effect of retained austenite in the material. Corrosion data (E_{corr} and I_{corr}) were extrapolated graphically using the Tafel method.

EIS (electrochemical impedance spectroscopy) tests were also performed using the same cell and electrolyte of the PDP at the value of the open circuit potential, and in a range of frequency $10^5–10^{-2}$ Hz with a perturbation amplitude of 10 mV, after 30 min and 24 h of immersion. An AMEL 2549 Potentiostat (Amel Electrochemistry S.r.l.) coupled with a Materials Instrument Spectrometer was employed. The fitting of experimental data was carried out using the software ZView (Scribner Associates Inc, Southern Pines, NC, USA) and the Randles circuit as the equivalent circuit. Both PDP and EIS tests were performed after 30 min of OCP stabilization. Each measure was repeated three times in order to assure statistical reproducibility.

3. Results and Discussion

3.1. Microstructural Characterization

The microstructure of the samples was analyzed along the cross section by electron microscopy.

A further increase in austenitizing temperature (900 °C), maintaining 350 °C of austempering temperature, led to carbide-free bainitic sheaves coarsening, as can be noted in Figure 5b.1,b.2. The microstructure consisted of a martensitic matrix, as before, but with a higher amount of bainite.

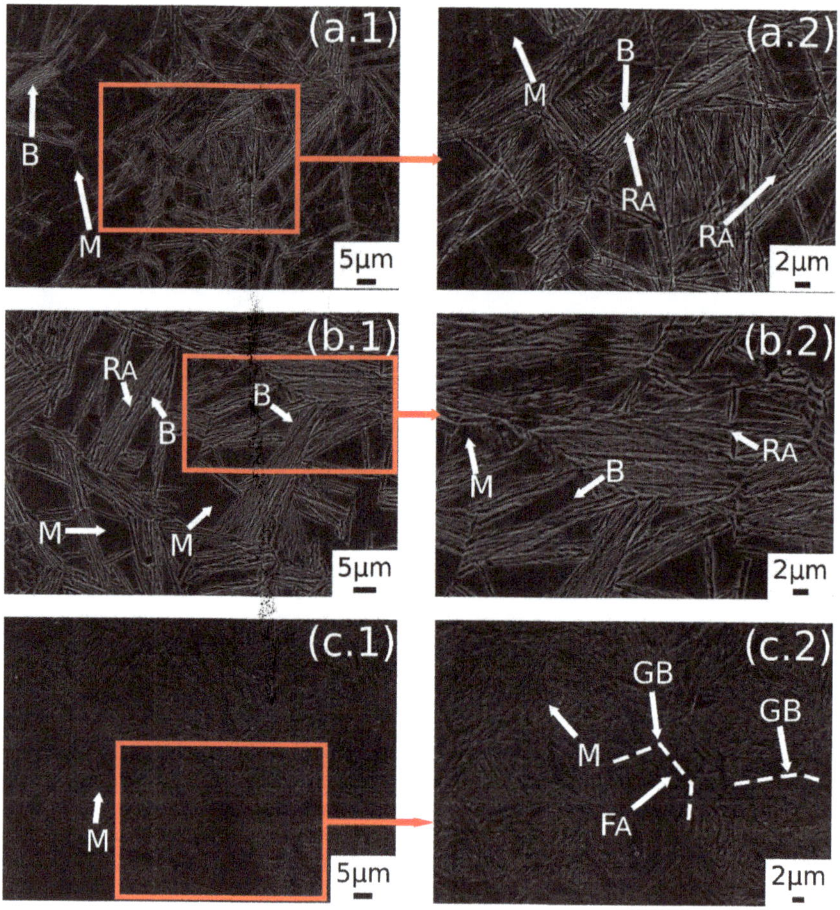

Figure 5. SE-SEM images of the surfaces of the samples: (**a.1**,**a.2**) 900 °C + 300 °C, (**b.1**,**b.2**) 900 °C + 350 °C, (**c.1**,**c.2**) 900 °C + 400 °C. (M martensite, F ferrite, R_A retained austenite, B carbide free bainite, GB prior austenite grain boundary, F_A allotriomorphic ferrite).

A prevalent martensitic microstructure with free-carbide bainite sheaves randomly oriented was observed in the sample austenitized at 900 °C for 30 min and soaked at 300 °C for 30 min, as shown in Figure 5a.1,a.2. A lower amount of bainite was found with decreasing the AT temperature. This microstructural evolution could be related to the effect of AT on the kinetic of bainitic transformation. In fact, higher austempering temperatures fasten the bainitic transformation kinetics [4]. The observed microstructural evolution agreed with the literature [1,2,34,35].

Austempering at 400 °C produced a microstructure consisting of martensite (Figure 5c.1,c.2) without bainite, due to austempering at a temperature superior to Bs, as demonstrated in [23]. In the image are also indicated the trace of prior austenite grain boundaries (GB) with allotriomorphic ferrite (F_A) [22]. R_{AB} and R_{AF} are recognizable also in the sample subjected to 900 °C austenitizing and 350 °C austempering (Figure 6). High-magnification micrographs allow for demonstrating that filmy austenite (R_{AF}) is located between the bainitic laths, while the R_{AB} are located at bainitic region boundaries.

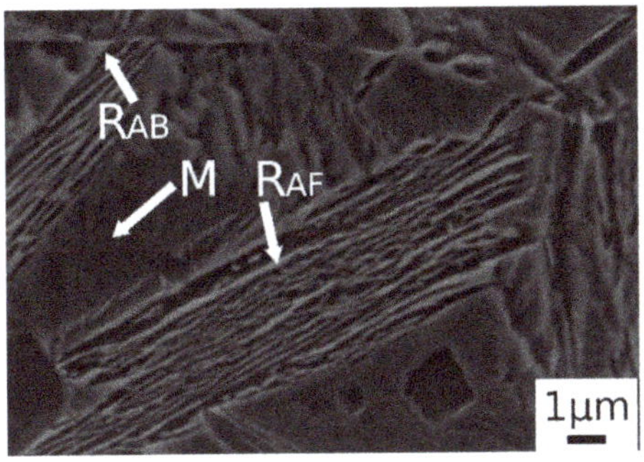

Figure 6. SE-SEM image at high magnification of the alloy after austenitizing at 900 °C (30 min) and austempering at 350 °C for 30 min. (M martensite, R_{AB} blocky retained austenite, R_{AF} filmy retained austenite).

3.2. X-ray Diffraction

X-ray diffraction patterns of austenitized samples at 780 °C, 830 °C, 850 °C are shown in Figure 7. In all the patterns are visible BCC iron peaks related to ferrite and bainite, body-centered tetragonal iron peaks related to martensite, and FCC iron peaks related to the presence of retained austenite. Displacement between ferrite and martensite peaks should explained by the high martensite carbon content and strong cell distortion [23,36].

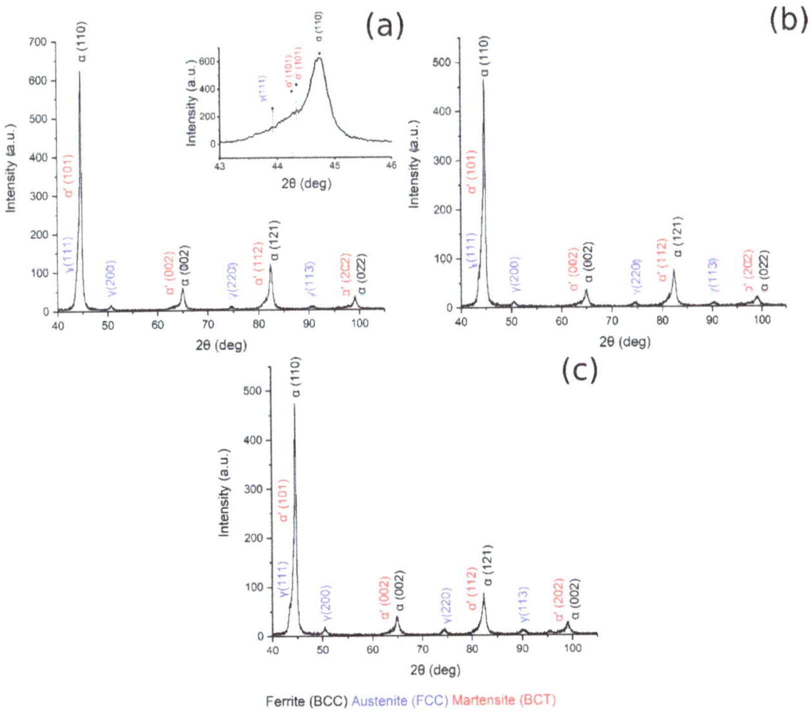

Figure 7. XRD pattern (**a**) 780 °C + 350 °C, in the box upon on the right can be found a zoom of the pattern in the range 43–46°, (**b**) 830 °C + 350 °C, (**c**) 850 °C + 350 °C.

Figure 8 shows the XRD pattern of the sample austenitized at 900 °C and soaked at 300, 350 and 400 °C for 30 min.

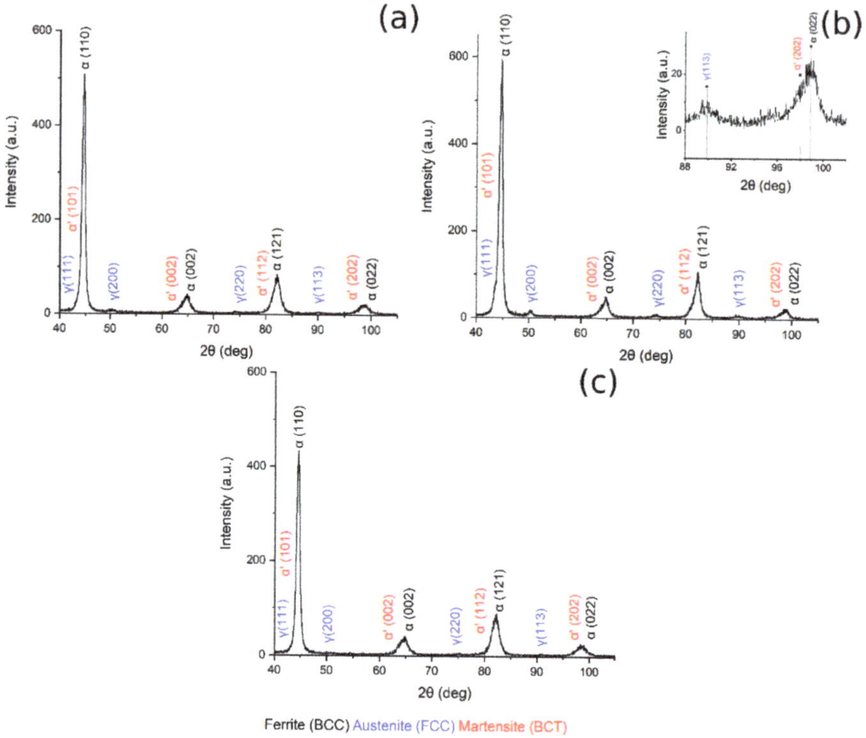

Figure 8. XRD pattern (**a**) 900 °C + 300 °C, (**b**) 900 °C + 350 °C, in the box upon on the right can be found a zoom of the pattern in the range 88–102°, (**c**) 900 °C + 400 °C.

Table 3 reports the volume fraction of the phases calculated by means of the Rietveld analysis performed with the Maud software on the XRD patterns (Figures 7 and 8). The results show that the microstructure mainly consisted of martensite and carbide free bainite with RA.

Table 3. Results of phase quantification performed by Rietveld analysis.

Heat Treatment	Retained Austenite (vol%)	Bainite/Ferrite (vol%)	Martensite (vol%)
780 (30 min)+ air-cooling (10 °C/s) + 350 (30 min) (40 °C/s) water-cooling	6.2 ± 0.3	45.8 ± 0.3	48.1 ± 0.3
830 (30 min)+ air-cooling (10 °C/s) + 350 (30 min) (40 °C/s) water-cooling	8.7 ± 0.3	36.4 ± 0.3	54.9 ± 0.3
850 (30 min)+ air-cooling (10 °C/s) + 350 (30 min) (40 °C/s) water-cooling	10.7 ± 0.3	40.2 ± 0.3	49.1 ± 0.3
900 (30 min)+ air cooling (10 °C/s) + 300 (30 min) (40 °C/s) water cooling	2.7 ± 0.3	43 ± 0.3	54.3 ± 0.3
900 (30 min)+ air-cooling (10 °C/s) + 350 (30 min) (40 °C/s) water-cooling	5.2 ± 0.3	24.4 ± 0.3	70.4 ± 0.3
900 (30 min)+ air-cooling (10 °C/s) + 400 (30 min) (40 °C/s) water-cooling	1.2 ± 0.3	42.8 ± 0.3	56 ± 0.3

By plotting the austenite volume fraction vs. the austenitizing temperature, it can be observed that (i) increasing the austenitizing temperature, from 780 °C to 850 °C, produced an increase in final Vγ, and that (ii) above 850 °C a decrease in the Vγ can be noted (Figure 9a).

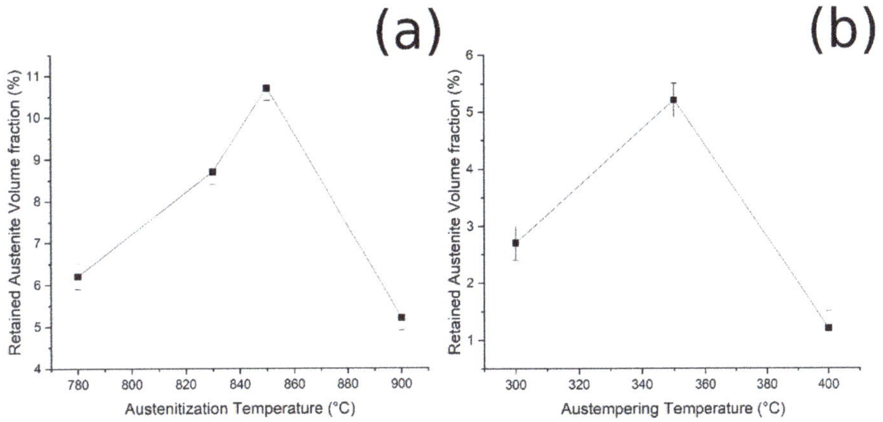

Figure 9. Relationship between (**a**) RA and austenitizing temperature, and (**b**) RA and austempering temperature (fixed austenitizing at 900 °C).

Furthermore, when the material was austenitized at 900 °C, the volume fraction of retained austenite increased from 300 °C to 350 °C of austempering, and then decreased (Figure 9b). These results could be explained by considering the bainitic transformation process and the kinetics: the higher the carbide-free bainite volume fraction, the higher is the amount of RA stabilized and retained at room temperature. The obtained results agree with the literature [2,17,27].

3.3. Corrosion Resistance

The corrosion resistance of the specimens was preliminarily measured by PDP tests, in a borate buffer solution, and the PDP curves are reported in Figures 10a and 11a. Values of corrosion potentials and current densities, graphically extrapolated from the curves, together with the corrosion rate, calculated with Equation (1), are reported in Table 4.

Table 4. Values of corrosion potential and corrosion current densities graphically extrapolated form Figures 10a and 11a. Corrosion rates calculated with Equation (1) are also reported in the last two columns.

Heat Treatment	E_{corr} (V)	I_{corr} (μA/cm^2)	Corrosion Rate (g/s × cm^2)	Corrosion Rate (mm/year)
780 (30 min)+ air-cooling (10 °C/s) + 350 (30 min) (40 °C/s) water-cooling	−0.63	3.5 ± 5%	1.01×10^{-9}	4.09×10^{-2} ± 5%
830 (30 min)+ air-cooling (10 °C/s) + 350 (30 min) (40 °C/s) water-cooling	−0.35	2.5 ± 5%	7.24×10^{-10}	2.92×10^{-2} ± 5%
850 (30 min)+ air-cooling (10 °C/s) + 350 (30 min) (40 °C/s) water-cooling	−0.47	2.3 ± 5%	6.66×10^{-10}	2.69×10^{-2} ± 5%
900 (30 min)+ air-cooling (10 °C/s) + 300 (30 min) (40 °C/s) water-cooling	−0.56	5.0 ± 5%	1.45×10^{-9}	5.84×10^{-2} ± 5%
900 (30 min)+ air-cooling (10 °C/s) + 350 (30 min) (40 °C/s) water-cooling	−0.48	4.6 ± 5%	1.33×10^{-9}	5.38×10^{-2} ± 5%
900 (30 min)+ air-cooling (10 °C/s) + 400 (30 min) (40 °C/s) water-cooling	−0.65	3.2 ± 5%	9.26×10^{-10}	3.74×10^{-2} ± 5%
Normalized 900 (30 min), air-cooling	−0.87	18.0 ± 5%	5.21×10^{-9}	2.10×10^{-1} ± 5%

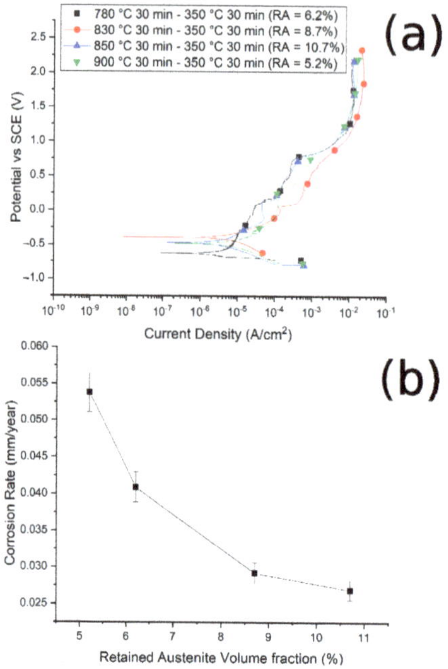

Figure 10. (**a**) Potentiodynamic polarization curves for the different samples austenitized at 780 °C, 830 °C, 850 °C, 900 °C and austempered at 350 °C. (**b**) Correlation between corrosion rate and RA volume fraction. Test performed in borate buffer solution.

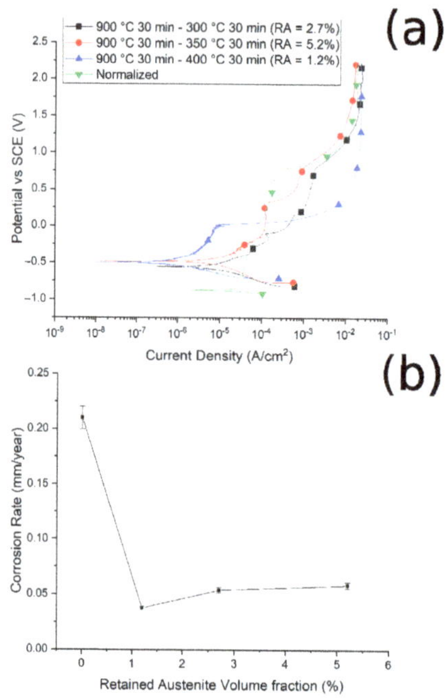

Figure 11. (**a**) Polarization curves of the samples austenitized at 900 °C and austempered at 300 °C, 350 °C, and 400 °C, and normalized at 900 °C. (**b**) Relationship between corrosion rate and volume fraction of retained austenite. Test performed in borate buffer solution.

It is possible to note that, as the austenitizing temperature and volume fraction of the retained austenite increased, the I_{corr} decreased, the corrosion rate (CR) decreased and the corrosion resistance increased, in agreement with Han et al. [30]. In Figures 10b and 11b are reported the trend of the corrosion rate (CT) with the RA volume fraction for the two studied routes. A clear correlation between the quantity of RA and the CR can be noted, especially in the case of route 1, with a decrease in the corrosion rate with the increase in the RA volume fraction. This phenomenon agrees with the results of Bignozzi et al. [27], who studied the effect of austenite on martensitic stainless steel's resistance. The effect of RA was explained by Hill et al. [28,37]: austenite was characterized by a lower amount of internal stress with respect to martensite. According to the authors, higher stresses led to a corrosion rate increase.

The results were confirmed by Hsu et al. [38], who studied the corrosion behavior of ADI cast iron in a 3.5% NaCl environment. According to their results, retained austenite, produced by austempering, acts as a corrosion inhibitor, lowering the material degradation with high efficiency.

Moreover, the results were confirmed in the research conducted by Yang et al. [39]. The authors in ref. [38] analyzed the difference in corrosion resistance between the quenching and tempering treatment (Q&T) and the quenching and partitioning (Q&P) of a medium carbon steel, in a sodium chloride solution. The investigated alloy, after Q&P, was characterized by a high amount of RA (~17.5%) with respect to the sample treated by Q&T, and this exhibited better corrosion properties both in short- and long-time immersion.

It can be noted that, at fixed austenitizing temperatures (900 °C), when increasing the austempering temperature, specimens showed better corrosion resistance, in agreement with the literature [40]. For this austenitizing temperature, the variations in the corrosion properties are smaller than in the case of route 1, as can be observed when comparing Figure 11b with Figure 10b, due to the lower variation in RA content between the various samples.

Considering the corrosion potentials (E_{corr}), an ennoblement in the corrosion potential can be observed for all the treated samples in comparison to the normalized sample (with no RA), confirming the increased corrosion performances of the samples that contained RA. In fact, an increase in the corrosion potential indicates and ennoblement of the tested material, and so an increase in the corrosion properties.

From the observation of the surface of the corroded samples (in Figure 12 is presented an example, but all the samples present the same surface after corrosion), it was possible to recognize mainly a uniform corrosion, with some localized corrosion sites, in accordance with the employed electrolyte. This allowed the application of the formula for corrosion rate (1) calculation reported in ASTM G102 [41],

$$CR = K_i \cdot \frac{I_{corr}}{\rho} EW \tag{1}$$

where:

(a) CR is the corrosion rate (mm/year);
(b) K_i is a constant equal to 3.27×10^{-3};
(c) I_{corr} is the current density ($\mu A/cm^2$);
(d) ρ is the material density, assumed equal to 7.8 g/cm^3;
(e) EW is the so-called equivalent weight, calculated according to (2),

$$EW = \frac{1}{\sum_{i=1}^{n} \frac{n_i \cdot f_i}{W_i}} \tag{2}$$

and n_i, f_i and W_i represent respectively the valence, the mass fraction and the atomic weight of the ith element.

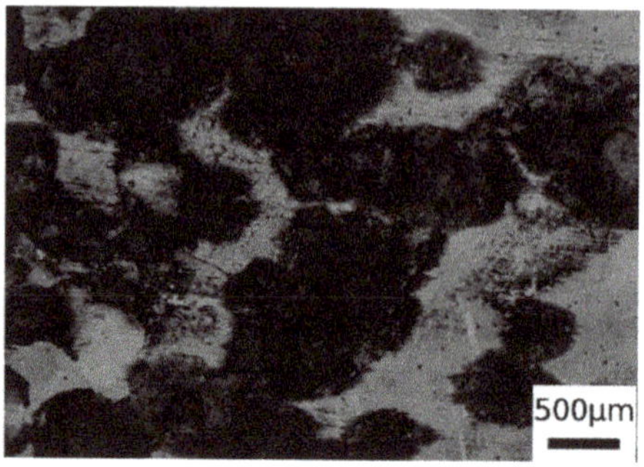

Figure 12. SEM micrographs surface of one sample after PDP tests (BSE mode) for the sample austenitized at 900 °C (30 min) and austempering at 350 °C for 30 min.

Further investigation on the corrosion behavior was performed with EIS tests on the samples with the best and the worst corrosion resistance, after 30 min and 24 h immersion in borate buffer solution.

The results, in term of Nyquist, Bode Modulus and Phase diagrams, are reported in Figure 13.

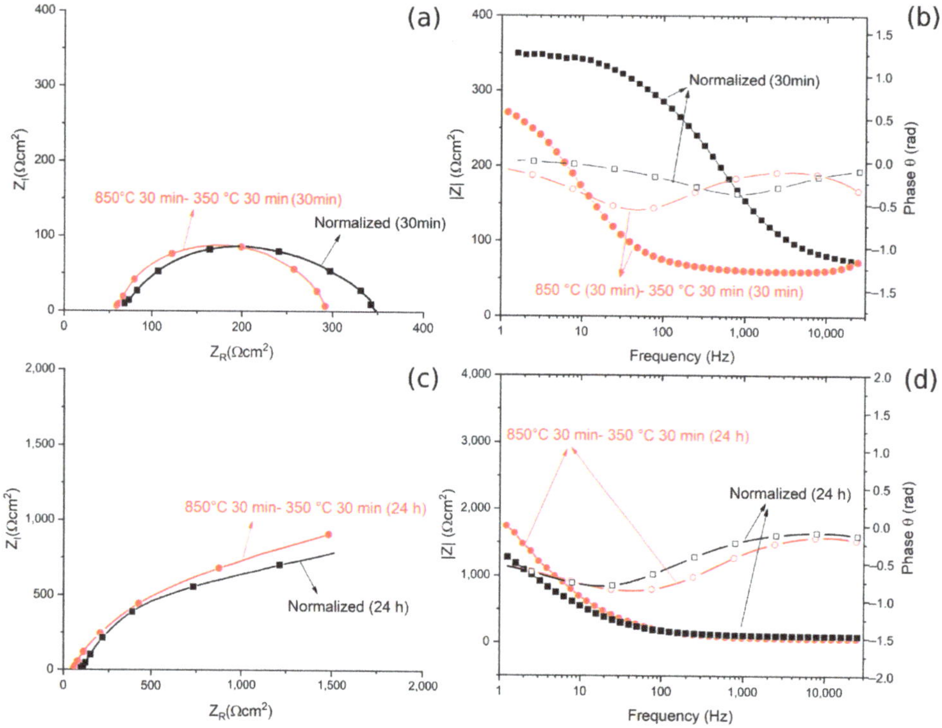

Figure 13. (**a**) Nyquist plot for the material austempered at 350 °C after austenitizing at 850 °C and in normalized conditions for 30 min immersion. (**b**) Bode modulus and phase plot for or the material austempered at 350 °C after austenitizing at 850 °C and in the normalized condition for 30 min immersion. (**c**) Nyquist plot for the material austempered at 350 °C after austenitizing at 850 °C, and in normalized condition for 24 h immersion. (**d**) Bode and phase plots for the material austempered at 350 °C after austenitizing at 850 °C and in normalized condition for 24 h immersion. All the tests were performed in borate buffer solution. Dots represent experimental data, lines the results of the fitting.

The obtained EIS data were fitted using the equivalent circuit shown in Figure 14, called Randles circuit, and the results are reported in Table 5. In this circuit, R_S represents the resistance of the electrolyte, R_O is the polarization resistance, which depends on the passive oxide film and is a measure of the corrosion resistance of the material, and CPE_O is a constant phase element, used instead of a capacitance because the measured capacitance is not ideal. The choice of the equivalent circuit is in accordance with the shape of the Bode plots, which showed a single time constant, and with the literature regarding natural oxide films on steels [42].

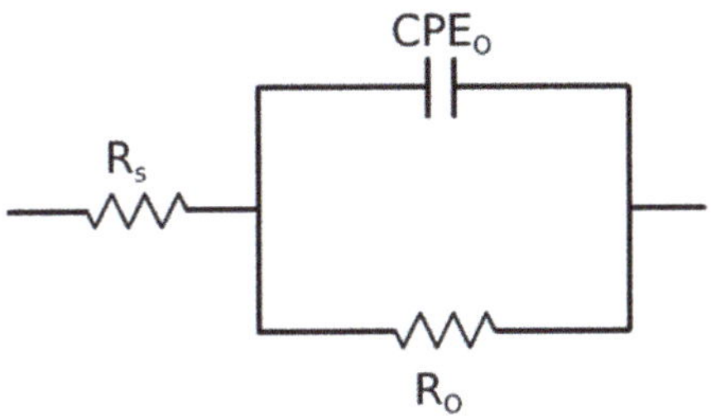

Figure 14. Equivalent circuit used for the fitting of EIS data.

Table 5. Results of the fitting of the experimental data with the equivalent circuit of Figure 14 for the material austempered at 350 °C after austenitizing at 850 °C and in normalized condition.

Equivalent Circuit Values	850 °C 30 min-350 °C 30 min (30 min)	Normalized (30 min)	850 °C 30 min-350 °C 30 min (24 h)	Normalized (24 h)
R_s (Ωcm^2)	57.81	54.40	86.38	86.40
Q_O ($F \times Hz^{1-n}$)	1.3×10^{-4}	1.2×10^{-4}	2.4×10^{-4}	2.8×10^{-4}
n_O	0.74	0.69	0.70	0.68
R_O (Ωcm^2)	286.90	290.70	3223	3210
X^2	1×10^{-3}	1×10^{-3}	1×10^{-3}	1×10^{-3}

The quality of the fitting of EIS data after both 30 min and 24 h of immersion was good, as confirmed by the low values of chi squared.

From the Nyquist plots a qualitative evaluation of the polarization resistance can be performed considering this as the real part of the impedance at low frequencies (or from the Bode modulus plot as the interception with the Y axis). This value of polarization resistance is a measure of the protective properties of the oxide coating formed on the steel.

After 30 min of immersion in the solution, no significant differences in terms of Z_R at low frequencies, and so in the polarization resistance, could be observed. This indicated that the different corrosion performance recorded in the PDP tests was not related to the characteristics of a passive layer (measured by EIS tests), but to the amount of retained austenite in the samples. In fact, RA, given its low amount of internal defects, reduces the amount of residual stress in the material, making it less susceptible to corrosion. This agreed with the results in the literature [26–28].

The data in Table 5 confirmed the results of the qualitative evaluation from the diagrams, in fact the values of R_o, that is the polarization resistance given by the oxide layer, are similar between the two samples. This allowed us to state that after 30 min of immersion, the volume fraction of RA did not influence the protective properties of the oxide film.

Similarly, after 24 h of immersion, no significant differences between the two specimens are observable in term of polarization resistance, R_o, both from the qualitative evaluation of the diagrams and from the analysis of the data in Table 5. Moreover, it is possible to observe that a long immersion period led to an increase in terms of R_o, due to the formation of a thicker passive layer, with semi conductive properties, as demonstrated by Hamadou [43] and Oblonsky [44]. The application of surface-enhanced Raman spectroscopy (SERS) on passive films in borate buffer solutions allowed Oblonsky, in ref. [44], to determine its composition: a mixture of amorphous $Fe(OH)_2$, γ-Fe_2O_3 and Fe_3O_4. It is permissible to think that a passive oxide film with a similar composition is formed on the surface of the steel examined in this work.

4. Conclusions

In this work a novel composition of high silicon steel has been studied. It was observed that:

- Austenitizing at 780 °C and austempering at 350 °C produced a dual-phase, ferritic-martensitic microstructure. By increasing the austempering temperature carbide-free bainite was formed, and, as the temperature increased, a coarsening of the bainitic sheaves was observed;
- Increasing the austenitizing temperature from 780 to 850 °C led to an increase in the retained austenite volume fraction, thanks to the increase in bainite amount, which favors carbon partitioning;
- At fixed austenitizing condition (900 °C), the bainite and RA amount increased with the austempering temperature up to 350 °C, whereas at 400 °C no bainite was found due to the absence of bainitic transformation, because 400 °C is above Bs;
- The corrosion resistance of the samples increased with the volume fraction of retained austenite;
- The variation in the corrosion properties, at 30 min of immersion in borate buffer solution, was not linked with variations in the protective properties of the oxide layer (Which did not change from EIS tests), but with the lower amount of residual stresses in the samples with higher amounts of retained austenite;
- The EIS tests, after 24 h of immersion in borate buffer solution, showed that the differences in the RA volume fraction did not produce differences in term of the polarization resistance of the oxide layer even after immersion. The polarization resistance, R_o, of the sample immersed for 24 h was higher than that of the sample after 30 min of immersion, due to the growth of a thicker passive film.

Author Contributions: Conceptualization, M.D., M.P. (Marina Polyakova) and K.B.; methodology, C.G. and L.P.; software, D.K. and M.F.; validation, D.K. and C.G.; formal analysis, M.F. and D.K.; investigation, A.G.S., M.F., M.P. (Mirko Pigato), L.P. and C.G.; resources, M.D., K.B. and M.P. (Marina Polyakova); data curation, M.F., A.G.S. and C.G.; writing—original draft preparation, M.F., L.P. and C.G.; writing—review and editing, L.P. and C.G.; visualization, M.F. and C.G.; supervision, K.B., M.P. (Marina Polyakova) and M.D.; project administration, M.D. and M.P. (Marina Polyakova); funding acquisition, M.D. and M.P. (Marina Polyakova). All authors have read and agreed to the published version of the manuscript.

Funding: The authors want to thank FONDAZIONE CARIPARO for the financial support with the visiting program TIP-STEP.

Institutional Review Board Statement: Not applicable.

Informed Consent Statement: Not applicable.

Data Availability Statement: The raw/processed data required to reproduce these findings cannot be shared at this time as the data also forms part of an ongoing study.

Conflicts of Interest: The authors declare no conflict of interest.

References

1. Putatunda, S.K.; Singar, A.V.; Tackett, R.; Lawes, G. Development of a high strength high toughness ausferritic steel. *Mater. Sci. Eng. A* **2009**, *513–514*, 329–339. [CrossRef]
2. Putatunda, S.K. Fracture toughness of a high carbon and high silicon steel. *Mater. Sci. Eng. A* **2001**, *297*, 31–43. [CrossRef]
3. Changlea, Z.; Shouhai, L.; Hanguanga, F.; Yinghua, L. Microstructure evolution and wear resistance of high silicon bainitic steel after austempering. *J. Mater. Res. Technol.* **2020**, *9*, 4826–4839. [CrossRef]

4. Fonstein, N. *Advanced High Strength Sheet Steels: Physical Metallurgy, Design, Processing, and Properties*; Springer: Cham, Switzerland, 2015; ISBN 9783319191652.
5. Qian, L.; Zhou, Q.; Zhang, F.; Meng, J.; Zhang, M.; Tian, Y. Microstructure and mechanical properties of a low carbon carbide-free bainitic steel co-alloyed with Al and Si. *Mater. Des.* **2012**, *39*, 264–268. [CrossRef]
6. Zhu, L.J.; Wu, D.; Zhao, X.M. Effect of silicon addition on recrystallization and phase transformation behavior of high-strength hot-rolled trip steel. *Acta Metall. Sin. Engl. Lett.* **2008**, *21*, 163–168. [CrossRef]
7. de Oliveira, P.G.B.; Mariani, F.E.; Casteletti, L.C.; Itman Filho, A.; Neto, A.L.; Totten, G.E. Boro-Austempering Treatment of High-Strength Bainitic Steels. *J. Mater. Eng. Perform.* **2020**, *29*, 3486–3493. [CrossRef]
8. Bahadur, S.; State, I.; Verhoeven, J.D.; State, I. Friction and wear behavior of high silicon bainitic structures in austempered cast iron and steel. *Wear* **1990**, *138*, 269–284.
9. Pezzato, L.; Gennari, C.; Chukin, D.; Toldo, M.; Sella, F.; Toniolo, M.; Zambon, A.; Brunelli, K.; Dabalà, M. Study of the effect of multiple tempering on the impact toughness of forged s690 structural steel. *Metals* **2020**, *10*, 507. [CrossRef]
10. Konstantinov, D.V.; Bzowski, K.; Korchunov, A.G.; Kuziak, R.; Pietrzyk, M.; Shiryaev, O.P. Multiscale Modelling of Structural and Phase Transformations in Steel During Drawing. *Vestn. Nosov Magnitogorsk State Tech. Univ.* **2016**, *14*, 90–98. [CrossRef]
11. Konstantinov, D.V.; Korchunov, A.G. Multi-Scale Computer Simulation of Metal Forming Processes. *Vestn. Nosov Magnitogorsk State Tech. Univ.* **2015**, *1*, 36–43.
12. Sherif, M.Y.; Mateo, C.G.; Sourmail, T.; Bhadeshia, H.K.D.H. Stability of retained austenite in TRIP-assisted steels. *Mater. Sci. Technol.* **2004**, *20*, 319–322. [CrossRef]
13. Pereloma, E.V.; Gazder, A.A.; Timokhina, I.B. Addressing retained austenite stability in advanced high strength steels. In Proceedings of the Materials Science Forum, Saint-Petersburg, Russia, 9–16 September 2013; Volume 738–739, pp. 212–216.
14. Podder, A.S.; Lonardelli, I.; Molinari, A.; Bhadeshia, H.K.D.H. Thermal stability of retained austenite in bainitic steel: An in situ study. *R. Soc. A Math. Phys. Eng. Sci.* **2011**, *467*, 3141–3156. [CrossRef]
15. Efremenko, V.G.; Hesse, O.; Friedrich, T.; Kunert, M.; Brykov, M.N.; Shimizu, K.; Zurnadzhy, V.I.; Šuchmann, P. Two-body abrasion resistance of high-carbon high-silicon steel: Metastable austenite vs nanostructured bainite. *Wear* **2019**, *418–419*, 24–35. [CrossRef]
16. Zhao, J.; Lv, B.; Zhang, F.; Yang, Z.; Qian, L.; Chen, C.; Long, X. Effects of austempering temperature on bainitic microstructure and mechanical properties of a high-C high-Si steel. *Mater. Sci. Eng. A* **2019**, *742*, 179–189. [CrossRef]
17. Changle, Z.; Hanguang, F.; Shengqiang, M.; Jian, L.; Yongping, L. Microstructure and properties of high-Si high-Mn bainitic steel after heat treatment. *Mater. Res. Express* **2019**, *6*. [CrossRef]
18. Palaksha, P.A.; Ravishankar, K.S. Influence of Austempering Heat Treatment on Microstructure and Mechanical Properties of Medium Carbon High Silicon Steel. *IOP Conf. Ser. Mater. Sci. Eng.* **2017**, *225*. [CrossRef]
19. Son, J.Y.; Kim, J.H.; Kim, W.B.; Ye, B.J. Effects of austempering conditions on the microstructures and mechanical properties in Fe-0.9%C-2.3%Si-0.3%Mn steel. *Met. Mater. Int.* **2010**, *16*, 357–361. [CrossRef]
20. Kumar, R.; Dwivedi, R.K.; Ahmed, S. Stability of Retained Austenite in Carbide Free Bainite during the Austempering Temperature and its Influence on Sliding Wear of High Silicon Steel. *Silicon* **2020**. [CrossRef]
21. Acharya, P.P.; Udupa, R.; Bhat, R. Microstructure and mechanical properties of austempered AISI 9255 high-silicon steel. *Mater. Sci. Technol.* **2018**, *34*, 355–365. [CrossRef]
22. Zhao, J.; Li, J.; Ji, H.; Wang, T. Effect of austenitising temperature on mechanical properties of nanostructured bainitic steel. *Materials* **2017**, *10*, 874. [CrossRef] [PubMed]
23. Franceschi, M.; Pezzato, L.; Gennari, C.; Fabrizi, A.; Polyakova, M.; Konstantinov, D.; Brunelli, K.; Dabalà, M. Effect of intercritical annealing and austempering on the microstructure and mechanical properties of a high silicon manganese steel. *Metals* **2020**, *10*, 1448. [CrossRef]
24. Lucio-Garcia, M.A.; Gonzalez-Rodriguez, J.G.; Casales, M.; Martinez, L.; Chacon-Nava, J.G.; Neri-Flores, M.A.; Martinez-Villafañe, A. Effect of heat treatment on H2S corrosion of a micro-alloyed C-Mn steel. *Corros. Sci.* **2009**, *51*, 2380–2386. [CrossRef]
25. Isfahany, A.N.; Saghafian, H.; Borhani, G. The effect of heat treatment on mechanical properties and corrosion behavior of AISI420 martensitic stainless steel. *J. Alloys Compd.* **2011**, *509*, 3931–3936. [CrossRef]
26. Pezzato, L.; Lago, M.; Brunelli, K.; Breda, M.; Calliari, I. Effect of the Heat Treatment on the Corrosion Resistance of Duplex Stainless Steels. *J. Mater. Eng. Perform.* **2018**, *27*, 3859–3868. [CrossRef]
27. Bignozzi, M.C.; Calcinelli, L.; Carati, M.; Ceschini, L.; Chiavari, C.; Masi, G.; Morri, A. Effect of Heat Treatment Conditions on Retained Austenite and Corrosion Resistance of the X190CrVMo20-4-1 Stainless Steel. *Met. Mater. Int.* **2019**, *26*, 1318–1328. [CrossRef]
28. Hill, H.; Raab, U.; Weber, S.; Theisen, W.; Wollmann, M.; Wagner, L. Influence of heat treatment on the performance characteristics of a plastic mold steel. *Steel Res. Int.* **2011**, *82*, 1290–1296. [CrossRef]
29. Candelária, A.F.; Pinedo, C.E. Influence of the heat treatment on the corrosion resistance of the martensitic stainless steel type AISI 420. *J. Mater. Sci. Lett.* **2003**, *22*, 1151–1153. [CrossRef]
30. Han, C.F.; Wang, Q.Q.; Sun, Y.F.; Li, J. Effects of Molybdenum on the Wear Resistance and Corrosion Resistance of Carbidic Austempered Ductile Iron. *Metallogr. Microstruct. Anal.* **2015**, *4*, 298–304. [CrossRef]
31. Modiano, S.; Fugivara, C.S.; Benedetti, A.V. Effect of citrate ions on the electrochemical behaviour of low-carbon steel in borate buffer solutions. *Corros. Sci.* **2004**, *46*, 529–545. [CrossRef]

32. Yuan, L.; Liu, Q.; Li, H.; Gao, B. Effect of Austenitizing Temperature on the Bainitic Transformation in a High-Carbon High-Silicon Steel. *Met. Sci. Heat Treat.* **2015**, *57*, 156–160. [CrossRef]
33. Podder, A.S.; Bhadeshia, H.K.D.H. Thermal Stability of Austenite Retained in Bainitic Steels. *Mater. Sci. Eng. A* **2010**, *527*, 2121–2128. [CrossRef]
34. Putatunda, S.K. Influence of austempering temperature on microstructure and fracture toughness of a high-carbon, high-silicon and high-manganese cast steel. *Mater. Des.* **2003**, *24*, 435–443. [CrossRef]
35. Moghaddam, P.V.; Hardell, J.; Vuorinen, E.; Prakash, B. Effect of retained austenite on adhesion-dominated wear of nanostructured carbide-free bainitic steel. *Tribol. Int.* **2020**, *150*, 106348. [CrossRef]
36. Luo, Q. A new XRD method to quantify plate and lath martensites of hardened medium-carbon steel. *J. Mater. Eng. Perform.* **2016**, *25*, 2170–2179. [CrossRef]
37. Hill, H.; Huth, S.; Weber, S.; Theisen, W. Corrosion properties of a plastic mould steel with special focus on the processing route. *Mater. Corros.* **2011**, *62*, 436–443. [CrossRef]
38. Hsu, C.H.; Chen, M.L. Corrosion behavior of nickel alloyed and austempered ductile irons in 3.5% sodium chloride. *Corros. Sci.* **2010**, *52*, 2945–2949. [CrossRef]
39. Yang, J.; Lu, Y.; Guo, Z.; Gu, J.; Gu, C. Corrosion behaviour of a quenched and partitioned medium carbon steel in 3.5 wt.% NaCl solution. *Corros. Sci.* **2018**, *130*, 64–75. [CrossRef]
40. Aithal, P.M.; Vijayan, V.; Surendranathan, A.O.; Udupa, K.R.; Samuel, K.G. Corrosion behavior of high and low temperature austempered ductile iron (ADI) in iron ore slurry. *Mater. Perform. Charact.* **2017**, *6*, 369–378. [CrossRef]
41. ASTM G102 Standard Practice for from Electrochemical Measurements. *Astm* **2015**, *89*, 1–7. [CrossRef]
42. Pezzato, L.; Brunelli, K.; Dolcet, P.; Dabalà, M. Plasma electrolytic oxidation coating produced on 39NiCrMo3 steel. *Surf. Coat. Technol.* **2016**, *307*, 73–80. [CrossRef]
43. Hamadou, L.; Kadri, A.; Benbrahim, N. Characterisation of passive films formed on low carbon steel in borate buffer solution (pH 9.2) by electrochemical impedance spectroscopy. *Appl. Surf. Sci.* **2005**, *252*, 1510–1519. [CrossRef]
44. Oblonsky, L.J. A surfave enhanced Raman spectroscopic study of the passive films formed in borate buffer of iron, nickel, chromium and stainless steel. *Science* **1995**, *37*, 17–41.

materials

Article

Influence of Chloride Concentration on Stress Corrosion Cracking and Crevice Corrosion of Austenitic Stainless Steel in Saline Environments

Chun-Ping Yeh *, Kun-Chao Tsai and Jiunn-Yuan Huang

Institute of Nuclear Energy Research (INER), 1000 Wenhua Rd., Longtan District, Taoyuan City 32546, Taiwan; tsaijohn@iner.gov.tw (K.-C.T.); jyhuang@iner.gov.tw (J.-Y.H.)

* Correspondence: cpyeh@iner.gov.tw; Tel.: +886-3-471-1400

Received: 27 October 2020; Accepted: 8 December 2020; Published: 10 December 2020

Abstract: Stainless steels are used as canister materials for interim storage of spent fuel. Crevice corrosion has proved to be a safety concern of 304L stainless steel spent fuel canisters, when exposed to the saline environments of coastal sites. To study the effects of chloride concentration and test duration on the crevice corrosion behavior, and the effect of relative humidity on the initiation of discrete SCC cracks, a test program was conducted on the 304L steel specimens sprayed with synthetic sea water of 3.5 wt.%. The salt-deposited specimens, wrapped up with a crevice former to form a crevice configuration, were then exposed to an environment at 45 °C with a pre-set 45%, 55%, and 70% relative humidity (RH), for 400 h and 10,000 h, respectively. The surface features and crack morphology of the tested 304L stainless-steel specimens were examined by energy-dispersive spectrometry (EDS) and electron back scatter diffraction (EBSD). For the specimens deposited with a chloride concentration of 1 g/m^2, no cracks were found in the corroded regions after 400-h exposure, whereas SCC cracks were observed with the specimens tested for 10,000 h at all three pre-set relative humidity. The specimens tested at the pre-set relative humidity 45% are characterized with discrete SCC cracks, but, on the other hand, those exposed to the environments of 55% and 70% relative humidity show SCC cracks of distinct features. From the results of 10,000-h tests, it is inferred that the chloride concentration threshold for SCC initiation of 304L stainless steel at 45 °C is between 0.1 g/m^2 and 1 g/m^2.

Keywords: chloride concentration; crevice corrosion; relative humidity; stainless steel; stress corrosion cracking

1. Introduction

Spent nuclear fuel, when discharged from reactors, is first stored in the spent fuel pool to cool down fuel temperature and radioactivity. Prior to final disposal, spent nuclear fuels are loaded and stored in the austenitic stainless-steel canister of the dry cask storage system. The stainless-steel canister is filled with helium gas to facilitate spent fuel decay heat removal by natural convection. Therefore, the surface of canister is exposed to ambient air. In Taiwan, the dry cask storage facilities for spent fuel are located in coastal regions. Sea salt could deposit on the canister surface by an airborne process or seawater spray. Stainless steels canisters are well-known to be susceptible to atmospheric localized corrosion in the presence of chloride ions [1–5]. The dry cask storage systems, as planned, are to operate for about 40–60 years [6]. Therefore, the integrity of the canisters must be maintained to ensure the spent fuel interim storage safety.

Stainless steels which with passive film on its surface have excellent corrosion resistance, and are often used in some aggressive environments. Nevertheless, austenitic stainless steels are inclined to suffer from localized corrosion, especially in the presence of crevice sites on the surfaces [7].

Crevice corrosion is a corrosion on metal surfaces at the gap between two surfaces, and the mass transfer process was restricted due to the narrow crevice configuration. It is inferred that crevice corrosion behaviors of stainless steel are associated with the evolution process of the passive film which inside crevice [8]. With increasing Cl^- ions concentration, the passive film is more unstable, and it leads to more severe crevice corrosion [9].

For crevice corrosion, by means of trapping chloride deposits on the surface of canister, it can be an essential factor of atmospherically-induced stress corrosion cracking (AISCC). The initiation and propagation of crevice corrosion are highly dependent on the temperature, aggressive ions (e.g., Cl^-) and crevice geometry. In the horizontal and vertical canister systems, the configuration of crevice is formed where the canister contacts with the support structure of storage module [10].

As for stainless steel, stress corrosion cracking (SCC) is susceptible in aggressive environments, for instance, the SCC caused by chlorides and particles of sea salt. At crevice sites, the increase of corrosion products volume could result in local stresses to prompt the happening of stress corrosion cracking. Moreover, Tani et al. reported that when crevice corrosion occurs, it will contribute to initiation of stress corrosion cracking on the surfaces beneath particles of sea salt [11].

Crevice corrosion behavior can be explained by two common mechanisms: Critical crevice solution (CCS) and IR drop mechanism [8]. As far as the critical crevice solution mechanism is concerned, it indicates that the depletion of oxygen within the crevice could cause the acidification of solution at crevice sites, which result in the breakdown of passive film and initiation of crevice corrosion. This theory accentuates the accumulation of aggressive ions (e.g., Cl^-) at crevice sites and then the depassivation followed by base metal dissolution [12,13]. As far as the IR drop mechanism is concerned, crevice corrosion takes place by potential transition from passive state to active state, when the value of IR exceeds a critical threshold. The IR drop involves the value of potential drop which is obtained by multiplication of the resistance of the electrolyte and the current at crevice sites [14,15]. Aggressive ions (e.g., Cl^-) encourage the crevice corrosion by raising the corrosion current, resulting in an increase of the IR value [16,17].

Scatigno and coworkers reported the influence of deposited salt on chloride-induced stress corrosion cracking in 304L steel at 90 °C with 70% RH. They suggested that atmospherically-induced stress corrosion cracking was occurred in the 1.7 g/m^2 $MgCl_2$ chloride concentration [18]. Shoji and coworkers presented a research for the atmospherically-induced stress corrosion cracking phenomenon with regard to relative humidity, temperature, and chloride ions, and demonstrated it happening in type 316L and 304L steels under ambient situations. It is inferred that $MgCl_2$ is the main sea-salt constituent for the promotion of low temperature atmospherically-induced stress corrosion cracking in type 316L and 304L steels [19]. For the corrosivity of chloride deposition on the canister surface under the given exposure ambient conditions, it was reported in the following descending order: $CaCl_2 > MgCl_2 > NaCl$. Apparently, it was controlled by the concentration of chloride on surfaces, due to the interaction of a given amount of salt with water vapor in the air [20]. In a summary of investigations on AISCC phenomena of stainless steels, Engelberg et al. summarized a susceptibility order for occurrence of AISCC of 304L steels. It was provided that the relative humidity ranges which atmospherically-induced stress corrosion cracking occurred in the 25 g/m^2 $MgCl_2$ chloride concentration, based on the experimental results of Shoji et al. [21]. Padovani et al. suggested that 316L stainless steel is susceptible to AISCC at 30 °C under $MgCl_2$ deposition [22]. Moreover, the critical RH for type 304L steel SCC initiation after 5000 h experiments with chloride concentration 1 g/m^2 is between 45% and 55% [23].

Masuda and coworkers illustrated the SCC behavior of the 304 stainless steel in terms of the pit growth, slip deformation and surface potential distribution. SCC tests were done at 28% RH and 343 K with $MgCl_2$ droplets on the specimen. Discontinuous cracks were often observed near the crack tip [24]. Based on the observations on the SCC phenomenon of 321 stainless steel in an $MgCl_2$ solution, Qiao and coworkers suggested that the interaction between discontinuous microcracks and main crack could enhance the effective stress intensity factor and promote crack coalescence, leading to mechanical fracture of the ligaments between the cracks in stainless steel [25].

The goal of the study is to evaluate the behavior of crevice corrosion for type 304L steel under different conditions of chloride concentrations, and test duration. Moreover, the effect of relative humidity on the initiation of discrete SCC cracks was also examined in this research to have a better understanding of the initiation of SCC at the sites of crevice corrosion under different relative humidity.

2. Experimental Procedure

Figure 1a,b present the dimensions of the crevice former and the test specimen and, respectively, and an assembled image of a test specimen and a crevice former is shown in Figure 1c. The thickness of the test specimen is 3 mm. The chemical compositions of 304L stainless steel used in this research are listed in Table 1. Specimens were manufactured from type 304L steel, and the material of crevice former was polytetrafluoroethylene (PTFE). Duplicate tests were conducted for each experimental condition. Before crevice corrosion test, the specimen surfaces were ground with 2000-grit abrasive paper.

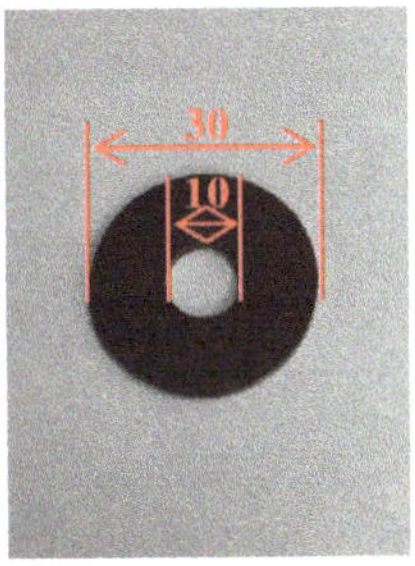

(a)

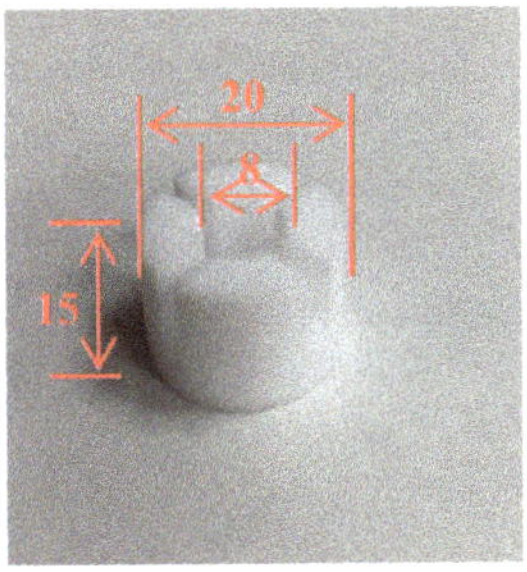

(b)

(c)

Figure 1. Image of the specimen and the crevice former used for crevice corrosion experiment. (**a**) Specimen, (**b**) crevice former, and (**c**) assembled image. (Unit: mm).

Table 1. Chemical compositions of 304L stainless steel used in this work.

Element	C	S	Si	Ni	Cr	Mn	Fe
wt.%	0.017	0.0290	0.450	9.000	18.000	1.540	Bal.

The chemical compositions of sea salt used in this study, based on ASTM D 1141-98 (13) Formula a, Table X1.1, Sec.6, are listed in Table 2. The specimens were firstly sprayed with 3.5 wt.% synthetic sea water and then dried on the hot plate at 60 °C for 15 min. After the specimens were dried, the mass change of the specimens was evaluated for the calculation of chloride concentration in g/m^2. The specimen chloride concentration was at one of the pre-determined levels of 0.1 and 1 g/m^2. The specimen and the crevice former were fixed with a M6 screw by applying a torque of 1.13 N m to form a crevice configuration. The constant-temperature/humidity chambers were kept at a combination of the ambient temperature of 45 °C with a RH = 45%, 55%, and 70%, for 400 h and 10,000 h, respectively. Samples for microstructural inspection were obtained through a metallographic process of sample preparation. The specimens were washed ultrasonically in deionized (DI) water and then dried cautiously and mounted in the resin. The mounted samples were polished by using Aluminum oxide powder. Surface features and microstructures of the tested specimens were examined with an optical microscope and a scanning electron microscope (SEM). The technique of energy-dispersive spectrometer (EDS) is widely used for providing semi-quantitative analysis of materials. In this study, the crack regions were selected for the EDS mapping analysis. The elemental distributions at crack regions were fully mapped by collecting a full EDS spectrum at each pixel for the area of the crack region. The technique of EDS was also used to determine the compositions of the corrosion bands. Furthermore, a SEM equipped with an electron backscatter diffraction (EBSD) detector was employed to examine the SCC features of the specimens. In addition, the kernel average misorientation (KAM) map for the tested specimens can be obtained by inputting the EBSD map into the software for further data processing. The crack morphology of the specimens was characterized with an SEM.

Table 2. Chemical compositions of sea salt used in this work.

Composition	NaCl	$MgCl_2$	Na_2SO_4	$CaCl_2$	KCl	$NaHCO_3$	KBr	$SrCl_2$	H_3BO_3	NaF
wt%	58.490	26.460	9.750	2.765	1.645	0.477	0.238	0.095	0.071	0.007

3. Results and Discussion

3.1. *Analysis of Surface Morphology*

Figures 2 and 3 are the macrographs of the 304L steel specimens with chloride concentration of 0.1 g/m^2 after testing for 400 and 10,000 h at three different relative humidity levels, respectively. A trace of crevice corrosion was discerned with the specimens after 400-h testing. Significant crevice corrosion was observed with the specimens tested for 10,000 h. As illustrated in Figures 2a–c and 3a–c, the rusted areas of both 400-h and 10,000-h tested specimens decrease with increasing the relative humidity from 45% RH to the least at 55% RH, but then increase at 70% RH; the rusted areas of the specimens tested for 10,000 h at 70% RH are clearly larger than those tested at 45% and 55% RH. Since the chloride concentration of 0.1 g/m^2 is much lower than those reported in the literature [19–21], the local chloride concentration may be reduced due to the higher relative humidity of 55% compared to the relative humidity of 45%, leading a limited concentration of chloride transported to the crevice sites of the specimens. As a result, it shows a tendency for a decrease in the corrosion effect when the humidity goes from 45% to 55%. However, due to a low chloride concentration of the specimens tested, it is necessary to have a higher relative humidity environment to facilitate the transportation of a sufficient amount of chlorine (Cl) to the crevice sites for the initiation of corrosion. When the humidity goes from 55% to 70%, the effect of higher relative humidity has an advantage over the effect of chlorine dilution, resulting in an increase in the rusted areas. It is hypothesized that the rusted areas are influenced by a synergy effect of relative humidity and chlorine dilution. Moreover, the deliquescence relative humidity (DRH) of NaCl at 45 °C is about 73% [26], which is close to the relative humidity of 70% used in this work. With the possible additional NaCl deliquescence, it is inferred that the corrosion effect may be enhanced, leading to the reduction of chlorine dilution effect at RH = 70%. Figures 4 and 5 respectively present the SEM micrographs of the specimens with chloride concentration of 0.1 g/m^2 after 400-h and 10,000-h testing at three different relative humidity levels. Rust on the specimens shows evidence of crevice corrosion induced by chloride. The morphologic features of crevice corrosion vary with relative humidity. Figure 4a,c reveal that there are some rust spots on the specimens tested for 400 h, and that shallow corrosion exists beneath the rust spots. Compared to Figure 4a,c for those exposed to 45% and 70% RH, the corroded areas of the specimens tested at 55% RH are smaller, Figure 4b. As with the 400-h tested specimens, Figure 4a,c, the specimens tested for 10,000 h show some rusted spots, beneath which there exists shallow corrosion, as exemplified in Figure 5a,c. And the corroded areas on the specimens exposed to 55% RH, Figure 5b, are smaller, relative to those shown in Figure 5a,c for the specimens tested at 45% and 70% RH. Moreover, no cracks were found with all the specimens tested for 400 and 10,000 h at three different relative humidity levels, Figures 4 and 5. In relative terms, the corroded areas on the 10,000-h tested specimens are larger than those on the corresponding specimens tested for 400 h.

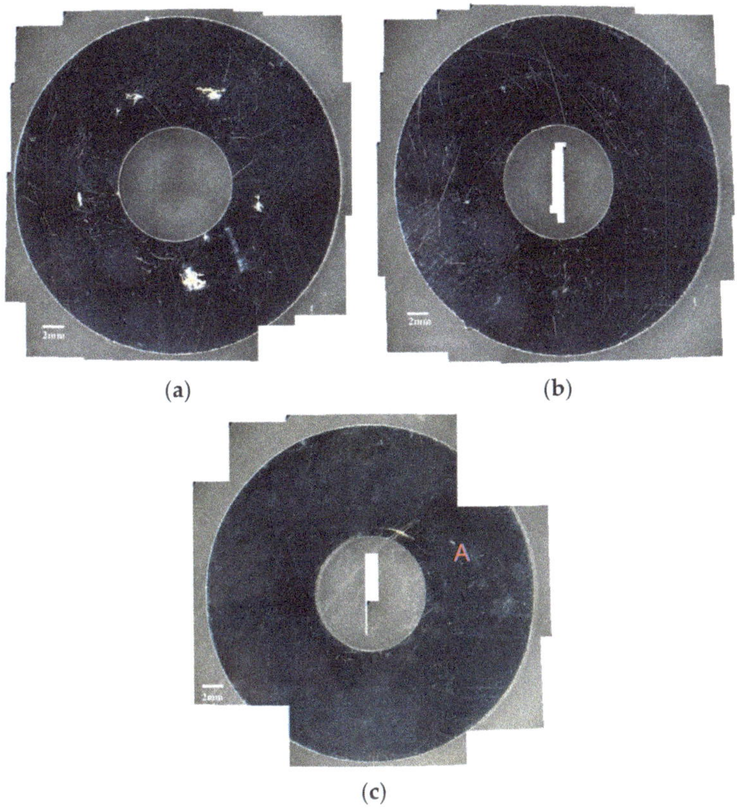

Figure 2. Macrographs of the corroded regions of the specimens with chloride concentration of 0.1 g/m^2 after 400-h testing at: (**a**) Relative humidity (RH) = 45%, (**b**) RH = 55%, and (**c**) RH = 70%.

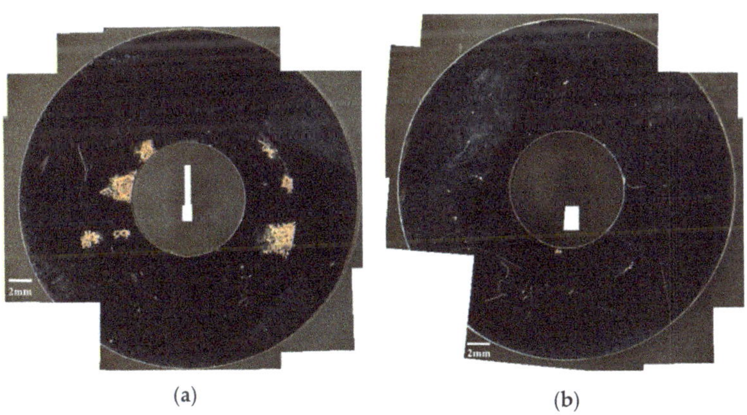

Figure 3. *Cont.*

(c)

Figure 3. Macrographs of the corroded regions of the specimens with chloride concentration of 0.1 g/m^2 after 10,000-h testing at: (**a**) RH = 45%, (**b**) RH = 55%, and (**c**) RH = 70%.

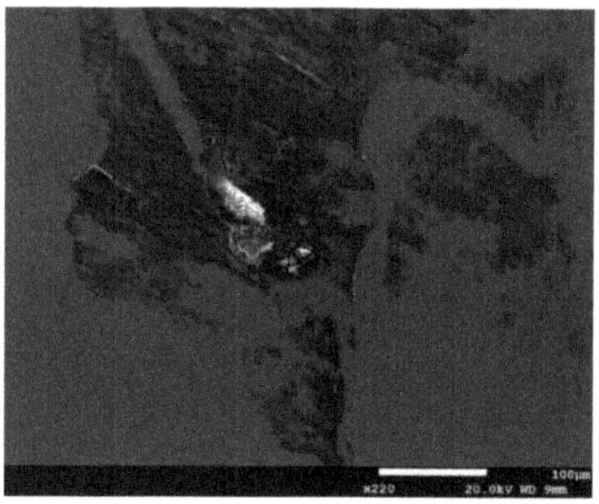

(a)

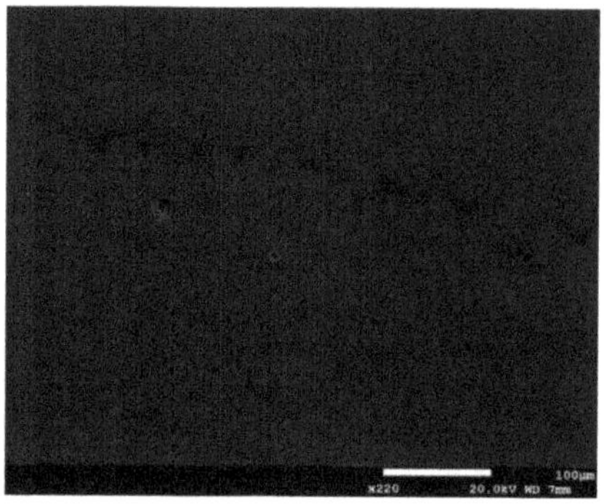

(b)

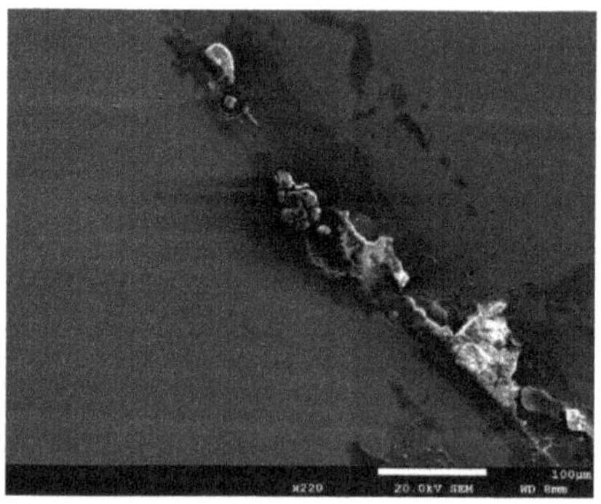

(c)

Figure 4. SEM micrographs of the specimens with chloride concentration of 0.1 g/m^2 after 400-h testing at: (**a**) RH = 45%, (**b**) RH = 55%, and (**c**) RH = 70%.

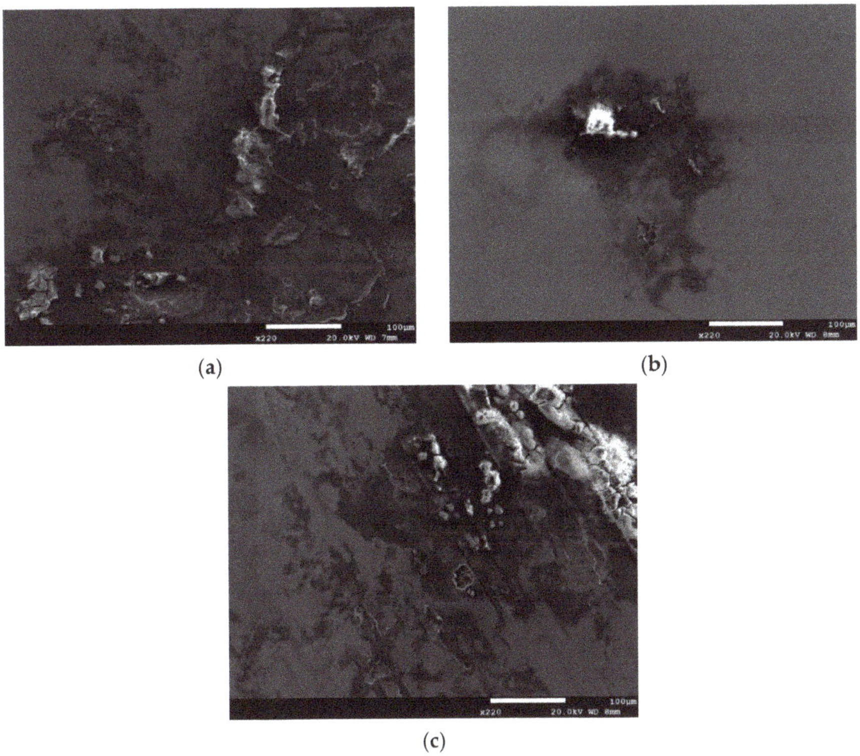

Figure 5. SEM micrographs of the specimens with chloride concentration of 0.1 g/m^2 after 10,000-h testing at: (**a**) RH = 45%, (**b**) RH = 55%, and (**c**) RH = 70%.

Figures 6 and 7 are the macrographs of the 304L austenitic stainless steel specimens deposited with 1 g/m^2 chloride concentration after 400-h and 10,000-h testing at three different relative humidity levels, respectively. Crevice corrosion was observed with the specimens after 400-h testing. The rusted areas on the specimens tested for 400 h decrease with increasing the relative humidity from 45% RH to a minimum at 55% RH, but then increase at RH = 70%, as shown in Figure 6. A comparison of Figure 7a–c illustrates that there is little difference in the rusted areas on the specimens after testing for 10,000 h under the RH = 45% and 55%, but that the areas of rusted increase significantly at 70% RH. Furthermore, for the 10,000-h tests, the specimens with 1 g/m^2 chloride deposit, Figure 7, show more serious corrosion at all three relative humidity levels than those with 0.1 g/m^2 chloride deposit, Figure 3. Figures 8 and 9 show the SEM micrographs of the specimens deposited with 1 g/m^2 chloride tested at different relative humidity levels for 400 and 10,000 h, respectively. As with the specimens with 0.1 g/m^2, rust on the specimens with 1 g/m^2 is evidence of crevice corrosion induced by chloride. Figure 8a,b reveal some rust spots present on specimens. It also illustrates shallow corrosion beneath rust spots. In addition to rust spots, Figure 8c illustrates larger corrosion pits on the specimens exposed to 70% RH, compared to the corrosion pits shown in Figure 8a,b for the specimens tested at 45% and 55% RH. No cracks were found with the specimens deposited with 1 g/m^2 chloride after testing for 400 h at three different relative humidity levels, as exemplified in Figure 8. The specimens after 10,000-h testing were severely corroded. The specimens exposed to 70% RH, Figure 9c, appear more severely corroded than those subjected to 45% and 55% RH of Figure 9a,b. Figure 9 shows more rust spots, shallow corrosion beneath the rust spots, and corrosion pits.

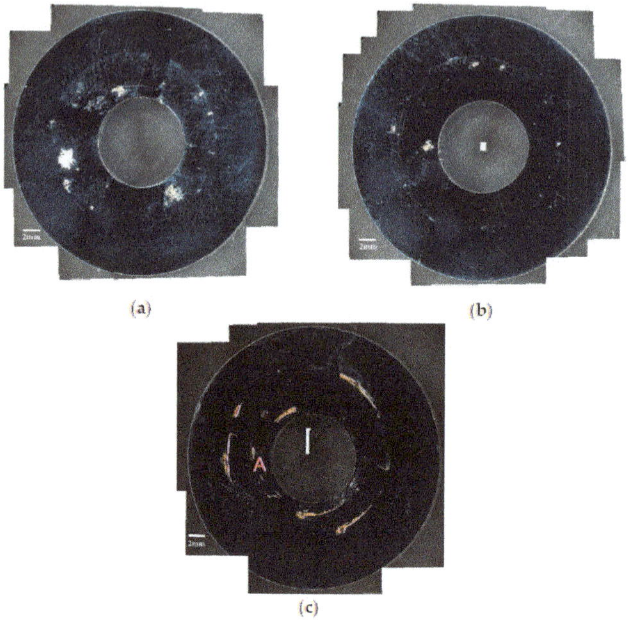

Figure 6. Macrographs of the specimens with 1 g/m^2 chloride concentration after 400-h testing at: (**a**) RH = 45%, (**b**) RH = 55%, and (**c**) RH = 70%.

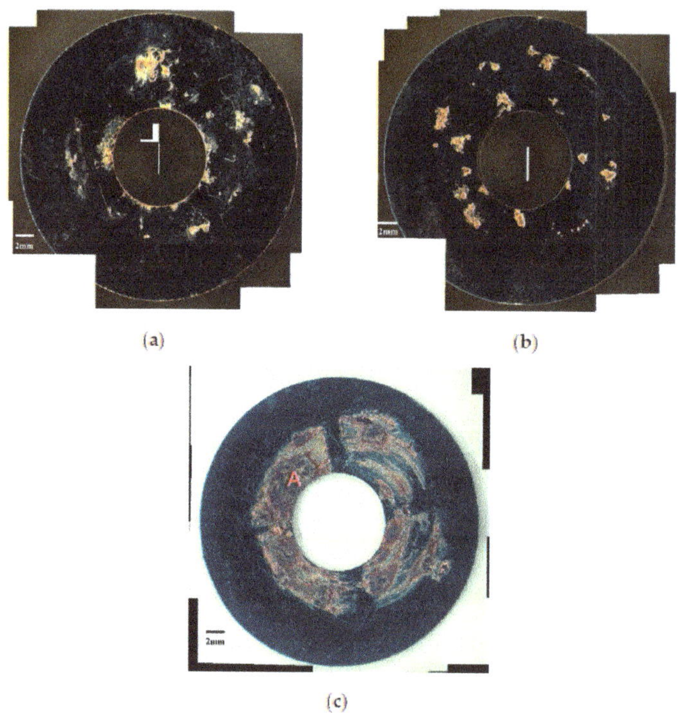

Figure 7. Macrographs of the specimens with 1 g/m^2 chloride concentration after 10,000-h testing at: (**a**) RH = 45%, (**b**) RH = 55%, and (**c**) RH = 70%.

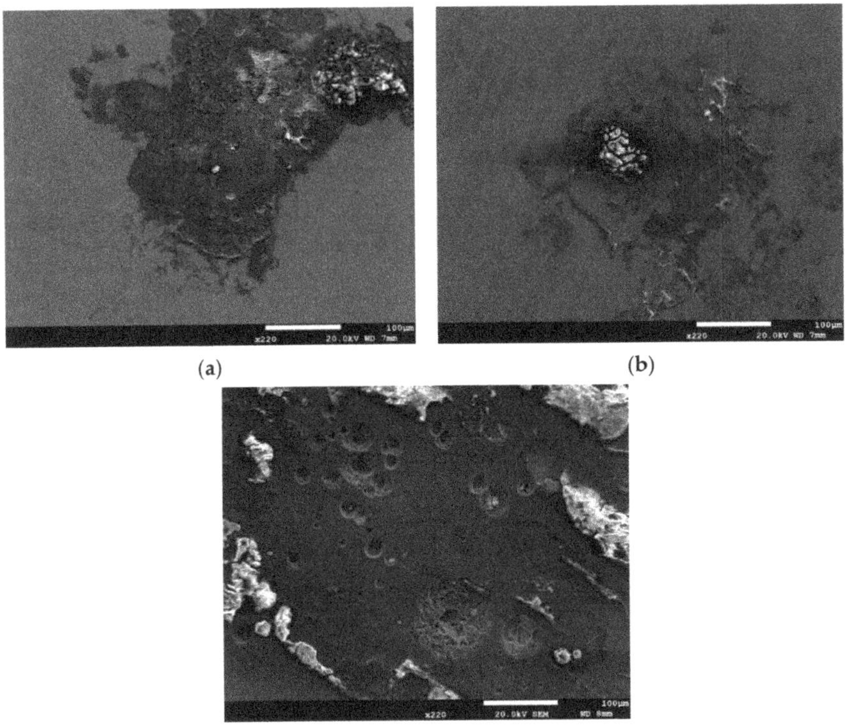

(**a**) (**b**) (**c**)

Figure 8. SEM micrographs of the specimens with 1 g/m^2 chloride concentration after 400-h testing at: (**a**) RH = 45%, (**b**) RH = 55%, and (**c**) RH = 70%.

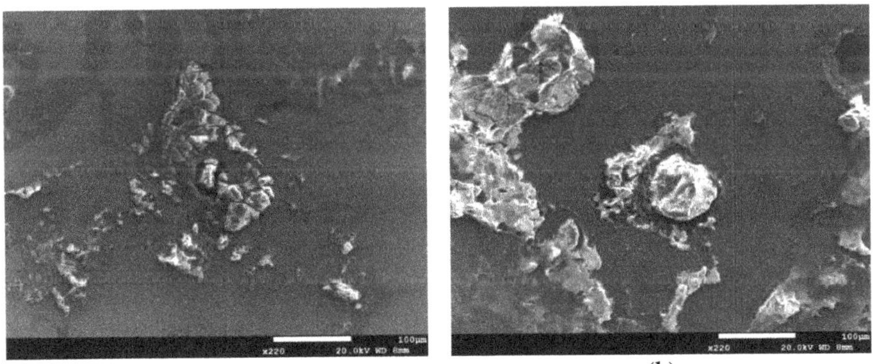

(**a**) (**b**)

Figure 9. *Cont.*

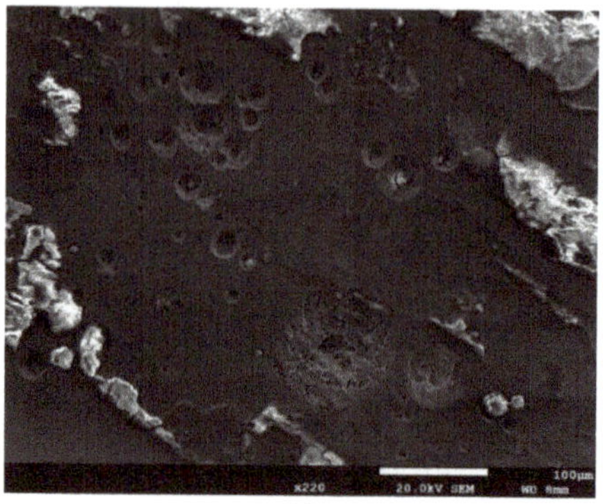

(c)

Figure 9. SEM micrographs of the specimens with 1 g/m^2 chloride concentration after 10,000-h testing at: (**a**) RH = 45%, (**b**) RH = 55%, and (**c**) RH = 70% with chloride concentration of 1 g/m^2.

SEM micrographs of Figure 10 reveal corrosion bands on the specimens with chloride concentration of 0.1 g/m^2 and 1 g/m^2, when exposed to 70% relative humidity for 400 and 10,000 h. Figure 10a–d present the SEM micrographs of corrosion bands of Figures 2, 3, 6 and 7 at point A, respectively. Corrosion bands are a unique corrosion feature of the specimens exposed to 70% relative humidity, not occurring with the specimens tested at 45% and 55% RH. Figure 10a,c show corrosion bands present on the 400-h tested specimens with chloride concentration of 0.1 g/m^2 and 1 g/m^2, respectively. For the specimens deposited with the same amounts of chloride, the number of corrosion bands on the specimen increases with increasing the test time, as illustrated by Figure 10a,b for the specimens deposited with 0.1 g/m^2 chloride and by Figure 10c,d for those with 1 g/m^2 chloride deposit after testing for 400 and 10,000 h, respectively. A comparison of Figure 10b,d further illustrates that the width of the corrosion band increases with increasing the chloride concentration of the specimen.

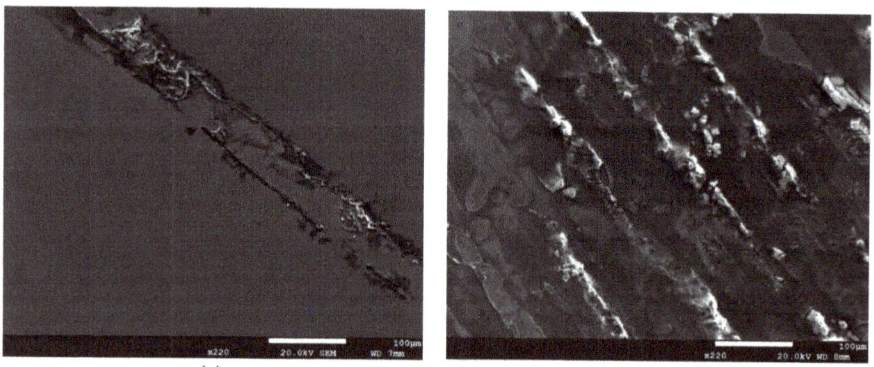

(a) (b)

Figure 10. *Cont.*

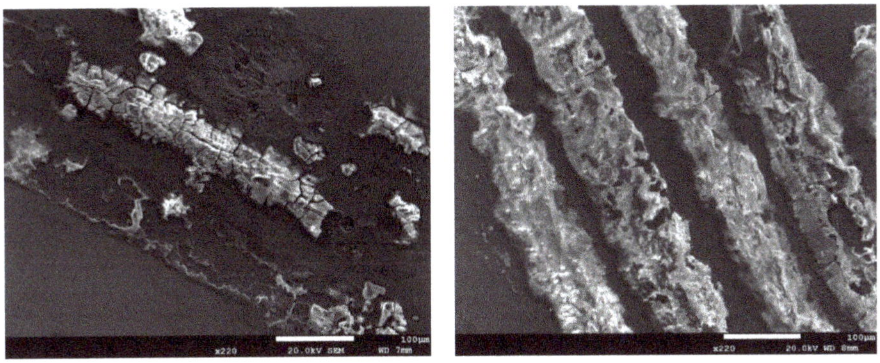

(**c**) (**d**)

Figure 10. SEM micrographs of corrosion bands of specimens deposited with: 0.1 g/m^2 chloride (**a**) after 400-h testing, (**b**) after 10,000-h testing and 1 g/m^2 chloride (**c**) after 400-h testing, and (**d**) after 10,000-h testing at 70% relative humidity.

3.2. *Analysis of EDS and EBSD*

Figure 11 depicts EDS examinations of corrosion bands on the specimen deposited with 1 g/m^2 chloride after testing for 10,000 h tests at 70% relative humidity. The EDS analysis results of corrosion bands are listed in Table 3. The chlorine and sulfur contents of Point A, the stainless-steel matrix, are too low to be detected. Points B, D, and E, located in the corrosion bands, were analyzed to have higher chlorine contents, and the oxide at Point C to contain a trace amount of chlorine. Furthermore, Points B, C, D, and E have higher S contents, compared to Point A, which may be accounted for by sulphates contained in synthetized sea water. Moreover, the weight % of Fe and O contents at Points B and D, located in the corrosion bands, were used to identify the possible empirical formula of corrosion products. It is hypothesized that the corrosion products may be $Fe(OH)_2$.

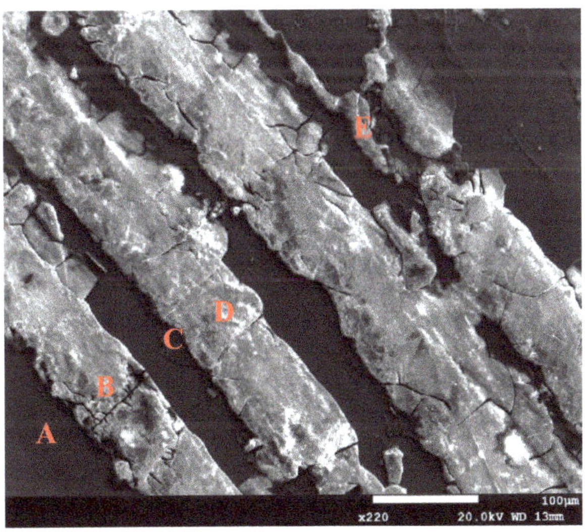

Figure 11. Energy-dispersive spectrometry (EDS) analysis of corrosion bands on the specimen with 1 g/m^2 chloride deposit after testing at 70% relative humidity for 10,000 h (EDS analysis results of Points A–E are listed in Table 3).

Table 3. EDS analysis results of corrosion bands (wt.%).

Location	O	Na	Mg	S	Cl	Ca	Cr	Mn	Fe	Ni
A	0.73	0.11	0.04	0.00	0.00	0.04	18.78	1.74	70.46	8.10
B	33.2	0.98	0.30	0.24	3.11	0.06	8.67	0.38	51.77	1.28
C	16.59	1.64	0.60	0.06	0.31	0.08	16.83	1.69	56.05	6.14
D	30.31	1.26	0.62	0.18	2.98	0.17	13.48	0.87	48.51	1.63
E	27.74	0.36	0.45	0.16	2.40	0.09	6.47	0.96	57.95	3.42

Cl^- ion is indispensable for the crevice corrosion initiation. The critical crevice solution mechanism ascribes the crevice corrosion initiation to the aggressive ions accumulation, particularly Cl^- at the crevice sites, resulting in the highly aggressive localized corrosion appearance that destroys the stainless steel passive film [27]. Points B and D have much higher Cl and O contents compared to Point C, which implies that the Cl^- ion may be accumulated more rapidly at those points to initiate crevice corrosion and form corrosion bands at the highest RH (RH = 70%). It is hypothesized that due to a low chloride concentration of the specimens tested, a higher relative humidity environment (RH = 70%) facilitate the transportation of chlorine (Cl) to a longer distance in straight lines to result in the linearity of the corrosion products. Therefore, corrosion bands were formed at local favorable sites and under favorable higher relative humidity environment conditions.

Figure 12 is the SEM micrographs revealing the morphology of stress corrosion cracking of the specimens deposited with 1 g/m^2 chloride after 10,000-h tests at different levels of relative humidity. Figure 12a shows discontinuous SCC cracks on the specimen exposed to 45% relative humidity, whereas Figure 12b,c demonstrate distinct SCC cracks on the specimens tested at RH = 55% and 70%. Moreover, Shoji et al. presented that $MgCl_2$ is the main sea-salt constituent for the promotion of low temperature atmospherically-induced stress corrosion cracking in type 316L and 304L steels. It was investigated that the AISCC phenomenon of specimens deposited with a chloride concentration of 25 g/m^2 by sea water [19]. Prosek et al. examined AISCC phenomenon of type 304 and 316L steel with a chloride concentration of 260 g/m^2 obtained by $MgCl_2$ droplets [20]. The chloride concentration of 1 g/m^2 used in this work is much lower than those reported in the literature [19–21]. Figure 12a shows short and shallow cracks on the specimens tested at 45% RH, as opposed to Figure 12b,c, long and deep cracks are present on those exposed to 55% and 70% RH. It is hypothesized that microcracks first nucleate discretely at favorable sites and, under favorable conditions, microcracks grow and then coalesce by breaking the ligaments between them into main cracks [24,25,28,29]. Based on the hypothesis, few short and shallow cracks observed on the specimens with chloride concentration of 1 g/m^2 at RH = 45%, the lowest RH of the three, could be accounted for by a limited amount of chloride transported to the crevice sites of the specimens. Due to a low chloride concentration of the specimens tested, it is necessary to have a higher relative humidity environment to facilitate the transportation of a sufficient amount of chlorine (Cl) to the crevice sites for crack nucleation. The chlorine (Cl) maps by EDS mapping, as shown in Figures 13c, 14c and 15c, are the evidence that could substantiate the aforementioned argument. It will be discussed in detail later on. However, no cracks were observed at three different relative humidity levels with the specimens deposited with chloride concentration of 0.1 g/m^2, as illustrated in Figure 5. It could be that the 0.1 g/m^2 chloride concentration was too low for the initiation of SCC in the specimens examined at the pre-set RH levels. From the results of this study, it is inferred that the chloride concentration threshold for the initiation of SCC for 304L steel at 45 °C is between 0.1 g/m^2 and 1 g/m^2. The increase of corrosion products volume could result in local stresses beneath the crevice former to prompt the happening of stress corrosion cracking. Moreover, crack length was observed to increase with the increasement of RH when comparing Figure 12b–c.

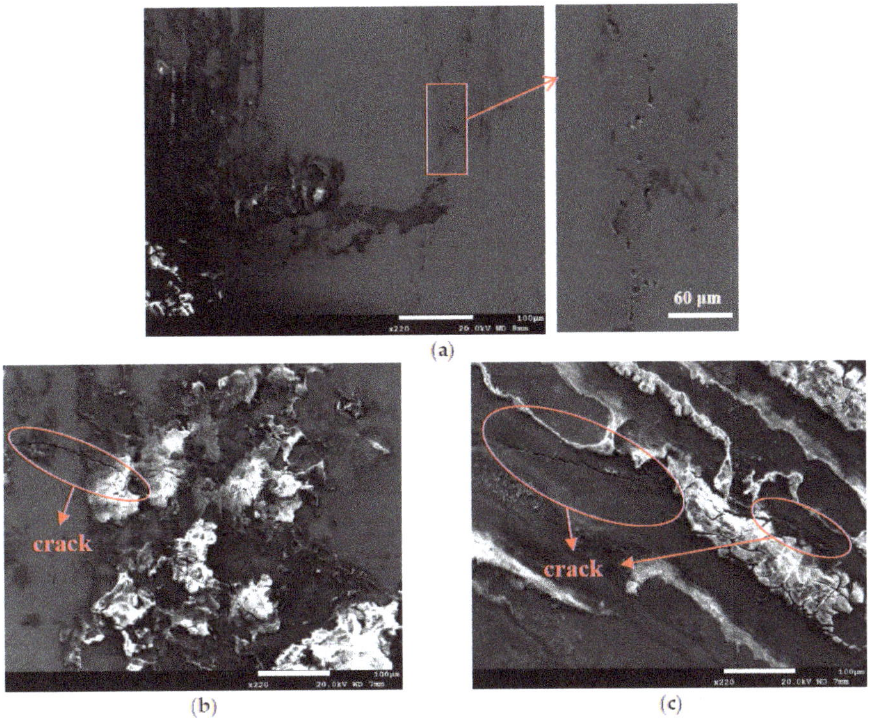

Figure 12. SEM morphology of stress corrosion cracks in the specimens with 1 g/m^2 chloride deposit after 10,000-h testing at: (**a**) RH = 45%, (**b**) RH = 55%, and (**c**) RH = 70%.

Figures 13–15 respectively present the EDS mapping for the crack regions of the specimens deposited with chloride concentration of 1 g/m^2 tested at RH = 45%, 55%, and 70% for 10,000 h. For the specimen exposed to 45% RH, the crack region is apparently enriched with oxygen, Figure 13b, but depleted with iron, Figure 13f. It may be explained that the corrosion products of iron were formed near the crack region, leading to the iron depletion at the crack region. Figure 13d,e respectively indicate enriched sulfur and manganese on the center and left bottom of the figure. The distribution of sulfur and manganese is only associated with the presence of non-metallic inclusions of MnS in the metal, and it have nothing to do with the crack. No indication of chlorine is discerned in the crack region of the specimen tested at 45% RH, as manifested in Figure 13c. For the specimens tested at 55% and 70% RH, it is found that more oxygen, chlorine and sulfur are enriched in the crack regions, as respectively demonstrated by Figures 14b–d and 15b–d, but that manganese, iron, are depleted, as manifested by Figure 14e,f and Figure 15e,f. It is also found that chromium is enriched near the crack regions, as respectively demonstrated by Figures 14g and 15g. Nickel is enriched near the crack regions, as manifested in Figure 15h. Moreover, a comparison of Figures 14c and 15c further shows that the concentration of chlorine in the crack region significantly increases with increasing the relative humidity from 55% to 70% RH.

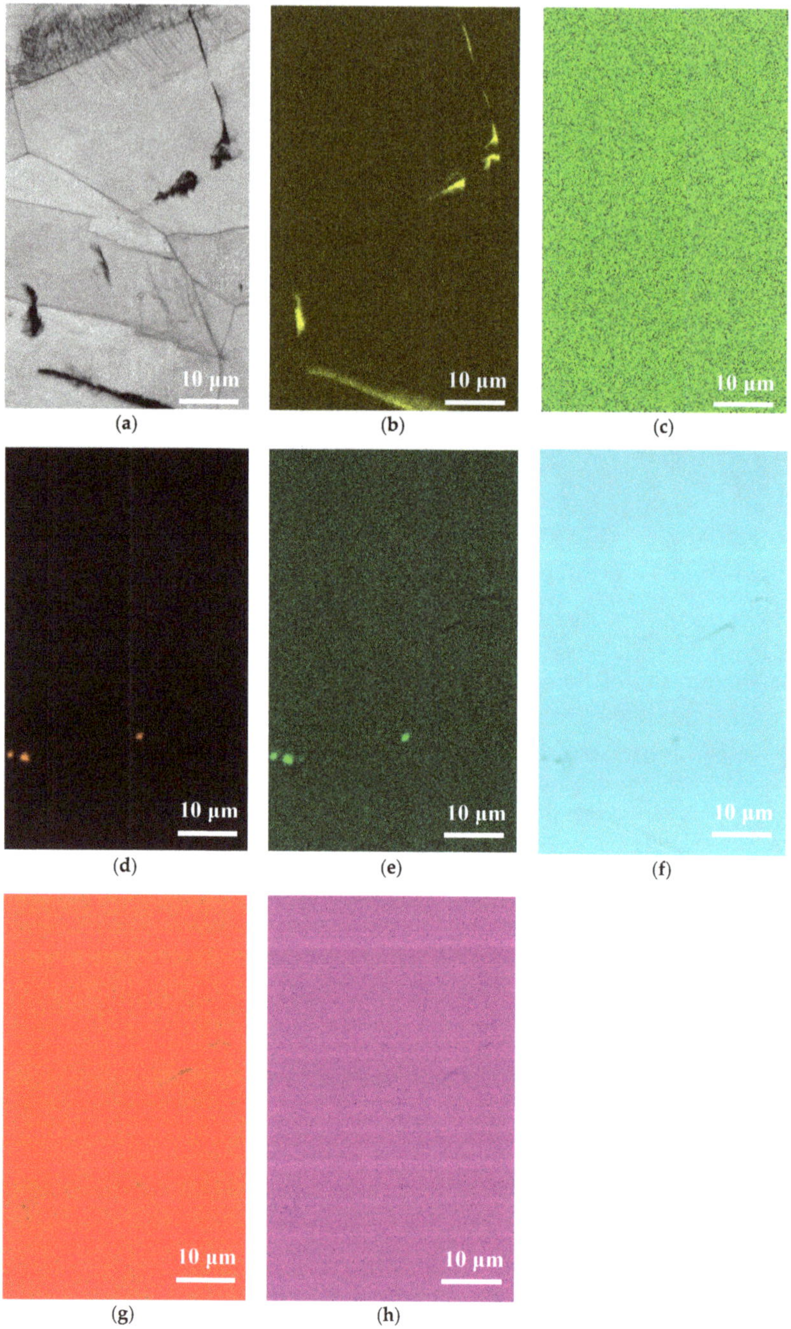

Figure 13. EDS mapping for the crack regions of the specimens with 1 g/m^2 chloride concentration tested for 10,000 h at RH = 45%: (**a**) Band contrast image, (**b**) O mapping, (**c**) Cl mapping, (**d**) S mapping, (**e**) Mn mapping, (**f**) Fe mapping, (**g**) Cr mapping, and (**h**) Ni mapping.

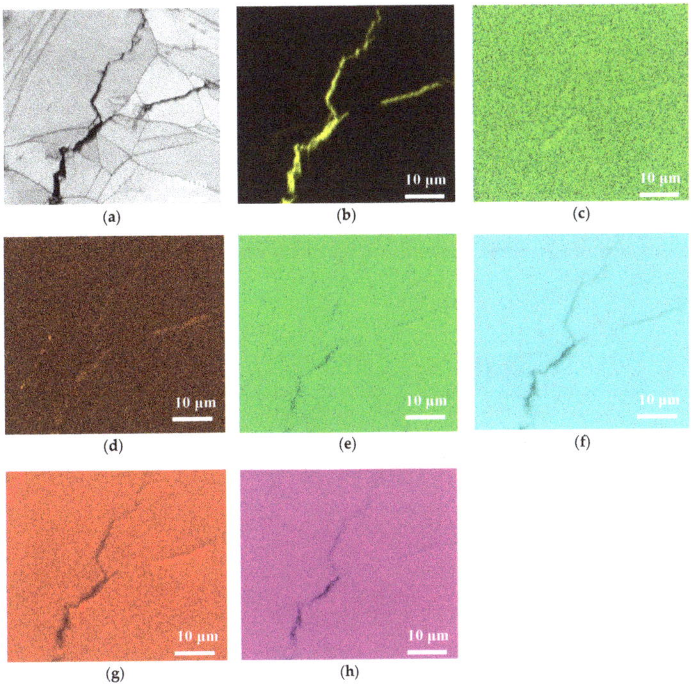

Figure 14. EDS mapping for the crack regions of the specimens with 1 g/m^2 chloride concentration tested for 10,000 h at RH = 55%: (a) Band contrast image, (b) O mapping, (c) Cl mapping, (d) S mapping, (e) Mn mapping, (f) Fe mapping, (g) Cr mapping, and (h) Ni mapping.

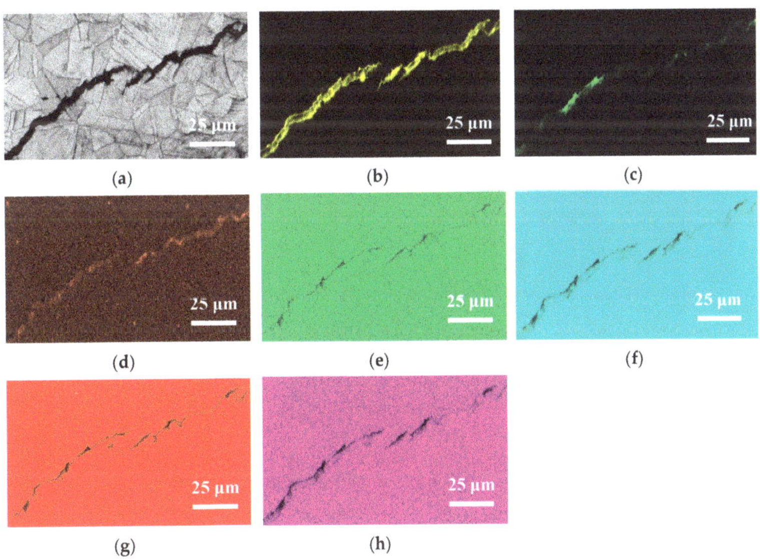

Figure 15. EDS mapping for the crack regions of the specimens with 1 g/m^2 chloride concentration tested for 10,000 h at RH = 70%: (a) Band contrast image, (b) O mapping, (c) Cl mapping, (d) S mapping, (e) Mn mapping, (f) Fe mapping, (g) Cr mapping, and (h) Ni mapping.

Figures 16–18 present the EBSD maps of the crack region of the specimen deposited with 1 g/m^2 chloride tested for 10,000 h at RH = 45%, 55% and 70%, respectively. Figure 16a,c, Figure 17a,c and Figure 18a,c demonstrate that the specimens were cracked by a trans-granular SCC (TGSCC) mode, which is in excellent agreement with the observations on TGSCC of stainless steel caused by chlorides at temperatures above 50 °C [11]. The kernel average misorientation (KAM) maps for specimens tested, Figures 16b, 17b and 18b, reveal high plastic strain around crack regions that could be related to plastic strain at the crack tip. The crack propagation was related to plastic strain at the crack tip. Although the discrete SCC cracks were only observed with specimens exposed to 45% RH, as shown in Figure 16, higher plastic strain was around crack region, and the crack mode of SCC is still TGSCC, similar to the continuous SCC cracks with specimens exposed to 55% and 70% RH, as respectively demonstrated by Figures 17 and 18.

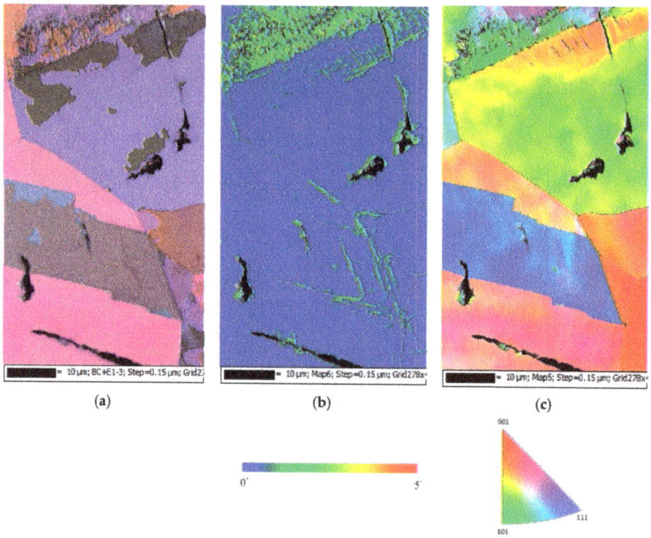

Figure 16. EBSD maps of the crack region of the specimen deposited with 1 g/m^2 chloride concentration tested for 10,000 h at RH = 45%: (**a**) Euler map, (**b**) kernel average misorientation (KAM) map, and (**c**) inverse pole figure (IPF) map.

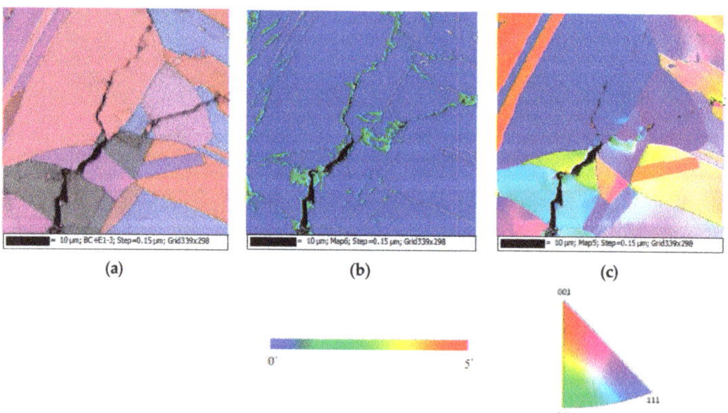

Figure 17. Electron backscatter diffraction (EBSD) maps of the crack region of the specimen deposited with 1 g/m^2 chloride concentration tested for 10,000 h at RH = 55%: (**a**) Euler map, (**b**) KAM map, and (**c**) IPF map.

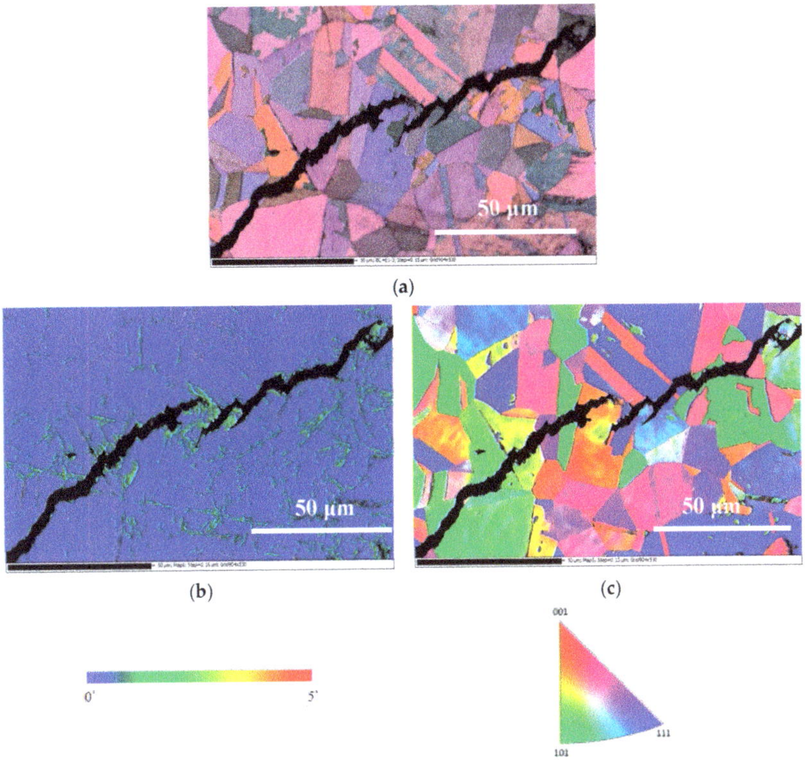

Figure 18. EBSD maps of the crack region of the specimen deposited with 1 g/m^2 chloride concentration tested for 10,000 h at relative humidity of 70%: (**a**) Euler map, (**b**) KAM map, and (**c**) IPF map.

4. Conclusions

304L steel crevice corrosion behaviors were investigated by testing the specimens at 45 °C under a combination of the chloride concentration of 0.1 and 1 g/m^2, the relative humidity of 45%, 55% and 70% and the test duration of 400 h and 10,000 h in this study. The effects of the chloride concentration and relative humidity on the corroded regions were evaluated and a chloride concentration threshold for the SCC initiation according to the test results was proposed. The chloride concentration of 0.1 and 1 g/m^2 used in this study differs from the concentration of salts in seawater. The conclusions of this study are given as below:

The specimens with chloride concentration of 0.1 g/m^2 had minimum corrosion areas at relative humidity of 55%, and there were no cracks at three different relative humidity levels. It may be that the concentration of 0.1 g/m^2 deposit chloride was too low for the initiation of SCC.

(1) The specimens with chloride concentration 1 g/m^2 tested for 400 h at 55% RH had minimum rusted areas. Nevertheless, there was little difference in the rusted areas of the specimens tested at 45% and 55% RH for 10,000 h, whereas the area of rust significantly increased when the RH increased to 70%. Some shallow corrosion beneath rust spots were found, but no cracks were observed in the corrosion region after 400-h testing, whereas SCC cracks were observed with the specimens tested for 10,000 h at all relative humidity levels. The discrete SCC cracks were observed with specimens exposed to 45% RH, but, on the other hand, distinct stress corrosion cracking cracks were found with those tested at 55% and 70% RH. This can be accounted for by a limited amount of chloride transported to the sites of crevice of the specimens with chloride

concentration of 1 g/m^2 exposed to 45% RH, the lowest RH. It may be necessary to have a higher RH environment to facilitate transportation of chloride to the crevice sites, because of the low chloride concentration of the specimens. The observations that the chlorine concentration at the crack regions increases with increasing the relative humidity give evidence to substantiate the aforementioned argument.

(2) The results of the 10,000-h tests of this research suggest that the chloride concentration threshold for the initiation of SCC in 304L stainless steel at 45 °C is between 0.1 g/m^2 and 1 g/m^2.

(3) Corrosion bands were exclusively observed on the specimens with chloride concentration of 0.1 g/m^2 and 1 g/m^2 tested at 70% RH. The width of corrosion band increased with increasing chloride concentration. Moreover, corrosion bands were analyzed to have higher chlorine contents.

(4) The 304L austenitic stainless steel specimens with chloride concentration of 1 g/m^2 tested at 45 °C were cracked with a trans-granular SCC mode, which was proved by the results of EBSD.

Author Contributions: Conceptualization, C.-P.Y. and K.-C.T.; formal analysis, C.-P.Y.; investigation, C.-P.Y.; project administration, J.-Y.H.; writing—original draft, C.-P.Y.; writing—review and editing, C.-P.Y. and J.-Y.H. All authors have read and agreed to the published version of the manuscript.

Funding: This research received no external funding.

Conflicts of Interest: The authors have no conflict of interest to declare.

References

1. Schoell, R.; Xi, L.; Zhao, Y.; Wu, X.; Yu, Z.; Kenesei, P.; Almer, J.; Shayer, Z.; Kaoumi, D. In situ synchrotron X-ray tomography of 304 stainless steels undergoing chloride-induced stress corrosion cracking. *Corros. Sci.* **2020**, *170*, 108687. [CrossRef]
2. Dastgerdi, A.A.; Brenna, A.; Ormellese, M.; Pedeferri, M.; Bolzoni, F. Experimental design to study the influence of temperature, pH, and chloride concentration on the pitting and crevice corrosion of UNS S30403 stainless steel. *Corros. Sci.* **2019**, *159*, 108160. [CrossRef]
3. Yeom, H.; Dabney, T.; Pocquette, N.; Ross, K.; Pfefferkorn, F.E.; Sridharan, K. Cold spray deposition of 304L stainless steel to mitigate chloride-induced stress corrosion cracking in canisters for used nuclear fuel storage. *J. Nucl. Mater.* **2020**, *538*, 152254. [CrossRef]
4. Wu, X. On residual stress analysis and microstructural evolution for stainless steel type 304 spent nuclear fuel canisters weld joint: Numerical and experimental studies. *J. Nucl. Mater.* **2020**, *534*, 152131. [CrossRef]
5. Guo, L.; Street, S.R.; Mohammed-Ali, H.B.; Ghahari, M.; Mi, N.; Glanvill, S.; Plessis, A.D.; Reinhard, C.; Rayment, T.; Davenport, A.J. The effect of relative humidity change on atmospheric pitting corrosion of stainless steel 304L. *Corros. Sci.* **2019**, *150*, 110–120. [CrossRef]
6. Saegusa, T.; Yagawa, G.; Aritomi, M. Topics of research and development on concrete cask storage of spent nuclear fuel. *Nucl. Eng. Des.* **2008**, *238*, 1168–1174. [CrossRef]
7. Aoyama, T.; Sugawara, Y.; Muto, I.; Hara, N. In situ monitoring of crevice corrosion morphology of type 316L stainless steel and repassivation behavior induced by sulfate ions. *Corros. Sci.* **2017**, *127*, 131–140. [CrossRef]
8. Li, Y.Z.; Wang, X.; Zhang, G.A. Corrosion behavior of 13Cr stainless steel under stress and crevice in 3.5wt.% NaCl solution. *Corros. Sci.* **2020**, *163*, 108290. [CrossRef]
9. Han, D.; Jiang, Y.; Shi, C.; Deng, B.; Li, J. Effect of temperature, chloride ion and pH on the crevice corrosion behavior of SAF 2205 duplex stainless steel in chloride solutions. *J. Mater. Sci.* **2012**, *47*, 1018–1025. [CrossRef]
10. Meyer, R.M.; Pardini, A.; Cuta, J.; Adkins, H.; Casella, A.; Qiao, A.; Larche, M.R.; Diaz, A.; Doctor, S.R. *NDE to Manage. Atmospheric SCC in Canister for Dry Storage of Spent Fuel: An Assessment*; PNNL-22495 401001060; Pacific Northwest National Laboratory: Richland, WA, USA, 1 September 2013.
11. Tani, J.I.; Mayuzumi, M.; Hara, N. Stress corrosion cracking of stainless-steel canister for concrete cask storage of spent fuel. *J. Nucl. Mater.* **2008**, *379*, 42–47. [CrossRef]
12. Brossia, C.; Kelly, R. Occluded solution chemistry control and the role of alloy sulfur on the initiation of crevice corrosion in type 304ss. *Corros. Sci.* **1998**, *40*, 1851–1871. [CrossRef]

13. Machuca, L.L.; Bailey, S.I.; Gubner, R.; Watkin, E.L.; Ginige, M.P.; Kaksonen, A.H.; Heidersbach, K. Effect of oxygen and biofilms on crevice corrosion of UNS S31803and UNS N08825 in natural seawater. *Corros. Sci.* **2013**, *67*, 242–255. [CrossRef]
14. Pickering, H.W. The significance of the local electrode potential within pits, crevices and cracks. *Corros. Sci.* **1989**, *29*, 325–341. [CrossRef]
15. Cai, B.P.; Liu, Y.H.; Tian, X.J.; Wang, F.; Li, H.; Ji, R.J. An experimental study of crevice corrosion behavior of 316L stainless steel in artificial seawater. *Corros. Sci.* **2010**, *52*, 3235–3242. [CrossRef]
16. Cho, K.; Pickering, H.W. The role of chloride ions in the IR& IR* criterion for crevice corrosion in iron. *J. Electrochem. Soc.* **1991**, *138*, L56–L58.
17. Shojaei, E.; Mirjalili, M.; Moayed, M.H. The influence of the crevice induced IR drop on polarization measurement of localized corrosion behavior of 316L stainless steel. *Corros. Sci.* **2019**, *156*, 96–105. [CrossRef]
18. Scatigno, G.G.; Dong, P.; Ryan, M.P.; Wenman, M.R. The effects of salt loading on chloride-induced stress corrosion cracking of 304L austenitic stainless steel under atmospheric conditions. *Materialia* **2019**, *8*, 100509. [CrossRef]
19. Shoji, S.; Ohnaka, N.; Furutani, Y.; Saitoh, T. Effects of relative humidity on atmospheric stress corrosion cracking of stainless steels. *Corros. Eng.* **1986**, *35*, 559–565. [CrossRef]
20. Prosek, T.; Iverson, A.; Taxsen, C.; Thierry, D. Low temperature on stress corrosion cracking of stainless steels in the atmosphere in the presence of chloride deposits. *Corrosion* **2009**, *65*, 105–117. [CrossRef]
21. Cook, A.B.; Lyon, S.B.; Stevens, N.P.C.; Gunther, M.; McFiggans, G.; Newman, R.C.; Engelberg, D.L. Assessing the risk of under-deposit chloride-induced stress corrosion cracking in austenitic stainless steel nuclear waste containers. *Corros. Eng. Sci. Technol.* **2014**, *49*, 529–534. [CrossRef]
22. Albores-Silca, O.E.; Charles, E.A.; Padovani, C. Effect of chloride deposition on stress corrosion cracking of 316L stainless steel used for intermediate level radioactive waste containers. *Corros. Eng. Sci. Technol.* **2011**, *46*, 124–128. [CrossRef]
23. Yeh, C.P.; Tsai, K.C.; Huang, J.Y. Effects of relative humidity on crevice corrosion behavior of 304L stainless-steel nuclear material in a chloride environment. *Metals* **2019**, *9*, 1185. [CrossRef]
24. Masuda, H. SKFM observation of SCC on SUS304 stainless steel. *Corros. Sci.* **2007**, *49*, 120–129. [CrossRef]
25. Qiao, L.J.; Gao, K.W.; Volinsky, A.A.; Li, X.Y. Discontinuous surface cracks during stress corrosion cracking of stainless steel single crystal. *Corros. Sci.* **2011**, *53*, 3509–3514. [CrossRef]
26. He, X.; Mintz, T.S.; Pabalan, R.; Miller, L.; Oberson, G. *Assessment of Stress Corrosion Cracking Susceptibility for Austenitic Stainless Steels Exposed to Atmospheric Chloride and Non-Chloride Salts*; NUREG/CR-7170; U.S. Nuclear Regulatory Commission: Montgomery, MD, USA, 2014.
27. Li, J.X.; Chu, W.Y.; Wang, Y.B.; Qiao, L.J. In situ TEM study of stress corrosion cracking of austenitic stainless steel. *Corros. Sci.* **2003**, *45*, 1355–1365. [CrossRef]
28. Zhu, L.K.; Yu, Y.; Li, J.X.; Qiao, L.J.; Volinsky, A.A. Stress corrosion cracking under low stress: Continuous or discontinuous cracks? *Corros. Sci.* **2014**, *80*, 350–358. [CrossRef]
29. Nash, B.K.; Kelly, R.G. Characterization of the crevice solution chemistry of 304stainless-steel. *Corros. Sci.* **1993**, *35*, 817–825. [CrossRef]

MDPI
St. Alban Anlage 66
4052 Basel
Switzerland
www.mdpi.com

Materials Editorial Office
E-mail: materials@mdpi.com
www.mdpi.com/journal/materials

www.ingramcontent.com/pod-product-compliance
Lightning Source LLC
Chambersburg PA
CBHW061551210926
51850CB00009B/2069
* 9 7 8 3 0 3 6 5 8 9 6 1 9 *